Contributions to Statistics

For further volumes:
http://www.springer.com/series/2912

Luigi Biggeri · Guido Ferrari
Editors

Price Indexes in Time and Space

Methods and Practice

Physica-Verlag

Editors
Professor Luigi Biggeri
University of Florence
Viale Morgagni 59
50134 Firenze
Italy
biggeri@ds.unifi.it

Professor Guido Ferrari
University of Florence
Viale Morgagni 59
50134 Firenze
Italy
ferrari@ds.unifi.it
and
Renmin University of China
59 Zhongguancun Street
Beijing 100872
China

ISBN 978-3-7908-2814-6 e-ISBN 978-3-7908-2140-6
DOI 10.1007/978-3-7908-2140-6
Springer Heidelberg Dordrecht London New York

Cover design: Integra Software Services Pvt. Ltd., Pondicherry

Printed on acid-free paper

Physica-Verlag is a brand of Springer
Springer is part of Springer Science+Business Media (www.springer.com)

Foreword

In his "Prime ricerche sulla rivoluzione dei prezzi in Firenze"* (1939), Giuseppe Parenti, by Fernand Braudel regarded as an author who "se classait, d'entrée de jeu et sans discussion possible, à la hauteur même d'Earl Jefferson Hamilton. . . ." begins his opening lines with a description/definition of the price revolution which took place in the XVI in Europe as "that extraordinary enhancement of all things that occurred in European countries around the second half of the XVI; revolution in the true meaning of the word, as not only, like any strong price increase, it modified the wealth distribution process and changed the relative position of the various social categories and of the different functions of the economic activity, but affected too, in a way that was not enough studied yet, the relative evolution of the various national economies, and finally, ..., certainly contributed to the birth, or at least to the dissemination, of the new naturalistic economic ideas, from which the economic science would have sprung". Definition that can be taken as the founding metaphor of this volume.

The ideal stimulus represented by Parenti's work may have opened the way to the now long standing tradition which links the research activity of the Department of Statistics of the University of Florence to price index numbers problems, concretized in the many works produced by its researchers and in the organization of the International Seminar on "Improving the Quality of Price Indices" held in Florence in 1995 under the joint auspices of the Department of Statistics itself and of Eurostat. This seminar can be viewed as a milestone for the research project on "Price Indexes in Time and Space" granted by the Italian Ministry of University (MIUR), which disembogued into the International Workshop on price indexes held in September 2008 at the Department of Statistics, and of which this volume constitutes the printed voice.

The work carried out by scholars, researchers, national and international organisms and institutions committed to the analysis of price index numbers theory and practice and price indexes production is too vast to be accounted for and discussed in this book. And, after all, such an exercise would go beyond its objectives.

---

*In "Studi di storia dei prezzi", cared of the Dipartimento di Statistica dell'Universitá di Firenze and of the "Fondation de la maison des sciences de l'homme, Paris", Ann Arbor, 1981.

We will therefore restrain ourselves to stress some points.

To begin with, a general remark is in order: both the interest of researchers and scholars and the attention and work of the statistical offices and organizations in the twentieth century was basically focussed on time indexes, more specifically on time price indexes and even more specifically, on Consumer Price Indexes (CPIs), particularly in the early stages of the journey.

Even Irving Fisher, in his classic volume "The Making of Index Numbers" (1922) discussed index numbers as synonymous of time (price) indexes, with no mention to possible space extension of the concept.

Furthermore, and to quote another prominent scholar and Nobel Prize winner, Ragnar Frisch in his fundamental article "Annual Survey of General Economic Theory: The Problem of Index Numbers" (Econometrica, 1936) claims "...will be confined to those (index numbers) whose object is to measure some sort of purchasing power". Here the expression "purchasing power" refers to time only.

Parallel to this main trend of thought, although shifted in time, and under the boost of very concrete motivations dictated by the needs for making international comparisons of income and Gross Domestic Product (GDP), Irving B. Kravis, Alan Heston and Robert Summers enlivened the United Nations International Comparison Project (ICP) in 1968 by publishing the first volume of the series on "International Comparisons of Real Product and Purchasing Power", reporting the research work the objective of which was to develop a comprehensive and reliable system of estimates of real GDP and the purchasing power of currencies based upon detailed price comparisons among countries.

This was probably the first time that the space purchasing power terminology and meaning, and therefore, the space CPI concept and the related purchasing power parity (PPP) definition appeared.

Official international statistical agencies, such as the United Nations Statistical Division (UNSD), the OECD, and Eurostat started working on the subject, in the framework of the ICP while continuing to be interested in, and producing time price indexes.

The National Statistical Offices (NSOs) did not follow that trend and continued to elaborate essentially time price indexes, namely time CPIs.

As a consequence, the two aspects of the same question continued to be treated and approached in parallel and their duality was somewhat, if not totally, ignored.

This volume intends to somehow bridge this gap, as is obvious from its title "Price Indexes in Time and Space".

If the measurement of time inflation and, subsequently as above said, of space price comparison has been a first and fundamental concern, other problems appeared downstage as well, claiming for their own adequate place.

Such is, firstly, the question of the space comparability of time CPIs, which has opened the way to the studies on harmonized price CPIs and their elaboration.

Secondly, great importance has increasingly been taken by price indexes other than the CPIs: wholesale, production, international trade price indexes and so on.

Again, noteworthy significance has been gained by price indexes utilized as National Accounts (NA) deflators, as well as those in the financial field.

This volume, and the underlying research project, reflect the above points. Indeed, the logic that has driven us has been that of stressing the close duality of time and space frames, the most advanced methodologies of statistical approach, also with extensive reference to the axiomatic approach and with emphasis on CPIs, both in time, basically as inflation measures and as time deflators of NA aggregates, and in space, again as (spatial) inflation measures or PPPs and as space GDP deflators, keeping in mind the problems of basket choice, weighting and integration-harmonization. All this, with a perspective as general as possible, which accounts for the highly relevant questions of the elaboration, use and validity of the sub-indexes, for the implications in the financial field and, last but not least, for the practical problems of construction and dissemination of the indexes. That is to say:

- the CPIs theory, the time-space background and the analysis of the time-space integration-harmonization;
- the space CPIs, the PPPs and the international comparisons of GDP;
- the time CPIs used as sub-indexes;
- the time indexes used as NA deflators;
- the price indexes in the financial field.

All the above confirms, we believe, that the subject of price index numbers retains its fascination and utility, despite the elapsing of time. If anything, it seems to strengthen all its virtues, due to the needs that the new theoretical and practical economic challenges entail.

As a matter of fact, a price index is a tool as simple as it is powerful and useful which does not cease to unfold its attractiveness and the many uses one can make of it.

It is the will to stress and emphasize once more the meaning and the effectiveness of price index numbers and to recover their whole potential in a comprehensive framework that has supported our endeavour and the related work.

The papers in this book deal with all the above topics in an effort to discuss them and afford some contribution to the theoretical debate as well as to the methodology and practice of elaboration.

An old Chinese saying warns: "you can dig a seventy-two-feet-deep well with hard work, but if you do not find water it is as if you had not worked at all".

We hope we found some water.

Firenze, December 2009

Luigi Biggeri
Guido Ferrari

Contents

Contributors

Bert M. Balk Rotterdam School of Management, Erasmus University, Rotterdam, and Statistics, Netherlands, bbalk@rsm.nl

Luigi Biggeri Istat, Rome, Italy; University of Firenze, Firenze, Italy, biggeri@ds.unifi.it

Carsten Boldsen Hansen United Nations Economic Commission for Europe, Geneva, Switzerland, carsten.hansen@unece.org

Isabella Carbonaro DET, University of Rome Tor Vergata, Italy, isabella.carbonaro@uniroma2.it

Margherita Carlucci Department of Economics, Sapienza University of Rome, Italy, margherita.carlucci@uniroma1.it

Giuseppe Cavaliere Dipartimento di Scienze Statistiche, Universita' di Bologna, Italy, giuseppe.cavaliere@unibo.it

Michele Costa Dipartimento di Scienze Statistiche, Universita' di Bologna, Italy, michele.costa@unibo.it

Luca De Angelis Dipartimento di Scienze Statistiche, Universita' di Bologna, Italy, l.deangelis@unibo.it

Rita De Carli ISTAT, Rome, Italy, decarli@istat.it

Marco Fattore Dipartimento di Metodi Quantitativi per le Scienze Economiche ed Aziendali, Università degli Studi di Milano - Bicocca, Via Bicocca degli Arcimboldi 1, 20126 – Milano, Italy, marco.fattore@unimib.it

Guido Ferrari Dipartimento di Statistica, Università di Firenze, Firenze, Italy, ferrari@ds.unifi.it, and Renmin University of China, PRC

K. Renuka Ganegodage School of Economics, The University of Queensland, St Lucia 4072, Australia, r.ganegodage@uq.edu.au

Giorgio Garau DEIR, University of Sassari, Sassari, Italy, giorgio@uniss.it

Javier Huerga European Central Bank, Frankfurt am Main, Germany, javier.huerga@ecb.int

José Mondéjar Jiménez Facultad de Ciencias Sociales de Cuenca, UCLM, Spain, Jose.Mondejar@uclm.es

Tiziana Laureti Department of Statistics and Mathematics for Economic Research, University of Naples "Parthenope", Naples, Italy, tiziana.laureti@uniparthenope.it

Patrizio Lecca Department of Economics, University of Strathclyde, Glasgow, UK, patrizio.lecca@gmail.com

Matteo M. Pelagatti Dipartimento di Statistica, Università degli Studi di Milano-Bicocca, Via Bicocca degli Arcimboldi, 8, I-20126, Milano, matteo.pelagatti@unimib.it

Alicia N. Rambaldi School of Economics, The University of Queensland, St Lucia 4072, Australia, a.rambaldi@uq.edu.au

D.S. Prasada Rao School of Economics, The University of Queensland, St Lucia 4072, Australia, d.rao@uq.edu.au

Raffaele Santioni Bank of Italy, Economic Research Unit, Rome, Italy, raffaele.santioni@bancaditalia.it

Lucia Schirru DEIR, University of Sassari, Sassari, Italy, luciaschirru@gmail.com

Paul Schreyer OECD Statistics Directorate, Paris Cedex 16, France, Paul.Schreyer@OECD.org

Jan de Haan Statistics Netherlands, Division of Macro-Economic Statistics and Dissemination, JM Voorburg, The Netherlands, j.dehaan@cbs.nl

Part I
Consumer Price Indexes Time-Space Integration

Are Integration and Comparison Between CPIs and PPPs Feasible?

Luigi Biggeri and Tiziana Laureti

1 Introduction

The importance of integration and comparison between the Consumer Price Indices (CPIs) and the Purchasing Power Parities (PPPs) has been widely discussed in literature (Heston, 1996; Rao, 2001a; ILO/IMF/OECD/UNECE/Eurostat & The World Bank, 2004; Ferrari, Laureti, & Mostacci, 2005), and recognised in two critical reviews of ICP (International Comparison Program) and PPP computation by international organisations (Castles, 1997; Ryten, 1998) as well.

A more integrated approach to CPI and PPP for household consumption is required in order to: (i) explore the feasibility of integrating the PPP activities with the streamlined activities of the National Statistical Offices (NSOs) for the compilation of CPIs; (ii) examine the relationship between the PPPs for international comparisons with the evolution of CPIs in the countries in question. Integration and comparison are very advantageous both among different countries and different areas or cities within a country (ILO/IMF/OECD/UNECE/Eurostat & The World Bank, 2004).

Over the last decades there has been very little harmonization of the activities and surveys of NSOs involved in both CPI and PPP work while the need for comparisons of CPIs and PPPs depends on the possibility of providing complete matrices of temporal-spatial price differences (ILO/IMF/OECD/UNECE/Eurostat & The World Bank, 2004) which can be used for a better comprehension of the factors which influence price levels and their changes in different countries.

Therefore the feasibility of integration and comparison between CPIs and PPPs is an important issue which we will deal with in this paper considering only household consumption aggregates and binary comparisons between two areas or countries.

Firstly, in Section 2 we will examine the integration issues considering the content of the different consumption baskets, which can be used for computing the CPIs in two countries and the PPPs between these countries, in order to verify the

---

L. Biggeri (✉)
Istat, Rome, Italy University of Firenze, Firenze, Italy
e-mail: biggeri@ds.unifi.it

L. Biggeri, G. Ferrari (eds.), *Price Indexes in Time and Space*, Contributions to Statistics, DOI 10.1007/978-3-7908-2140-6_1, © Springer-Verlag Berlin Heidelberg 2010

overlapping of the baskets, to identify a basis for integrating the price and expenditure share data for the CPI and PPP computation and to compare these results in a consistent space-time comparison of consumer prices. The potential problems and benefits that may arise from developing an integrated approach to collect the necessary information are also specified.

However, the integration approach may be hampered by using the "identity products principle" which is commonly applied for the calculation of PPPs and can seriously influence the representativeness of the PPP product list of the consumption baskets in different countries or regions within a country, and negatively affect the comparisons between PPPs and CPIs. For these reasons it is also advisable to include less comparable products in the PPP baskets.

Section 3 illustrates a simple statistical approach for investigating the advantage of broadening the definition of comparability in order to include additional products in the PPP calculation, in terms of coverage and representativeness of the computed PPPs and to evaluate the importance of different factors which affect the results of the computations.

Regarding the comparison between CPIs and PPPs it is also important to examine how the changes in consumer price levels over time in the two countries (computed by the CPIs) affect the movements over time of the PPPs calculated for household consumption.

It is not possible to totally integrate and link the commonly computed CPIs and PPPs and to carry out a direct comparison for the time being, because these indices differ in the basket of products and services in question and in the formulae used.

Section 4 illustrates a methodological approach based on the decomposition of the formulae in order to approximately evaluate the economic factors which explain the divergences between the CPIs of the two countries from time $t-1$ to time t, and the movement of the PPPs concerning the two countries in the same period.

Finally, the concluding remarks in Section 5 explain how to carry out the integration of data collection and increase the comparability of CPIs and PPPs and then underline the usefulness of the methods suggested. Lastly, a huge organisational and costly effort by the NSOs is required in order to obtain the amount of data to be collected and estimated at least in a benchmark year for achieving the desired results.

2 Integration of CPIs and PPPs

CPIs and PPPs share conceptual similarities. CPIs measure changes in price levels of products and services over time within a country, whereas PPPs measure differences in price levels across countries or regions within a country. Therefore, CPIs and PPPs refer respectively to time and spatial dimension of price differences. However, the results obtained are different according to the baskets of goods and services considered and formulae used.

In order to analyse the possible integration of CPI and PPP activities, Rao (2001a) discussed the issue of optimizing the flow of data from CPI to PPP and presented a figure of the intersection of price data sets at a national level of a generic country, in order to verify the comparisons of sets of products and services between CPI and PPP lists within the country.

Bearing in mind the aims of this paper and considering two different countries, we are interested both in the integration of the price data collection for calculating the two indices and in the comparison between CPIs and the change in the level of PPPs. Therefore, also the CPIs of the two countries should be comparable.

In the following sub-sections, we will analyse the comparison of the CPI baskets of products in the two countries in question, then the comparison of the different baskets used for calculating the CPIs and PPPs in the two countries, and finally the potential problems and benefits involved in developing an integrated approach for collecting the required information.

2.1 The Comparison of CPI Baskets in the Two Countries

With the aim of comparing the items included in the CPI baskets of two countries, it may be necessary to divide the products and services included in the baskets into two parts: non-comparable and comparable items (with at least a minimum degree of comparability). In this way it is possible to verify the degree of overlapping of the sets of elementary items (products and services) representative of the elementary household expenditure aggregate, included in the consumption baskets used for the CPI calculations. The items priced in different countries could be identical or quite different depending on the heterogeneity level of the two countries concerning the population's consumption behaviour.

Considering Fig. 1, where the CPI baskets of the two countries l and J are represented, composed by N_j and N_l items respectively, it is clear that there are fewer problems in finding an overlapping area when fairly similar or homogeneous countries are being compared in terms of consumption markets and behaviour. In this case, it is possible that $N_j = N_l$ i.e. the total number of products included in the two CPI baskets of each country in question could be identical. Moreover the characteristics of the products chosen for computing the CPIs and the elementary expenditure

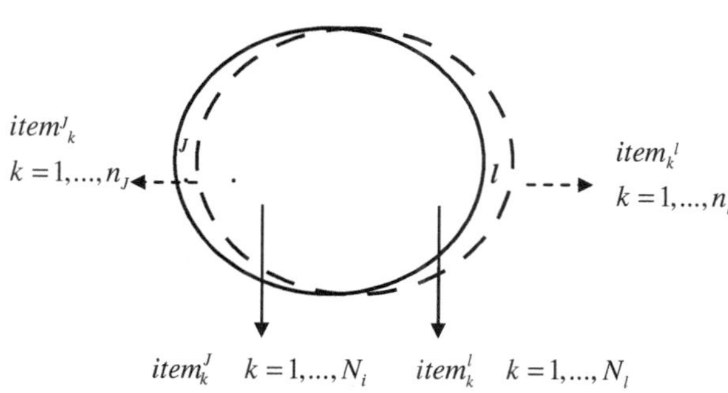

Fig. 1 Comparison of CPI baskets in the two countries, j and l

aggregates could be similar in the two countries. On the other hand, when comparisons involve countries that are fairly heterogeneous, the overlapping area will decrease.

The problem of identifying identical or similar products in the two countries can be related to the different number of items whose prices are to be collected for computing the CPIs in the two countries ($N_j \neq N_l$). Moreover, the definition and the identification of the elementary aggregates and products in the basket, and in particular the methods and practices used for price data collection, can greatly differ in the two countries according to the local situation of consumption, the differences in the consumer markets, the statistical infrastructures and the available resources.

However, even if the number of the products is the same ($N_j = N_l$), the physical and economic characteristics of the products and services which are used for calculating the CPIs can be different in the two countries due to the different patterns of consumption.

Therefore, the outer sets in Fig. 1 consist of n_j and n_l products and services (or groups of products and services) which are *typical or characteristic* regarding the consumption behaviour in country j and l respectively. These items should be considered separately in the outer sets since they have different price determining characteristics or technical parameters, and cannot be used directly for calculating comparable CPIs in the two countries.

It is worth noting that the above theoretical framework for comparing different consumption baskets is not applied from a practical point of view because the NSO of a certain country when computing national CPIs does not usually consider the comparability of the items included in that country's consumption basket with those included in the consumption basket of the other country in question.

The main components of CPIs are the data on prices of a large range of products and services *representative* of the consumption baskets of households and the information on weights associated with the various product categories reflecting the importance attached to different items.

The collection of prices and the expenditure weights are based on a classification of goods and services obtained by using a standard system such as the Classification of Individual Consumption according to Purpose (COICOP), or similar national classifications. The lowest level of product classification at which expenditure weights are available is used for identifying the *elementary aggregate indices* to be progressively aggregated to the total household expenditure level in order to obtain the general total CPI.

Within the elementary aggregate, *considered as strata sample*, the sample items to be included in the CPI computation are chosen considering the criteria of representativeness in terms both of the importance of all the products included in the elementary aggregate concerning consumption expenditure and their evolution of price changes over time. The elementary price index is computed using only price data, meaning that the index is estimated without using any weights within the elementary aggregate.

In this context, it is obvious that the items included in the CPI baskets of two countries can be quite different and it is not easy to compare these CPIs if no specification of the characteristics of products and services is given in order to harmonise the computation of the CPIs.

For this purpose the European HICPs (Harmonised Indices of Consumer Prices) are computed (ILO/IMF/OECD/UNECE/Eurostat & The World Bank, 2004, Annex 1) to measure inflation on a comparable basis taking into account differences in national definitions. They are based on the prices of goods and services available for purchase in the economic territory of each EU Member State for the purpose of directly satisfying consumer needs. The definitions of prices to be collected and of the groups of products and services to be considered are harmonised and agreed on.

The European HICPs are classified according to the four-digit categories or subcategories of the COICOP-HICP, which is the classification that has been adapted to the needs of HICPs, in order to have groups of products that are *approximately comparable* in terms of the specific items which must satisfy the same groups of consumers' need.

HICPs must also be based on appropriate sampling procedures, taking into account the national diversity of products and prices and among other things they illustrate what national consumer price indices have in common among the various countries. Three important sampling dimensions are considered: the item dimension, the outlet dimension and the regional dimension.

Therefore, the comparability criteria used in the HICPs is quite "weak" in terms of comparability of single products since in the HICP calculation the *representativeness criteria* is the most important aspect.

2.2 The Comparison of CPI and PPI Baskets in Two Countries

Considering the above theoretical comparison of the CPI baskets in two countries (Fig. 1), the comparison between CPI and PPP baskets can only refer to the overlapping set of items, and in this case products are defined according to the need of computing adequate PPPs at elementary level. The computation of PPPs and the feasibility of the integration between CPI and PPP activities require the evaluation of the degree of comparability of the products in order to measure the price differential between the two countries in question and the corresponding definition of the representative products used for computing the elementary price indices for CPI estimation. For this purpose Fig. 1 can be modified as shown in Fig. 2.

By following the definitions of comparability and representativeness discussed in Biggeri, De Carli and Laureti (2008), we must underline that for computing PPPs the shaded overlapping area $\varpi_{j,l}$ includes only n_{jl} *identical products* with the same characteristics and are therefore strictly comparable but with different systems of weights in the two countries (j and l). The prices of these products can be and are usually used for calculating PPPs between the two countries.

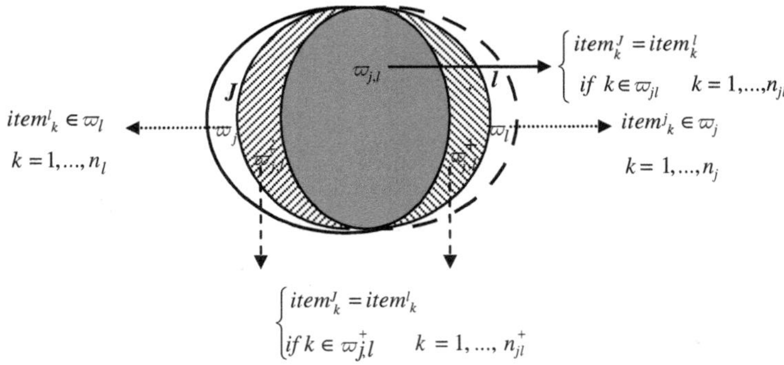

Fig. 2 Comparison of CPI and PPP baskets in two countries

The PPPs are computed at level of Basic Heading (BH) which consists of a fairly homogeneous group of items showing a low dispersion of price ratios. The basic heading level is normally the lowest level of aggregation for which expenditure data are available; therefore the PPPs at this level are computed without using weights for the individual items (Hill, 1997). The basic heading level may be considered similar to the elementary level used in CPI calculation. For the aggregation of price evolution and price differences above the elementary level or basic heading level, the expenditure share weights are common requirements for both CPIs and PPPs (Balk, 1996, 2001; Diewert, 1993).

However, the choice of the items (products and services) to be included in the BH follows different criteria (OECD-Eurostat, 2006; World Bank, 2007).

The main principle used in PPP computation in developing a product list requires a selection of "identical products" for the two countries. Identical products ensure that there are no quality issues in the measurement of the PPPs and the results only provide a measure of price differences. However, this is the most contentious issue in constructing PPPs, because the use of the identity principle can have serious implications for the representativeness of the product list of the consumption baskets in different countries.[1]

Therefore, referring to Fig. 2 and considering the BHs, the degree of the representativeness of items in the overlapping area $\varpi_{j,l}$ can be different in the two countries compared. In fact, since the patterns of consumption can greatly differ in these two countries, products that are representative and easily found in country j may not be easily found in l, due to differences in supply conditions, income levels, taste, climate, customs, etc. From a practical point of view it is evident that the strict comparability of products, obtained through a detailed specification, leads to PPPs for which it is possible to measure pure price differences. At the same time, this strict

[1] There are several operational procedures used by international organizations in order to deal with these problems (see among others, Kravis, Kenessey, & Heston, 1975, and more recently Rao, 2001b)

comparability will lower the degree of coverage in terms of products considered and of the general representativeness of a given product in different countries, (and even within a country); therefore the real consumption basket of these countries can be inadequately represented. In this case, the overall accuracy and reliability of the calculated PPPs will be affected.

The two overlapping sets of goods and services marked $\varpi_{j,l}^{+}$ contain n_{jl}^{+} less comparable items, whose prices are used for the computation of CPIs but usually not for PPP calculation. However, they could also be used for calculating PPPs by using a broader definition of comparability or by applying adjustments for quality differences.

The inclusion of the less comparable products in countries j and l for the computation of PPPs will increase the degree of coverage and probably the degree of representativeness of the comparison. However the calculated PPPs may correspond to different products, thus reflecting both pure price differences and the different representativeness of the selected products in the different countries.

As already mentioned, the outer sets marked ϖ_j and ϖ_l consist of some goods and services (n_j and n_l) which are typical (or characteristic) of the consumption behaviour in countries j and l respectively. Two products included in the outer areas cannot be considered comparable for PPP purposes, even if we use a broader definition of comparability, because consumers may be willing to pay more for one product than another.[2] Moreover, these products may not be on sale in one of the two countries and vice-versa.

The number of typical products in each country is usually different ($n_j \neq n_l$) although in some cases it can be the same in both countries ($n_j = n_l$). It is clear that the higher the number of typical products, the larger the outer areas will be.

As shown in Fig. 2 the total number of the items in the CPI basket in each country is obtained as the sum of the items included in the different subsets of products classified according to the imputed degree of comparability. For example, in country j the total number of products priced for CPI calculations is expressed as $N_j = n_{jl} + n_{jl}^{+} + n_j$. Similarly, the number of items in country l is expressed as $N_l = n_{jl} + n_{jl}^{+} + n_l$.

2.3 Problems and Benefits Involved in Developing an Integrated Approach for the Collection of the Necessary Information for CPIs and PPPs

The calculation of PPPs and standard CPIs is based on similar data requirements. From a practical point of view at present it clear that the definitions of the products

[2]When a characteristic is price determining the absence or presence of that characteristic will affect the price that consumers are prepared to pay for the product. There are several examples of price determining characteristics (Word Bank, ICP handbook, 2007). For example, the possession, or absence, of air conditioning will usually affect the price of an automobile. Consumers in most countries will pay more to obtain it. The size of a packet of rice is price determining as consumers will pay more for a kilo than half a kilo.

to be used for PPP computation may be quite different from the definitions of the products used for the computation of CPIs and in any case the price data collection follows different criteria for the computation of the two indices. Even if we refer to the overlapping area of identical products of CPI and PPP baskets (as in Fig. 2), the same item in the two baskets can be considered identical in theory but not in practice, since the definition of the products and services in the CPI computation is not usually well specified in terms of their characteristics.

There are several problems concerning the integration of data collection[3] for both CPIs and PPPs. On one hand it is necessary to evaluate the comparability of products and identify the identical products in the two countries, meaning that we must verify the characteristics and the quality of the products chosen for the CPIs and PPPs. On the other hand, it is also essential to verify whether the products priced in different countries are "representative" of their consumption within the basic headings or not. A related problem is whether the coverage of the products priced is adequate[4] concerning the basic heading to which they belong.

There are two different approaches for verifying these conditions, which do not necessarily exclude each other.

The first approach consists in analysing the definition of each item used for each CPI elementary aggregate and comparing it with the similar item used for the BH in the PPP computation. This analysis has been implemented in some experiments in various countries (see for example, Bretell and Gardiner, 2002; Wingfield, Fenwick & Smith, 2005; Aten, 2005, 2006; Melser & Hill, 2005) and also in Italy in order to compute the PPPs at regional level within the country (De Carli, 2008). The results indicate that these analyses can be very difficult and time-consuming to implement. It is often necessary to review the definitions of the items whose prices are collected for the CPIs while in other cases it is necessary to implement specific surveys (for example for clothing, footwear and furniture) in order to obtain adequate price data for PPPs which are coherent with the identity product principle. Moreover, if it is not possible to find the items with the same strictly comparable definitions, methods of spatial quality adjustment must be used in order to compare the products and services of the two countries.

This approach may guarantee the strict comparability of more items included both in CPIs and PPPs, but does not provide any information on the representativeness and coverage of the PPP item list, which represents the consumption baskets in different countries. In order to solve this problem, it is necessary to collect data on expenditure weights for the products and services belonging to each elementary and BH aggregate. However at present the elementary aggregates and BHs are the lowest level aggregate for which expenditure data are available. Therefore, in order to carry out specific analyses to assess the degree of representativeness and coverage

[3] As far as some useful initiatives that could provide a framework for a practical integrated approach for the integration of PPP and CPI work are concerned we refer to the suggestions of other mentioned authors and, in particular, to the ILO manual which mentions two core strategies to do it: the "Use of characteristics approach" and "linking approach to international comparisons".

[4] These issues are currently being researched, and Rao (2001b) offer a modified approach that attached weights proportional to coverage and representativeness.

of PPPs the expenditure weight data within the elementary aggregates should be collected or evaluated, at least in a benchmark year.

The second approach focuses on the "reconciliation" of the definitions of products in the PPP and CPI baskets, using a broader definition of comparability for the computation of the PPPs (Krijnse-Locker, 1984). In this way a larger number of items included in the CPI baskets become comparable with those considered in other countries and can be used for the computation of new enlarged PPPs, thus achieving a higher level of comparability between CPIs and PPPs.

In our opinion, it is necessary to go beyond the criteria of identical products currently used for computing PPPs, because the cost of living could be misinterpreted if the comparison is based on two identical products which satisfy the same consumer need but are more frequently purchased in one country than the other and vice versa.

In order to compare the levels of expenditure between two countries for a specific basket of an elementary aggregate which can fulfil specific consumer demand, it is better to refer to the most frequently purchased products in each country since even if they are not strictly comparable they will certainly represent the products purchased in these countries. However, it is important that these products are purchased by consumers in order to satisfy the same specific needs.

The use of a broader definition of comparability might be achieved by using the Structured Product Descriptions (SPDs) suggested by the ICP Global office of the World Bank and used in the 2005 ICP (Diewert, 2008). In fact, SPDs provide the framework for selecting the representative items to be priced. These price movements, taken together, can supply a good estimate of the overall change in prices for the group of similar products as a whole. When completing a SPD, collectors are identifying a specific product with all its relevant characteristics and distinguishing it from the other products in the same elementary aggregate. These product characteristics may be used to specify a particular product to be included in the calculation of CPIs and PPPs.

These product characteristics were used to specify a particular product to be included in the calculation PPPs. The SPDs could also be used for collecting product prices in order to construct CPIs thus obtaining a harmonized framework to carry out the comparison among countries. Apart from increasing the number of items to be included in the computation, the above mentioned analyses can improve the representativeness of the consumption basket of the countries examined. However, disaggregated data concerning expenditure weights within the elementary aggregates are necessary for evaluating improvements in representativeness and coverage of the PPPs and in comparability between CPIs and PPPs[5].

A successful integration of PPP activity with the CPI compilation depends on to what extent these two activities can be based on a common pool of data and information available at a national level and at a territorial level within a country.

Nevertheless, concerning data collection it is clear that for achieving the integration of PPP computation with CPI activities an increased amount of information

---

[5]For the international comparisons the integration work also requires the harmonisation of the definitions and classifications of products and of the methods for the collection of data.

to be collected and processed during the construction of the CPI is required and therefore a lot of extra work is necessary (see also Rao, 2001a; ILO/IMF/OECD/UNECE/Eurostat & The World Bank, 2004; Ferrari, Laureti & Mostacci, 2005). So it is clear that the NSOs and organizations involved in CPI construction must believe that PPP computation and any results from PPPs are a natural extension of current CPI activities, and produce much more important statistical information on which possible economic analysis can be performed.

However, the NSOs must be aware that the integration activities could also result in tangible benefits as many authors have underlined. In short the potential benefits are:

- increased coverage in terms of products and share of the household expenditure for the PPPs;
- improved quality of the PPP estimations in terms of the representativeness of the consumption baskets of the countries involved;
- increased coherence between the results of PPP and CPI calculations;
- possibility for computing the PPPs at reduced intervals of time, taking into account the high frequency of collection of data for CPI purposes, thus overcoming the difficulties linked to the use of CPIs for the temporal updating of the PPPs;
- improved research on methods for quality adjustment in order to make more comparable similar products, which could enable us to verify the quality changes over time and quality differences across countries.

One more important advantage is the possible development of PPPs across different cities and/or regions within countries. In this case it is easier to make a reliable comparison and a successful integration of the PPPs and CPIs between two regions within a country because the level of homogeneity concerning consumer behaviour is usually higher and the definitions of all the products are more similar.

However, NSOs must evaluate some of the above mentioned benefits, especially concerning the coverage and the quality improvement of the PPP estimations in order to decide whether it is better to implement the integration of the CPIs and PPPs and more importantly if less comparable products should be included in the PPP computation. Moreover, NSOs must assess the pros and cons in terms of comparability between CPIs and PPPs. In the following sections we will suggest some statistical methods for carrying out these evaluations.

3 A Methodological Approach for Deciding whether to Include Less Comparable but More Representative Products in the PPP Calculation

3.1 Inclusion of Less Comparable Products vs Identical Products

In order to understand to what extent it is profitable to include less comparable products in the list for the computation of spatial indices, and in particular when they come from the CPI calculation, we need a method for measuring the effect caused by their inclusion in the PPP calculation.

On one hand, the inclusion of less comparable products should increase both the coverage referring to the share of the household expenditure of each set of products and the representativeness of PPPs referring to the values of the PPPs concerning different sets of products, on the other hand by doing so the degree of comparability of the same products will decrease. Therefore, there is a sort of trade-off between the concept of representativeness and comparability.

In order to select the right number of products we will propose a simple method based on the calculation of three different indices referring to three different sets of products, already illustrated in Fig. 2 and represented by the overlapping areas $\varpi_{j,l}$ (shaded), the $\varpi^+_{j,l}$ (striped) and the outer areas ϖ_j and ϖ_l.

By following Biggeri, De Carli & Laureti (2008) the three spatial indices are calculated as ratios of the weighted geometric mean prices of the three sets of products of the two countries in question. The first two indices are called *Average Prices' Parities (APPs)* to differentiate them from the currently computed PPPs, and the third is called the *Characteristicity Index (CI),* because it measures the influence of typical products of the country's basket on spatial comparisons.

It is worth noting that all these spatial indices can be calculated by using country j or country l as the reference country thus obtaining APPs comparing country j to country l or vice-versa. Below only indices, calculated considering country l as reference country, are shown since the results are similar but opposite.

By only considering the strictly comparable products, which are those included as identical products in the overlapping area $\varpi_{j,l}$ in Fig. 2, we will calculate the following spatial index:

$$APP^{\varpi_{jl}}_{l,j} = \frac{\prod_{k=1}^{n_{jl}} \left(p^j_k\right)^{w^j_k}}{\prod_{k=1}^{n_{jl}} \left(p^l_k\right)^{w^l_k}} \tag{1}$$

where p^j_k (w^j_k) denotes the price (weight, as share of expenditure) of item specification k in country j, p^l_k (w^l_k) is the price (weight) of item specification k in country l, $\sum_{k=1}^{n_{jl}} w^j_k = \sum_{k=1}^{n_{jl}} w^l_k = 1$ and n_{ij} is the number of identical items priced in both countries.

Considering only the less comparable products, which are contained in the two striped areas $\varpi^+_{j,l}$ in Fig. 2, a second index $APP^{\varpi^+_{jl}}_{l,j}$ can be computed as a ratio of the *weighted geometric mean prices:*

$$APP^{\varpi^+_{jl}}_{l,j} = \frac{\prod_{k=1}^{n^+_j} \left({}^+p^j_k\right)^{{}^+w^j_k}}{\prod_{k=1}^{n^+_l} \left({}^+p^l_k\right)^{{}^+w^l_k}} \tag{2}$$

where ${}^{+}p_k^j$ (${}^{+}w_k^j$) and ${}^{+}p_k^l$ (${}^{+}w_k^l$) are the price (weight) of the less comparable products priced in country j and l respectively and $\sum_{k=1}^{n_j^+} {}^{+}w_k^j = \sum_{k=1}^{n_l^+} {}^{+}w_k^l = 1$.

Finally, by considering the typical products we can calculate the CI as the ratio between the weighted geometric average prices of the typical products in the two countries in question, included in the outer areas ϖ_j and ϖ_l:

$$CI_{l,j} = \frac{\prod_{k=1}^{n_j} \left({}^{*}p_k^j\right)^{{}^{*}w_k^j}}{\prod_{k=1}^{n_l} \left({}^{*}p_k^l\right)^{{}^{*}w_k^l}} \tag{3}$$

where ${}^{*}p_k^j$ (${}^{*}w_k^j$) and ${}^{*}p_k^l$ (${}^{*}w_k^l$) are the price (weight) of the characteristic products in country j and l respectively and $\sum_{k=1}^{n_j} {}^{*}w_k^j = \sum_{k=1}^{n_l} {}^{*}w_k^l = 1$.

After having calculated these indices we must check their values so assess if they are equal or different to 1 and then compare the results.

For example if $APP_{l,j}^{\varpi_{jl}^+}$ is equal to 1 it seems that the inclusion of less comparable products does not add further information to the comparison of the level of prices between the two countries compared to the information from the $APP^{\varpi_{jl}}$ index although it increases the coverage. When as usual the $APP_{l,j}^{\varpi_{jl}^+}$ is different to 1, the inclusion of less comparable products shows a different behaviour of the prices for these products in the two countries.

Having established that the second APP index, $APP_{l,j}^{\varpi_{jl}^+}$, is different to 1 we must compare it to the $APP_{l,j}^{\varpi_{jl}}$ index.

By comparing the values of $APP_{l,j}^{\varpi_{jl}^+}$ and $APP_{l,j}^{\varpi_{jl}}$ we can assess whether the computation of the APP for the less comparable products adds further information to the comparison of the level of prices between the two countries.

If the two indices are equal to one another it would be advantageous to compute PPPs by using less comparable products because in this way the representativeness and coverage of the computed PPPs are improved.

When the two indices differ we should evaluate the degree of divergence and the trade-off between comparability and representativity. Moreover in order to include the right number of products we must consider to what extent the two different sets of products weigh on the total household consumption expenditure in both countries. This can be done by considering the total share of expenditure for those products included in the different countries. The cost for applying quality adjustment methods should also be considered when deciding whether to include less comparable products.

The value of the index $APP_{l,j}^{\varpi_{jl}^{+}}$ can be much higher than that of $APP_{l,j}^{\varpi_{jl}}$ meaning that the calculation of spatial indices based only on identical products do not fully represent the consumption baskets of these countries.

The characteristicity index is not useful for deciding whether to include other products in the PPP calculation since the products on which this index is based are so different and typical of each country that they cannot be considered when comparing the price level of the two countries. On the other hand, if we are aware of the value of the $CI_{l,j}$ and the corresponding weight in terms of consumer expenditure concerning typical products we can evaluate the loss in terms of the overall representativeness (*characteristicity effect*). If the typical products of each country weigh heavily, a direct comparison between the two countries in question would be impossible.

3.2 *Interpretation of the Factors Influencing the PPPs Based on Products with Different Degree of Comparability*

Although the information obtained from the computation and comparison of the three indices is sufficient for deciding the number of products to be included in the computation of the PPPs, this evaluation can be improved by using a decomposition technique.

In fact, we can suggest an interesting decomposition of the first two indices which can be used to assess the importance of the different factors that affect the value of binary spatial indices.

Considering for example the $APP_{l,j}^{\varpi_{jl}^{+}}$, calculated referring to the less comparable products, the following decomposition is obtained:

$$APP_{l,j}^{\varpi_{jl}^{+}} = \prod_{k=1}^{n_{jl}^{+}} \left(\frac{{}^{+}p_k^j}{{}^{+}p_k^l} \right)^{w_k^l} \cdot \prod_{k=1}^{n_{jl}^{+}} \left({}^{+}p_k^j \right)^{w_k^j - w_k^l} \tag{4}$$

The first factor on the right hand side of (4) represents the *Pure Price Effect* (PPE), corresponding to a bilateral PPP, using a weighted Jevons index with weights of country l. The *Weight Effect* (WE), expressed by the second factor on the right hand side of (4), concerns the impact of the difference in consumption pattern in the two countries in question. When the products have a similar degree of representativeness concerning consumer behaviour, the difference in the weights corresponding to the item k is close to zero.

By introducing the variables $\alpha_k = \ln\left(\frac{p_k^j}{p_k^l}\right)$ and $c_k^{l,j} = \left(w_k^j - w_k^l\right)$, which express the logarithm of price ratio concerning item k in country j and l and the difference between the corresponding expenditure weights, respectively (where $k = 1, \ldots, n_{jl}^{+}$), after simple algebra, formula (4) can be equivalently expressed as:

$$APP_{l,j}^{\varpi_{jl}^{+}} = \exp(\overline{\alpha}) \exp\left(n_{ij} \cdot s_{\alpha} \cdot s_{w_k^l} \cdot R_{w_k^l,\alpha}\right) \times \exp\left(n_{ij} \cdot s_{\ln P^j} \cdot s_{c_k^{l,j}} \cdot R_{\ln P^j, c_k^{l,j}}\right) \tag{4bis}$$

Thus it is possible to identify the factors that influence the *Pure Price Effect,* that is $s_{w_k^l}$ the standard deviation of the weighing system of the base country l, s_α, the standard deviation of the logarithm of the price ratios, $R_{w_k^l,\alpha}$ the linear correlation coefficient between the log price ratios and the weights of country l . It is worth noting that $\exp(\overline{\alpha}) = \prod_{k=1}^{n} \left(\frac{p_k^l}{p_k^j}\right)$ is the unweighted geometric mean of the price ratios between country j and l. This index is the Jevons index which is the best estimator when the log - distribution of price changes is Normal. Therefore, the spatial index and the evaluation of the degree of the influence of the factors in which it is decomposed depends on the shape of the distribution of the ratio between the prices of the products in the baskets of the two countries. As the distribution of the ratios of the price levels in two countries may vary according to the choice of the reference country, the influence of the shape of the distribution on the spatial indices could cause problems and therefore further analyses may be required.

Similarly, the *Weight Effect* is influenced by $s_{\ln P^j}$, the standard deviations of prices of country j,$s_{c_k^{l,j}}$, the standard deviation of the difference between the weights in the two countries compared $c_k^{j,l} = \left(w_k^l - w_k^j\right)$ and $R_{\ln P^j, c_k^{l,j}}$, the linear correlation coefficient between the prices and the differences in the corresponding weights.

Although it is possible to obtain similar decomposition forms as already mentioned (considering country j as the reference country) that give two estimations of the effects which differ slightly, the most important aspect is that we obtain statistical measures (standard deviation, central tendency and correlation coefficient) concerning the variability of price changes and the consumers' behaviour in the two countries which can be interpretable from a statistical and economic point of view.

On the other hand, the symmetric treatment of countries can be achieved for the pure price effect and for the weight effect by using a geometric mean of the indices and then by applying a geometric average to the results therefore obtaining Törnqvist indices. Considering the PPE calculated by using less comparable products we can state that:

$$^{T}PPE_{l,j} = \sqrt{\prod_{k=1}^{n_{jl}^{+}} \left(\frac{{}^{+}p_k^j}{{}^{+}p_k^l}\right)^{w_k^l} \cdot \prod_{k=1}^{n_{jl}^{+}} \left(\frac{{}^{+}p_k^j}{{}^{+}p_k^l}\right)^{w_k^j}}$$

$$^{T}PPE_{j,l} = \sqrt{\prod_{k=1}^{n_{jl}^{+}} \left(\frac{{}^{+}p_k^l}{{}^{+}p_k^j}\right)^{w_k^j} \cdot \prod_{k=1}^{n_{jl}^{+}} \left(\frac{{}^{+}p_k^l}{{}^{+}p_k^j}\right)^{w_k^l}}$$

where $^{T}PPE_{l,j} = \frac{1}{^{T}PPE_{j,l}}$

4 Comparison Between the Computed CPIs and PPPs

Considering the comparison between CPIs and PPPs and referring to household consumption it is important to examine how the changes in levels of consumer prices over time in the two countries (computed by the CPIs) affect the movements over time of the PPPs calculated for the household consumptions.

As already stated, it is not possible to totally integrate and link the currently computed CPIs and PPPs.

Although CPIs are conceptually very similar to PPPs, since their aim is to measure price level differences over time and across space respectively, the formulae used in the calculations are quite different.

We suggest comparing the CPIs between two countries by considering the Laspeyres type index, which is the formula generally used by most NSOs for the construction of CPIs and comparing the PPPs over time by using the formulae presented above. Following this procedure it is not possible to carry out a direct comparison since the two price indices differ in the formula used and in the basket of goods and services to be included in the calculation. Nevertheless the comparison can be carried out by decomposing the two different formulae used in time and space comparisons. By following this decomposition method, considering country l as the reference country, we can compare CPIs across space, thus measuring and interpreting the factors which explain the divergences between the CPIs of the two countries from time $t-1$ to time t:

$$ {}_{t-1}\mathbf{P}_t^j - {}_{t-1}\mathbf{P}_t^l = \sum_{k=1}^{n} {}_{t-1}P_{k,t}^j \cdot {}_{t-1}w_k^j - \sum_{k=1}^{n} {}_{t-1}P_{k,t}^l \cdot {}_{t-1}w_k^l $$

where ${}_{t-1}P_{k,t}^j = \frac{p_{t,k}^j}{p_{t-1,k}^j}$ and ${}_{t-1}P_{k,t}^l = \frac{p_{t,k}^l}{p_{t-1,k}^l}$ are elementary price indices in area j and l respectively, ${}_{t-1}w_k^j$ and ${}_{t-1}w_k^l$ are the weights, expressed by expenditure shares on commodity or service k in the base period $t-1$, relating to country j and country l, and $\sum_k {}_{t-1}w_k^j = \sum_k {}_{t-1}w_k^j = 1$.

On the other hand, by using a decomposition approach and considering country l as the reference country, we can compare APPs (and in a similar way PPEs or PPPs) over time in order to understand the influencing factors, which refer to the variations from time $T-1$ to time T of the APPs comparing the price levels of two countries calculated at time $t-1$ and time t:

$$ \frac{APP^T{}_{l,j}}{APP^{T-1}{}_{l,j}} = \frac{\prod_{k=1}^{n_{jl}} \left(p_k^{j,T}\right)^{w_k^{l,T}} \Big/ \prod_{k=1}^{n_{jl}} \left(p_k^{l,T}\right)^{w_k^{l,T}}}{\prod_{k=1}^{n_{jl}} \left(p_k^{j,T-1}\right)^{w_k^{l,T-1}} \Big/ \prod_{k=1}^{n_{jl}} \left(p_k^{l,T-1}\right)^{w_k^{l,T-1}}} $$

4.1 Comparing CPIs Across Space

Considering CPIs calculated in fairly homogeneous countries (for example at territorial level across different areas in the same country) it is reasonable to assume that the products purchased are the same (number and characteristics) in the two areas compared.[6]

By using the decomposition methods, suggested in (Biggeri and Giommi, 1987; Biggeri, Brunetti & Laureti, 2008) the divergences between the CPI for country j, ${}_{t-1}\mathbf{P}_t^j$, and the CPI for country l as reference country, ${}_{t-1}\mathbf{P}_t^l$, for each aggregation level, can be decomposed as follows[7] :

$$ {}_{t-1}\mathbf{P}_t^j - {}_{t-1}\mathbf{P}_t^l = \sum_k {}_{t-1}w_k^l \left({}_{t-1}P_{k,t}^j - {}_{t-1}P_{k,t}^l \right) + \sum_k {}_{t-1}P_{k,t}^j \cdot \left({}_{t-1}w_k^j - {}_{t-1}w_k^l \right) \quad (5) $$

It is clear that a divergence emerging from a comparison between the CPIs referring to the two countries depends on two main factors:

- the different evolution of the prices of the products and services (elementary price index effect), which is expressed by the first factor on the right hand side.
- the differences regarding the behaviour of consumers in their purchases, that is on the share of the expenditure devoted to the different products and services (weight effect).

By introducing the variables $\delta_k = \left({}_rP_{k,t}^j - {}_rP_{k,t}^l \right)$and $d_k = \left({}_rw_k^j - {}_rw_k^l \right)$, which express the difference between the sets of elementary price indices in country j and l and the differences between the expenditure weights respectively, after applying simple algebra, we can identify the various factors which influence the two effects[8]:

$$ {}_r\mathbf{P}_t^j - {}_r\mathbf{P}_t^l = \left[\overline{\delta} + \left(n \cdot s_{w^l} \cdot s_\delta \cdot R_{w^l,\delta} \right) \right] + \left(n \cdot s_{P^j} \cdot s_d \cdot R_{P^j,d} \right) \quad (5\text{bis}) $$

The first factor on the right hand side of (5 bis), that is the elementary price effect $\left[\overline{\delta} + \left(n \cdot s_{w^l} \cdot s_\delta \cdot R_{w^l,\delta} \right) \right]$, is influenced by $\overline{\delta} = \frac{1}{n} \sum_k \delta_k$ the distance between the centres of the two distributions of elementary price indices, s_{w^l}, the standard deviation of the weights, by s_δ the standard deviation of the elementary price index

[6] If this hypothesis is not satisfied the results of the decomposition are approximate.

[7] We must underline that by applying similar procedures and considering area j as reference area, after some simple algebra, we can obtain four different forms of the decomposition of the CPI differences, that give two estimations of the effects which differ slightly (see Biggeri et al. 2008). In actual fact a unique measure of the difference could be achieved but it is irrelevant to the aim of this paper so we will leave this issue for further development.

[8] Once again by applying similar procedure we can obtain similar decomposition forms by changing the reference country. However the results may differ slightly.

differences and by $R_{w^l,\delta}$, the linear correlation coefficient between the weighting system of the index in the country l and the difference in the elementary indices in the two countries.

From an economic point of view the factor $\overline{\delta} = \frac{1}{n}\sum_k {}_rP^j_{tk} - \frac{1}{n}\sum_k {}_rP^l_{tk} = {}_r\overline{P}^j_t - {}_r\overline{P}^l_t$ plays an important role both in determining the "price effect" and influencing the overall difference between the two CPIs considered.

This factor expresses the differences between the unweighted arithmetic mean of the period r to t price relatives, which is the Carli index and can be considered the best estimator when the price change distribution is Normal (Roger, 2000). Therefore, the divergence between the CPIs of two different countries, and the evaluation of the degree of the influence of the factors in which it is decomposed, depends on the shape of the distribution of the two elementary indices compared. Departures from normality can arise from either kurtosis, skewness, or a combination of both.

When the price change distribution is negatively skewed in country j and symmetric or positively skewed in country l, for example, the overall difference between the CPIs will be mainly influenced by the "weight effect" and in particular by the correlation between price changes and weights. In this case the value of $\overline{\delta}$ is negative and this shows that the products which are more widely consumed and therefore have a higher expenditure weight, experience a major increase in price and this has a rising effect on the aggregate price index.

On the other hand, the "weight effect" $\left(n \cdot s_{Pj} \cdot s_d \cdot R_{Pj,d}\right)$ is determined by similar factors, that is by s_{Pj}, the standard deviations of elementary indices of country j, by s_d, the standard deviation of the difference between the weights in the two countries compared and $R_{Pj,d}$, the linear correlation coefficient between the elementary price indices and the difference in the corresponding weights.

Once again statistical measures (standard deviation, central tendency and correlation coefficient) are obtained which express the different characteristics of the price change distributions and the consumers' behaviour in the two countries. By understanding how the dispersion of the elementary price index distribution affects the difference between CPIs at territorial level we get an important insight of the behaviour of consumers and the process of inflation.

4.2 Comparing APPs Over Time

The movement in the spatial price indices referring to countries j and l from time $T-1$ to time T is expressed by:

$$\frac{APP^T{}_{l,j}}{APP^{T-1}{}_{l,j}} = \frac{\prod_{k=1}^{n_{jl}} \left(p_k^{j,T}\right)^{w_k^{l,T}} \Big/ \prod_{k=1}^{n_{jl}} \left(p_k^{l,T}\right)^{w_k^{l,T}}}{\prod_{k=1}^{n_{jl}} \left(p_k^{j,T-1}\right)^{w_k^{l,T-1}} \Big/ \prod_{k=1}^{n_{jl}} \left(p_k^{l,T-1}\right)^{w_k^{l,T-1}}} \tag{6}$$

By taking the natural logarithms and by adding and subtracting the ratio between the two hybrid means (for countries j and l) obtained for each country by using the weights of the base period $T-1$ and the price of the corresponding country at time T, we can state that:

$$\begin{aligned}
\ln\left(\frac{APP^{T}{}_{l,j}}{APP^{T-1}{}_{l,j}}\right) &= \left[\sum_{k=1}^{n_{jl}} w_k^{j,T}\ln\left(p_k^{j,T}\right) - \sum_{k=1}^{n_{jl}} w_k^{l,T}\ln\left(p_k^{l,T}\right)\right] \\
&- \left[\sum_{k=1}^{n_{jl}} w_k^{j,T-1}\ln\left(p_k^{j,T-1}\right) - \sum_{k=1}^{n_{jl}} w_k^{l,T-1}\ln\left(p_k^{l,T-1}\right)\right] + \\
&+ \left[\sum_{k=1}^{n_{jl}} w_k^{j,T-1}\ln\left(p_k^{j,T}\right) - \sum_{k=1}^{n_{jl}} w_k^{l,T-1}\ln\left(p_k^{l}\right)\right] \\
&- \left[\sum_{k=1}^{n_{jl}} w_k^{j,T-1}\ln\left(p_k^{j,T}\right) - \sum_{k=1}^{n_{jl}} w_k^{l,T-1}\ln\left(p_k^{l}\right)\right]
\end{aligned}$$

By applying the exponential function, after simple algebra the above decomposition can be equivalently expressed as:

$$\frac{APP^{T}{}_{l,j}}{APP^{T-1}{}_{l,j}} = \frac{\prod_{k=1}^{n_{jl}}\left(p_k^{j,T}/p_k^{j,T-1}\right)^{w_k^{j,T-1}}}{\prod_{k=1}^{n_{jl}}\left(p_k^{l,T}/p_k^{l,T-1}\right)^{w_k^{l,T-1}}} \cdot \frac{\prod_{k=1}^{n_{jl}}\left(p_k^{j,T}\right)^{w_k^{j,T}-w_k^{j,T-1}}}{\prod_{k=1}^{n_{jl}}\left(p_k^{l,T}\right)^{w_k^{l,T}-w_k^{l,T-1}}} \tag{7}$$

The first product on the right hand side of (7) represents the divergence between the movement in price changes from time $T-1$ to time T in the two countries compared *(price effect)*.

In fact, this factor is the ratio between two consumer price indices in the two countries j and l, calculated following the weighted Jevons index using the weights of each country.

From an economic point of view, by examining this formula it is possible to understand the link between temporal and spatial variations of consumer prices and to measure to what extent the changes in price levels over time, measured by the CPIs, influence the changes over time of the spatial indices computed at time T and $T-1$.

The second product, which refers to the *weight effect (WE)*, is related to the impact of the difference in consumer behaviour in the two countries.

In order to improve our interpretation of the influence of the various factors, we suggest a further decomposition of the price and weight effect in formula (7).

Considering the price effect in the APP movement from time $T-1$ to time T, after applying simple algebra, we obtain the following decomposition:

$$\frac{\prod_{k=1}^{n_{jl}}\left(p_k^{j,T}/p_k^{j,T-1}\right)^{w_k^{j,T-1}}}{\prod_{k=1}^{n_{jl}}\left(p_k^{l,T}/p_k^{l,T-1}\right)^{w_k^{l,T-1}}} = \prod_{k=1}^{n_{jl}}\left(\frac{p_k^{j,T}/p_k^{j,T-1}}{p_k^{l,T}/p_k^{l,T-1}}\right)^{w_k^{l,T-1}} \cdot \prod_{k=1}^{n_{jl}}\left(\frac{p_k^{j,T}}{p_k^{j,T-1}}\right)^{w_k^{j,T-1}-w_k^{l,T-1}} \tag{8}$$

In order to obtain an equivalent decomposition form of (8), which highlights various statistical measure concerning the log- price ratios and weights in the two countries in question, we denote $\eta_k = \left[\ln\left(\frac{p_k^{j,T}}{p_k^{j,T-1}}\right) - \ln\left(\frac{p_k^{l,T}}{p_k^{l,T-1}}\right)\right]$ the differences between the set of logarithms of elementary price indices in countries j and l, and $v_k = \left(w_k^{j,T-1} - w_k^{l,T-1}\right)$ the difference between expenditure weights in the two countries at time $T-1$ decomposition (8) is equivalently expressed as:

$$\frac{\prod_{k=1}^{n_{jl}}\left(p_k^{j,T}/p_k^{j,T-1}\right)^{w_k^{j,T-1}}}{\prod_{k=1}^{n_{jl}}\left(p_k^{l,T}/p_k^{l,T-1}\right)^{w_k^{l,T-1}}} = \exp\left(n_{ij}\cdot s_{w_k^{l,T-1}}\cdot s_\eta \cdot R_{w_k^{l,T-1},\eta} + \overline{\eta}\right) \cdot \exp\left(n_{ij}\cdot s_{\ln\frac{p_k^{j,T}}{p_k^{j,T-1}}}\cdot s_v \cdot R_{\ln\frac{p_k^{j,T}}{p_k^{j,T-1}},v}\right) \tag{8bis}$$

Regarding the weight effect in the APP movements, a similar procedure can be applied in order to obtain the following decomposition form:

$$\frac{\prod_{k=1}^{n_{jl}}\left(p_k^{j,T}\right)^{w_k^{j,T}-w_k^{j,T-1}}}{\prod_{k=1}^{n_{jl}}\left(p_k^{l,T}\right)^{w_k^{l,T}-w_k^{l,T-1}}} = \prod_{k=1}^{n_{jl}}\left(\frac{p_k^{j,T}}{p_k^{l,T}}\right)^{w_k^{j,T}-w_k^{j,T-1}} \cdot \prod_{k=1}^{n_{jl}}\left(p_k^{l,T}\right)^{\left[\left(w_k^{j,T}-w_k^{j,T-1}\right)-\left(w_k^{l,T}-w_k^{l,T-1}\right)\right]} \tag{9}$$

By expressing $g_k = \left(w_k^{j,T} - w_k^{j,T-1}\right)$ the difference between expenditure weights in country j over the period from $T-1$ to T and $h_k = \left[\left(w_k^{j,T} - w_k^{j,T-1}\right) - \left(w_k^{l,T} - w_k^{l,T-1}\right)\right]$ the difference between the movements in the weighting system in country j and l respectively, we obtain:

$$\frac{\prod_{k=1}^{n_{jl}}\left(p_k^{j,T}\right)^{w_k^{j,T}-w_k^{j,T-1}}}{\prod_{k=1}^{n_{jl}}\left(p_k^{l,T}\right)^{w_k^{l,T}-w_k^{l,T-1}}} = \exp\left(n_{ij}\cdot s_{\ln\frac{p_k^{j,T}}{p_k^{l,T}}}\cdot s_g \cdot R_{\ln\frac{p_k^{j,T}}{p_k^{l,T}},g}\right)\cdot\exp\left(n_{ij}\cdot s_{\ln p_k^{l,T}}\cdot s_h \cdot R_{\ln p_k^{l,T},h}\right) \tag{9bis}$$

As clearly shown in the formulae above the factors affecting price and expenditure share changes are once again statistical measures concerning the characteristics of price and weight distributions.

It is worth noting that similar decompositions can be obtained for the PPP and WE indices.

In short, the comparisons of the CPIs across countries and of the PPPs over time show that the differences and ratios can be decomposed into two components which refer to the prices and to the system of weights respectively. The two components are affected by the same factors which can be described by using statistical measures concerning the distributions of price changes and of the consumption expenditure shares in the two countries. The smaller the share of typical products the higher the levels of accuracy and reliability of the comparisons (of the CPIs across countries and of the PPPs over time). If the formulae used for the CPIs and PPPs were the same (for example Jevons type formula) the comparisons and the links would be exact.

5 Concluding Remarks

It is difficult and expensive to integrate price data collection for computing CPIs and PPPs that allow a coherent comparison between CPIs and the change of PPPs level at the same time. From a theoretical and practical point of view this paper presents a more integrated approach for the computation of CPIs and PPPs, which involves inter-temporal and inter-country comparisons of consumer prices.

Regarding the debated issue of integration, we have illustrated the feasibility of integrating PPP and CPI activities undertaken by NSOs. At the same time we have underlined the benefits and examined the problems which arise when developing an integrated approach for the collection of the required information.

We have pointed out that the identity product principle used in the computation of the PPPs does not guarantee the representativeness of the chosen basket for both the countries involved in the comparison. A broader definition of the comparability of products increases the comparability between CPIs and PPPs and gives possibilities and advantages which may require more data (i.e. the expenditure write weight data within the elementary aggregates should be collected or evaluated, at least in a reference year) in order to verify if these advantages truly exist. We have suggested a simple statistical method for investigating the advantage of broadening the definition of comparability thus including additional products in the PPP calculation. This can be done by computing binary spatial indices for the comparison of consumer price levels between the two countries, through the ratio of the weighted geometric mean prices of the two countries, and then by decomposing them according to the different sub-sets of products with various degrees of comparability (strictly comparable, less comparable and characteristic or typical of each country). Besides, the computed spatial indices could be decomposed to evaluate the importance of the different factors that affect their value - price effect, consumption basket (weighting) effect and characteristicity effect. Whenever the NSOs collect all the necessary information on prices and weights at a product level within the elementary aggregate to compute these indices at least in one reference period, we can obtain valuable information which is useful for deciding the optimal number and the specific characteristic of the items to be included in the computation of the PPPs.

Concerning the results of the comparisons between CPIs and PPPs, at present it is not possible to fully integrate and link the commonly computed CPIs and PPPs. However, it is possible to obtain some insight on the factors that affect the differences between them. In fact, the formulae can be decomposed in order to approximately measure the factors (essentially due to the evolution of prices and to the share of consumption expenditure concerning the different products and services) which explain the divergences between the CPIs of the two countries from time $t-1$ to time t, and the movement of the PPPs concerning the two countries in the same period. These decompositions give us a better understanding of the links between the CPIs and PPPs from an economic point of view. However, as shown in the last part of this paper, the links between the CPIs and PPPs and their explanation would only be exact if weighted geometric means were also used for the computations of CPIs.

In conclusion, the integration and comparison between CPIs and PPPs are certainly feasible and methods can be found for achieving our objective even if it is not always easy, and above all it is necessary to carry out more research on these issues at an international level in order to agree on a broader definition of comparability of products for the computation of the PPPs by using the analyses that we have suggested in this paper.

References

Aten, B. (2005). Report on Interarea Price Levels, 2003, Working Paper No. 2005-11, November 30, 2005 Bureau of Economic Analysis, 1–56.

Aten, B. (2006). Interarea Price Levels: An Experimental Methodology, Monthly Labor Review, September, 129, 9, 47–61.

Balk, B. M. (1996). A comparison of ten methods for multilateral international price comparisons. *Journal of Official Statistics, 12*, 199–222.

Balk, B. M. (2001). *Aggregation methods in international comparisons: What have we learned?* Joint World Bank-OECD Seminar on Purchasing Power Parities, Washington, DC.

Biggeri, L., & Giommi, A. (1987). *On the accuracy and precision of the consumer price indices. Methods and applications to evaluate the influence of the sampling of households.* Proceedings of the 46th Session of the ISI (International Statistical Institute), Book 2, Tokyo, 137–154.

Biggeri, L., Brunetti, A., & Laureti, T. (2008, May 8–9). *The interpretation of the divergences between CPIs at territorial level: Evidence from Italy.* Paper presented at the Joint UNECE/ILO meeting on Consumer Price Indices, Geneva.

Biggeri, L., De Carli, R., & Laureti, T. (2008, May 8–9). *The interpretation of the PPPs: A method for measuring the factors that affect the comparisons and the integration with the CPI work at regional level.* Paper presented at the Joint UNECE/ILO meeting on Consumer Price Indices, Geneva.

Bretell, S., & Gardiner, B. (2002, August 27–31). *The development of a system of European regional purchasing power parities.* Dortmund: European Regional Science Association Congress.

Castles, I. (1997). *The OECD-EUROSTAT PPP program: Review of practice and procedures*, Paris: OECD.

De Carli, R. (2008, May 8–9). *An experiment to calculate PPPs at regional level in Italy: Procedure adopted and analyses of the results*. Paper presented at the Joint UNECE/ILO meeting on Consumer Price Indices, Geneva.

Diewert, W. E. (1993). *Test approaches to international comparisons, Essays in index number theory* (Vol. I), W. E. Diewert & A. O. Nakamura (Eds.), Amsterdam: North-Holland.

Diewert, W. E. (2008). New Methodology for Linking Regional PPPS, ICP Bulletin, 5, 2, August, 1, 10–21.

Ferrari, G., Laureti, T. & Mostacci, F. (2005, November). Time-space harmonization of consumer price indexes in Euro-Zone countries. *International Advances in Economic Research, 11*(4), 2005.

Heston, A. W. (1996). Some problems in item Price Comparisons with Special Reference to uses of CPI Prices in Estimating Spatial Heading Parities. In Improving the Quality of Price Indices: CPI and PPP, EUROSTAT, Proceedings of the International Seminar held in Florence, December 18–20, 1995.

Hill, R. (1997). A Taxonomy of multilateral methods for making international comparisons of prices and quantities. *Review of Income and Wealth, 43*(1), 49–69.

ILO/IMF/OECD/UNECE/Eurostat & The World Bank. (2004). *Consumer Price Index Manual: Theory and Practice*, Peter Hill (ed.), Geneva: International Labour Office.

Kravis, I. B., Kenessey, Z., & Heston, A. W. (1975). *A system of international comparisons of gross product and purchasing power*. Baltimore: Johns Hopkins University Press.

Krijnse Locker, H. K. (1984). On the estimation of purchasing power parities on the basic heading level. *Review of Income and Wealth, 30*(2), 135–152.

OECD-Eurostat. (2006). *Methodological manual on purchasing power parities*, Luxembourg. An electronic version of the manual can be found at the web site of Eurostat.

Melser, D., & Hill, R. (2005). *Developing a Methodology for Constructing Spatial Cost of Living Indexes*. A paper of statistics New Zealand.

Rao, D. S. Prasada (2001a). *Integration of CPI and PPP: Methodological issues, feasibility and recommendations*. Joint World Bank-OECD Seminar on PPP, Washington, DC.

Rao, D. S. Prasada (2001b). *Weighted EKS and generalised CPD methods for aggregation at basic heading level and above basic heading level*. Joint World Bank-OECD Seminar on PPP, Washington, DC.

Roger, S. (2000). *Relative prices, inflation and core inflation* (International Monetary Fund (IMF) Staff Working Paper WP/00/58). Washington, DC: IMF.

Ryten, J. (1998). *The Evaluation of the International Comparison Project (ICP)*. Washington, DC: IMF.

Wingfield, D., Fenwick, D., & Smith, K. (2005, February). Relative regional consumer price levels in 2004, Office for National Statistics, Economic Trends 615.

World Bank. (2007). *International Comparison Program Handbook*, it is possible to consult the up-dated version of the Handbook in the web site of the World Bank.

Retrospective Approximations of Superlative Price Indexes for Years Where Expenditure Data Is Unavailable

Jan de Haan, Bert M. Balk, and Carsten Boldsen Hansen

1 Introduction

The computation of superlative price index numbers is hampered by the fact that this requires quantity or expenditure data for current periods, whereas such data usually become available with considerable time lags. Statistical agencies may want to inform the public about the substitution bias of their Consumer Price Index (CPI) by calculating superlative price index numbers retrospectively. In countries that revise the CPI weights every year, calculating annual superlative index numbers retrospectively is a simple exercise. However, the majority of countries revise the weights less frequently, and their CPIs are typically Lowe indexes where the expenditure weights are fixed for several years. Though detailed expenditure data is lacking for the years in between consecutive weight-reference years, the question can be asked: would it be possible to interpolate superlative price indexes for such intermediate years? This is indeed the case: using a Constant Elasticity of Substitution (CES) framework, a superlative price index can be approximated once we have estimated the elasticity of substitution. The Lloyd–Moulton price index does not make use of current-period expenditure data, so it is even possible to approximate a superlative index in real time and extrapolate the time series.

In this paper we present several alternative methods which make use of all the available data and approximate a consistent time series of annual superlative price index numbers. The idea – which may be appealing to statistical agencies that are reluctant to rely on the assumptions underlying the CES theory – is to approximate the expenditure shares relating to the intermediate years by using linear combinations of expenditure shares from the weight-reference years. Our main aim

J. de Haan (✉)
Statistics Netherlands, Division of Macro-Economic Statistics and Dissemination, JM Voorburg, The Netherlands
e-mail: j.dehaan@cbs.nl

The views expressed in this chapter are those of the authors and do not necessarily reflect the views of their organizations. We thank Martin B. Larsen (Statistics Denmark) for preparing the data set and David Vellekoop for research assistance.

L. Biggeri, G. Ferrari (eds.), *Price Indexes in Time and Space*, Contributions to Statistics, DOI 10.1007/978-3-7908-2140-6_2, © Springer-Verlag Berlin Heidelberg 2010

is to clarify some issues that arise when approximating retrospectively a superlative price index. Furthermore, we argue that the methods applied by a number of researchers could be improved.

By way of introducing the subject, in Sect. 2 we recall that the Fisher and Törnqvist indexes are instances of a general class of superlative indexes and show how the Lloyd–Moulton index fits in. In Sect. 3 we describe our approximation methods as well as the Lloyd–Moulton method. The approximations are generalized in Sect. 4 to the case when the price reference period differs from the weight reference period (as with Lowe CPIs). Section 5 extends the analysis to three or more weight-reference years and suggests chain linking during these years. Section 6 provides an illustrative example based on data that have been used for compiling the Danish CPI. Section 7 concludes.

2 Superlative and Lloyd–Moulton Price Indexes

The Quadratic Mean (QM) of order r price index was defined by Diewert (1976) as

$$P_{\mathrm{QM}}^{0t}(r) \equiv \left[\frac{\sum_i s_i^0 \left(p_i^t/p_i^0\right)^{r/2}}{\sum_i s_i^t \left(p_i^t/p_i^0\right)^{-r/2}} \right]^{1/r} = \left\{ \left[\sum_i s_i^0 \left(p_i^t/p_i^0\right)^{r/2} \right]^{2/r} \left[\sum_i s_i^t \left(p_i^t/p_i^0\right)^{-r/2} \right]^{-2/r} \right\}^{1/2} \quad (r \neq 0), \tag{1}$$

where p_i^0, p_i^t and s_i^0, s_i^t denote the price and expenditure share of commodity i in base period 0 and current or comparison period t ($t > 0$), respectively. It is a superlative index. By setting $r = 2(1-\sigma)$ expression (1) becomes

$$P_{\mathrm{QM}}^{0t}(2(1-\sigma)) = \left\{ \left[\sum_i s_i^0 \left(p_i^t/p_i^0\right)^{1-\sigma} \right]^{1/(1-\sigma)} \times \left[\sum_i s_i^t \left(p_i^t/p_i^0\right)^{-(1-\sigma)} \right]^{-1/(1-\sigma)} \right\}^{1/2}, \tag{2}$$

which is the geometric mean of the Lloyd (1975)–Moulton (1996) price index

$$P_{\mathrm{LM}}^{0t}(\sigma) = \left[\sum_i s_i^0 \left(p_i^t/p_i^0\right)^{1-\sigma} \right]^{1/(1-\sigma)} \tag{3}$$

and its "current weight (CW) counterpart"

$$P_{\mathrm{CW}}^{0t}(\sigma) = \left[\sum_i s_i^t \left(p_i^t/p_i^0\right)^{-(1-\sigma)}\right]^{-1/(1-\sigma)}. \tag{4}$$

The price index $P_{\mathrm{LM}}^{0t}(\sigma)$ monotonically decreases and $P_{\mathrm{CW}}^{0t}(\sigma)$ monotonically increases as σ increases, which implies that there exists a unique value σ^{0t} such that[1]

$$P_{\mathrm{LM}}^{0t}\left(\sigma^{0t}\right) = P_{\mathrm{CW}}^{0t}\left(\sigma^{0t}\right) = P_{\mathrm{QM}}^{0t}\left(2\left(1-\sigma^{0t}\right)\right). \tag{5}$$

Thus for $\sigma = \sigma^{0t}$ the Lloyd–Moulton index becomes superlative. The drawback of (5) is of course that, unless σ^{0t} happens to be constant over time, we would be using different superlative index number formulas for different periods.

For $\sigma = 0$ the Lloyd–Moulton price index and its CW counterpart reduce to the Laspeyres and Paasche price indexes, respectively, and the QM index reduces to the Fisher price index

$$P_{\mathrm{F}}^{0t} = \left\{\left[\sum_i s_i^0 \left(p_i^t/p_i^0\right)\right]\left[\sum_i s_i^t \left(p_i^0/p_i^t\right)\right]^{-1}\right\}^{1/2} = \left\{\left[\frac{\sum_i p_i^t q_i^0}{\sum_i p_i^0 q_i^0}\right]\left[\frac{\sum_i p_i^t q_i^t}{\sum_i p_i^0 q_i^t}\right]\right\}^{1/2}, \tag{6}$$

where q_i^0 and q_i^t denote the quantities consumed or purchased in periods 0 and t, respectively. If we replace the arithmetic averages of the price relatives in Eq. (6) by corresponding geometric averages then we obtain the Törnqvist index

$$P_{\mathrm{T}}^{0t} = \left\{\left[\prod_i \left(p_i^t/p_i^0\right)^{s_i^0}\right]\left[\prod_i \left(p_i^0/p_i^t\right)^{s_i^t}\right]^{-1}\right\}^{1/2} = \prod_i \left(p_i^t/p_i^0\right)^{(s_i^0+s_i^t)/2}. \tag{7}$$

As a matter of fact this index would also be obtained if in expression (2) the arithmetic averages were replaced by geometric averages, so this "trick" is independent of σ. Notice that the QM index is not defined for $\sigma = 1$. It can be shown that for $\sigma \to 1$ $P_{\mathrm{LM}}^{0t}(\sigma)$ and $P_{\mathrm{CW}}^{0t}(\sigma)$ tend to the Geometric Laspeyres and Paasche price indexes, so that $P_{\mathrm{QM}}^{0t}(2(1-\sigma))$ tends to the Törnqvist price index.

[1] The Lloyd–Moulton index is a generalized mean of order $1-\sigma$, which is strictly increasing in $1-\sigma$ and thus strictly decreasing in σ. Its "current-weight counterpart" can be written as the inverse of a generalized mean of order $1-\sigma$ and is thus strictly increasing in σ. The solution σ^{0t} must be obtained by some numerical method.

In empirical studies, particularly when the price and quantity data exhibit smooth trends, the differences between Fisher or Törnqvist index numbers are often negligible. This seems to corroborate Diewert's (1978, p. 884) finding that "all superlative indexes closely approximate each other".[2]

3 Approximating Superlative Price Indexes

We now consider two distant years 0 and T and one or more intermediate years (years in between 0 and T). Suppose that price data for all $t = 0, \ldots, T$ are known, and expenditure shares for years 0 and T, but that the expenditures shares for the intermediate years are unavailable. This will be the case for countries that do not annually revise their weights in the CPI but instead revise them, say, every three to five years. Suppose 0 and T are those weight-reference or benchmark years. Obviously, superlative price indexes cannot be computed for intermediate years $t = 1, \ldots, T-1$. The problem addressed here is how to approximate, or interpolate, superlative price indexes, for example P_{F}^{0t} or P_{T}^{0t}, given the lack of expenditure data.

3.1 Using Lloyd–Moulton Price Indexes

A first possibility would be estimating Lloyd–Moulton price index numbers. As shown by Eq. (5), for $\sigma = \sigma^{0t}$ the Lloyd–Moulton formula produces a superlative index going from 0 to t. Due to the unavailability of data, we cannot compute σ^{0t}, but we can compute σ^{0T} and then assume that $\sigma^{0t} \cong \sigma^{0T}$ for $t = 1, \ldots, T-1$. That is, we assume σ^{0t} (which makes the Lloyd–Moulton index equal to the CW index) to be constant over time. Since the Lloyd–Moulton index $P_{\mathrm{LM}}^{0T}(\sigma^{0T})$ will be numerically close to Fisher or Törnqvist indexes, we could also compute the value of σ for which $P_{\mathrm{LM}}^{0T}(\sigma)$ is equal to P_{F}^{0T} or P_{T}^{0T}. The last method was used by Shapiro and Wilcox (1997) and is suggested in the international CPI Manual (ILO et al., 2004).

Assuming constancy of the parameter σ is consistent with a Constant Elasticity of Substitution (CES) framework in which σ figures as the elasticity of substitution, which is assumed to be the same for all pairs of commodities. Balk (2000) proposed a two-level, nested CES approach: at the upper aggregation level there is a fixed set of product groups (strata, or elementary aggregates), whereas at the lower level (that is, within the strata or elementary aggregates) the set of commodities is

[2] Yet, not all the superlative price indexes are necessarily numerically similar. The problem is that "as the parameter r increases in absolute value, the superlative price (quantity) index number formula becomes increasingly sensitive to outliers in the price-relatives (quantity-relatives) distribution" (Hill, 2006, p. 38). Anyway, for small absolute values of r, which is the usual case, we do expect small numerical differences between different superlative indexes.

allowed to change over time.[3] An interesting result is that the value of the elasticity of substitution should be less than 1 at the upper level but greater than 1 at the lower level. In this chapter we are dealing with the upper level, hence expect a value of σ less than 1. The estimated value will depend on the actual aggregation level. Shapiro and Wilcox (1997), who employed a US data set consisting of 9,108 item-area strata, found that a value of 0.7 generated price index numbers very similar to those computed with the Törnqvist formula.

3.2 Using Estimated Expenditure Shares

A more statistically-oriented approach to approximating a superlative price index is the following.[4] Suppose that expenditure shares exhibit reasonably smooth trends. If year t is close to year 0 we would expect s_i^t to be close to s_i^0; moving from benchmark year 0 to benchmark year T the expenditure share s_i^t will move toward s_i^T. This suggests that we approximate s_i^t by a moving linear combination of s_i^0 and s_i^T:

$$\hat{s}_i^t = \left[ts_i^T + (T-t)s_i^0\right]/T = (t/T)s_i^T + (1-t/T)s_i^0 \qquad (t = 0,1,\ldots,T), \qquad (8)$$

which is a weighted mean of s_i^0 and s_i^T with t/T and $1-t/T$ as weights. The limiting cases of (8) are $\hat{s}_i^t = s_i^0$ and $\hat{s}_i^t = s_i^T$. Thus it is rather natural to approximate the Fisher price index for year t by

$$\hat{P}_{\mathrm{F}}^{0t} = \left\{\left[\sum_i s_i^0\left(p_i^t/p_i^0\right)\right]\left[\sum_i \hat{s}_i^t\left(p_i^0/p_i^t\right)\right]^{-1}\right\}^{1/2} \qquad (9)$$

and the Törnqvist price index by

$$\hat{P}_{\mathrm{T}}^{0t} = \left\{\left[\prod_i \left(p_i^t/p_i^0\right)^{s_i^0}\right]\left[\prod_i \left(p_i^0/p_i^t\right)^{\hat{s}_i^t}\right]^{-1}\right\}^{1/2} = \prod_i \left(p_i^t/p_i^0\right)^{(s_i^0+\hat{s}_i^t)/2}. \qquad (10)$$

Of course for $t = T$ we have $\hat{P}_{\mathrm{F}}^{0T} = P_{\mathrm{F}}^{0T}$ and $\hat{P}_{\mathrm{T}}^{0T} = P_{\mathrm{T}}^{0T}$.

A related alternative approach goes as follows. The average shares $(s_i^0 + \hat{s}_i^t)/2$ in (10) can be written as

$$\left(s_i^0 + \hat{s}_i^t\right)/2 = (1-t/2T)s_i^0 + (t/2T)s_i^T, \qquad (11)$$

[3]Balk (2000) addressed substitution effects as well as the treatment of new and disappearing goods in a nested CES price index. See De Haan (2005), Melser (2006) and Ivancic (2007) for empirical evidence on these topics at the lower aggregation level.

[4]This subsection draws heavily from an unpublished paper by Balk (1990a).

so that expression (10) becomes

$$\hat{P}_{\mathrm{T}}^{0t} = \left[\prod_i \left(p_i^t/p_i^0\right)^{s_i^0}\right]^{1-t/2T} \left[\prod_i \left(p_i^t/p_i^0\right)^{s_i^T}\right]^{t/2T}. \tag{12}$$

Substituting the identity

$$\prod_i \left(p_i^t/p_i^0\right)^{s_i^T} = \prod_i (p_i^t/p_i^T)^{s_i^T} \left[\prod_i \left(p_i^0/p_i^T\right)^{s_i^T}\right]^{-1} \tag{13}$$

into expression (12), we obtain

$$\hat{P}_{\mathrm{T}}^{0t} = \left[\prod_i \left(p_i^t/p_i^0\right)^{s_i^0}\right]^{1-t/2T} \left[\frac{\prod_i (p_i^t/p_i^T)^{s_i^T}}{\prod_i (p_i^0/p_i^T)^{s_i^T}}\right]^{t/2T}. \tag{14}$$

Now, replacing the geometric averages in expression (14) by arithmetic averages, using $s_i^\tau = p_i^\tau q_i^\tau / \sum_i p_i^\tau q_i^\tau \, (\tau = 0, T)$ and doing some rearranging, we define the Quasi Fisher (QF) index:

$$\begin{aligned} \hat{P}_{\mathrm{QF}}^{0t} &\equiv \left[\frac{\sum_i p_i^t q_i^0}{\sum_i p_i^0 q_i^0}\right]^{1-t/2T} \left[\frac{\sum_i p_i^t q_i^T}{\sum_i p_i^0 q_i^T}\right]^{t/2T} \\ &= \left[\sum_i s_i^0 \left(p_i^t/p_i^0\right)\right]^{1-t/2T} \left[\sum_i s_i^{T*} \left(p_i^t/p_i^0\right)\right]^{t/2T}, \end{aligned} \tag{15}$$

where $s_i^{T*} = (p_i^0/p_i^T)p_i^T q_i^T / \sum_i (p_i^0/p_i^T)p_i^T q_i^T = (p_i^0/p_i^T)s_i^T / \sum_i (p_i^0/p_i^T)s_i^T$ are price backdated expenditure shares. Expression (15) is a weighted geometric average of a Laspeyres price index and a Lowe price index. Since its limiting values are $\hat{P}_{\mathrm{QF}}^{00} = 1$ and $\hat{P}_{\mathrm{QF}}^{0T} = P_{\mathrm{F}}^{0T}$, $\hat{P}_{\mathrm{QF}}^{0t}$ should be seen as an approximation of the Fisher index.

Triplett (1989a) proposed the following approximation formula, called the Time-series Generalized Fisher Ideal (TGFI) index:

$$\begin{aligned} \hat{P}_{\mathrm{TGFI}}^{0t} &\equiv \left[\frac{\sum_i p_i^t q_i^0}{\sum_i p_i^0 q_i^0}\right]^{1/2} \left[\frac{\sum_i p_i^t q_i^T}{\sum_i p_i^0 q_i^T}\right]^{1/2} \\ &= \left[\sum_i s_i^0 \left(p_i^t/p_i^0\right)\right]^{1/2} \left[\sum_i s_i^{T*} \left(p_i^t/p_i^0\right)\right]^{1/2}. \end{aligned} \tag{16}$$

Whereas in (15) the exponents depend on t, they are fixed at the value 1/2 in (16).[5] Note that $1 - t/2T > 1/2$ and $t/2T < 1/2$ for $0 < t < T$. This suggests that the TGFI index is biased in the sense that it places too less weight on the first component, the Laspeyres index, and too much weight on the second component, the Lowe index. Note also that the Lowe index in (16) should be an approximation of the Paasche index if $\hat{P}^{0t}_{\mathrm{TGFI}}$ is meant to approximate the Fisher index. Under normal circumstances the Paasche index will be less than the Laspeyres. Thus, the TGFI index most likely understates the Fisher index if the Lowe index in (16), which has been given too much weight, will be less than the Paasche. Such a situation is likely to happen if long-run trends in relative price changes exist. To illustrate this point, suppose the price change between the benchmark years, p_i^T/p_i^0, of commodity i is relatively large. When relative price changes are persistent, the price change between year 0 and year t, p_i^t/p_i^0, is also relatively large as is the price change between year t and year T, p_i^t/p_i^T. Due to substitution effects we then expect $q_i^T < q_i^t$ for such commodities, and hence the Lowe index to be less than the Paasche index.

4 Lowe CPIs and Approximate Superlative Price Indexes

In practice it takes some time to compile the weighting scheme of a CPI. Consequently, the typical CPI is a Lowe index instead of a Laspeyres index, based on quantities or expenditure shares pertaining to one or more years preceding the index reference period. Let 0 be the quantity reference year and $b\,(b > 0)$ the price reference year. The Lowe CPI going from year b to year $t\,(t > b)$ with quantity reference year 0 is then

$$P_{\mathrm{L}}^{bt} = \frac{\sum_i p_i^t q_i^0}{\sum_i p_i^b q_i^0} = \sum_i s_i^{0*}\left(p_i^t/p_i^b\right), \tag{17}$$

where $s_i^{0*} = (p_i^b/p_i^0)p_i^0 q_i^0 / \sum_i (p_i^b/p_i^0)p_i^0 q_i^0 = (p_i^b/p_i^0)s_i^0 / \sum_i (p_i^b/p_i^0)s_i^0$ are price updated expenditure shares.[6] Now we show how the approximations of a superlative

[5] All indexes are "forward looking"; they are going from benchmark year 0 to year t. Retrospectively we could also calculate "backward looking" indexes going from benchmark year T to year t; see also Triplett (1989b). Given some approximation method, the product of the forward looking index going from 0 to t and the inverse of the backward looking index going from T to t will in general not be equal to the index going from 0 to T.

[6] In some countries the CPI is a Young index, which results from replacing the price-updated expenditure shares in (17) by the actual shares of period 0. A Young index can be interpreted in different ways. For example, statistical agencies computing a Young index may target at a Laspeyres index with price and weight reference period b and use the period 0 shares as estimates of the period b shares. Essentially they assume that the elasticity of substitution between any two commodities is equal to 1. In Denmark the Young CPI is interpreted as an approximation of the (superlative) Walsh

index described in Sect. 3.2 can be generalized to the situation in which $0 < b < t$ $(t = 0,1,\ldots,T)$.

First, similar to expression (8) the expenditure shares pertaining to period b are approximated by

$$\hat{s}_i^b = (b/T)s_i^T + (1 - b/T)s_i^0. \tag{18}$$

Then, natural approximations of the Fisher and Törnqvist indexes, P_{F}^{bt} and P_{T}^{bt}, are

$$\hat{P}_{\mathrm{F}}^{bt} = \left\{ \left[\sum_i \hat{s}_i^b \left(p_i^t / p_i^b \right) \right] \left[\sum_i \hat{s}_i^t \left(p_i^b / p_i^t \right) \right]^{-1} \right\}^{1/2}, \tag{19}$$

$$\hat{P}_{\mathrm{T}}^{bt} = \prod_i \left(p_i^t / p_i^b \right)^{(\hat{s}_i^b + \hat{s}_i^t)/2}, \tag{20}$$

respectively.

Second, the average shares in (20) are equal to

$$\left(\hat{s}_i^b + \hat{s}_i^t \right)/2 = [1 - (b+t)/2T]s_i^0 + [(b+t)/2T]s_i^T, \tag{21}$$

so that (20) becomes

$$\hat{P}_{\mathrm{T}}^{bt} = \left[\prod_i \left(p_i^t / p_i^b \right)^{s_i^0} \right]^{1-(b+t)/2T} \left[\prod_i \left(p_i^t / p_i^b \right)^{s_i^T} \right]^{(b+t)/2T}. \tag{22}$$

Substituting

$$\prod_i \left(p_i^t / p_i^b \right)^{s_i^0} = \prod_i \left(p_i^t / p_i^0 \right)^{s_i^0} \left[\prod_i \left(p_i^b / p_i^0 \right)^{s_i^0} \right]^{-1} \tag{23}$$

and

$$\prod_i \left(p_i^t / p_i^b \right)^{s_i^T} = \prod_i \left(p_i^t / p_i^T \right)^{s_i^T} \left[\prod_i \left(p_i^b / p_i^T \right)^{s_i^T} \right]^{-1} \tag{24}$$

into (22), we obtain

---

index (Boldsen Hansen, 2006, 2007). In Sect. 6 we present evidence for Denmark on the value of the elasticity of substitution.

$$\hat{P}_{\mathrm{T}}^{bt} = \left[\frac{\prod_i (p_i^t/p_i^0)^{s_i^0}}{\prod_i (p_i^b/p_i^0)^{s_i^0}}\right]^{1-(b+t)/2T} \left[\frac{\prod_i (p_i^t/p_i^T)^{s_i^T}}{\prod_i (p_i^b/p_i^T)^{s_i^T}}\right]^{(b+t)/2T} . \tag{25}$$

Then, replacing all four geometric averages in (25) by arithmetic averages gives rise to the alternative (Quasi Fisher) approximation:

$$\begin{aligned}\hat{P}_{\mathrm{QF}}^{bt} &\equiv \left[\frac{\sum_i p_i^t q_i^0}{\sum_i p_i^b q_i^0}\right]^{1-(b+t)/2T} \left[\frac{\sum_i p_i^t q_i^T}{\sum_i p_i^b q_i^T}\right]^{(b+t)/2T} \\ &= \left[\sum_i s_i^{0*}\left(p_i^t/p_i^b\right)\right]^{1-(b+t)/2T} \left[\sum_i \tilde{s}_i^{T*}\left(p_i^t/p_i^b\right)\right]^{(b+t)/2T} .\end{aligned} \tag{26}$$

$\hat{P}_{\mathrm{QF}}^{bt}$ is a weighted geometric mean of the Lowe index (17) and a Lowe index based on price backdated shares $\tilde{s}_i^{T*} = (p_i^b/p_i^T)s_i^T / \sum_i (p_i^b/p_i^T)s_i^T$. Note that these shares differ from those in expression (15). However, (26) reduces to (15) for $b = 0$.

To obtain a price index going from b to t one could also use (15) two times and divide $\hat{P}_{\mathrm{QF}}^{0t}$ by $\hat{P}_{\mathrm{QF}}^{0b}$. However, the resulting index cannot be called an approximation of a superlative index and will most likely differ from $\hat{P}_{\mathrm{QF}}^{bt}$.

Finally, the unweighted counterpart to (26) is

$$\begin{aligned}\hat{P}_{\mathrm{TGFI}}^{bt} &\equiv \left[\frac{\sum_i p_i^t q_i^0}{\sum_i p_i^b q_i^0}\right]^{1/2} \left[\frac{\sum_i p_i^t q_i^T}{\sum_i p_i^b q_i^T}\right]^{1/2} \\ &= \left[\sum_i s_i^{0*}\left(p_i^t/p_i^b\right)\right]^{1/2} \left[\sum_i \tilde{s}_i^{T*}\left(p_i^t/p_i^b\right)\right]^{1/2} .\end{aligned} \tag{27}$$

Nimmo, Hayes, and Pike (2007) retrospectively approximated quarterly Fisher price index numbers for New Zealand from 2002 to 2006 but seem to have used a slightly adjusted version of expression (27).[7] Their first weight-reference period is not a calender year but a broken year 2000/2001, and these weights were price updated to the June 2002 quarter. Most interestingly, Nimmo et al. (2007) extended their time series beyond the second weight-reference period (2003/2004) to the June

[7]Unfortunately their description is not entirely clear on this point. Quarterly index numbers can indeed be calculated using (27). However, the computation of quarterly or monthly (approximate) superlative indexes may not be very useful as seasonality disturbs their interpretation. Diewert (2000) discusses the problems faced when constructing annual superlative index numbers using monthly price data. The US Bureau of Labor Statistics publishes an experimental monthly superlative index (see Cage et al., 2003).

2006 quarter.[8] In an annual framework, this implies extrapolating the time series by applying (27) to years $t > T$, which yields

$$\hat{P}^{bt}_{\text{TGFI}} = \left[\frac{\sum_i p_i^T q_i^0}{\sum_i p_i^b q_i^0}\right]^{1/2} \left[\frac{\sum_i p_i^T q_i^T}{\sum_i p_i^b q_i^T}\right]^{1/2} \left[\frac{\sum_i p_i^t q_i^0}{\sum_i p_i^T q_i^0}\right]^{1/2} \left[\frac{\sum_i p_i^t q_i^T}{\sum_i p_i^T q_i^T}\right]^{1/2}$$
$$= \hat{P}^{bT}_{\text{TGFI}} \left[\frac{\sum_i p_i^t q_i^0}{\sum_i p_i^T q_i^0}\right]^{1/2} \left[\frac{\sum_i p_i^t q_i^T}{\sum_i p_i^T q_i^T}\right]^{1/2} \tag{28}$$
$$= \hat{P}^{bT}_{\text{TGFI}} \left\{\left[\sum_i \tilde{s}_i^{0*} (p_i^t/p_i^T)\right]\left[\sum_i s_i^T (p_i^t/p_i^T)\right]\right\}^{1/2},$$

where $\tilde{s}_i^{0*} = (p_i^T/p_i^0)s_i^0 / \sum_i (p_i^T/p_i^0)s_i^0$ are price updated shares of benchmark year 0. The right-hand side of (28) is the product of two factors: $\hat{P}^{bT}_{\text{TGFI}}$, given by expression (27) for benchmark year T, and the unweighted geometric mean of two price indexes going from year T to year t, a Lowe index (based on year 0 quantities) and a Laspeyres index.[9] The first factor is an *unweighted* mean of two Lowe indexes instead of the preferred weighted mean. The second factor will generally be a biased measure of price change between year T and year t. It is obvious that $\hat{P}^{bt}_{\text{TGFI}}$, given by (28), cannot be considered as an approximation of a superlative index.

5 Three or More Benchmark Years

Except for the Lloyd–Moulton approach, a time series of approximate superlative index numbers between two benchmark years (weight-reference years) can only be extended when expenditure data from a third benchmark year become available. Suppose we have expenditure data of three benchmark years: 0, T^1 and T^2. Extending expression (8), expenditure shares for intermediate years t ($t = 0, \ldots, T^2$) are approximated as

$$\hat{s}_i^t = \left(t/T^1\right) s_i^{T^1} + \left(1 - t/T^1\right) s_i^0 \qquad \text{for} \qquad 0 \le t \le T^1; \tag{29a}$$

$$\hat{s}_i^t = \left[\left(t - T^1\right) / \left(T^2 - T^1\right)\right] s_i^{T^2} + \left[1 - \left(t - T^1\right) / \left(T^2 - T^1\right)\right] s_i^{T^1} \qquad \text{for} \quad T^1 < t \le T^2. \tag{29b}$$

[8] It should be mentioned that "in some cases adjustments were made to reflect quantity changes since 2003/2004" (Nimmo et al., 2007, p. 4).

[9] For a discussion of the substitution bias of a Lowe price index, see Balk and Diewert (2004).

There are two main scenarios for extending the time series to year T^2: (1) a direct index going from 0 to T^2, or (2) a chained index as the product of a direct index going from 0 to T^1 and a direct index going from T^1 to T^2.

Let us start with the first scenario. Using expressions (29a) and (29b) the natural approximations, given by (9) and (10), are still valid. The alternative approximation becomes a little bit more complicated. For $T^1 < t \leq T^2$ the average of year 0 and year t expenditure shares can be written as

$$\begin{aligned}\left(s_i^0 + \hat{s}_i^t\right)/2 = & s_i^0/2 + \left\{\left[1 - \left(t - T^1\right)/\left(T^2 - T^1\right)\right] s_i^{T^1}\right. \\ & \left. + \left[\left(t - T^1\right)/\left(T^2 - T^1\right)\right] s_i^{T^2}\right\}/2.\end{aligned} \tag{30}$$

A similar line of reasoning as in Sect. 3.2 then yields the following approximation of the Fisher index for $T^1 < t \leq T^2$:

$$\begin{aligned}\hat{P}^{0t}_{\mathrm{QF}} &= \left[\frac{\sum_i p_i^t q_i^0}{\sum_i p_i^0 q_i^0}\right]^{1/2} \left\{\left[\frac{\sum_i p_i^t q_i^{T^1}}{\sum_i p_i^0 q_i^{T^1}}\right]^{1-\frac{t-T^1}{T^2-T^1}} \left[\frac{\sum_i p_i^t q_i^{T^2}}{\sum_i p_i^0 q_i^{T^2}}\right]^{\frac{t-T^1}{T^2-T^1}}\right\}^{1/2} \\ &= \left[\sum_i s_i^0 \left(p_i^t/p_i^0\right)\right]^{1/2} \left\{\left[\sum_i s_i^{T^{1*}} \left(p_i^t/p_i^0\right)\right]^{1-\frac{t-T^1}{T^2-T^1}} \left[\sum_i s_i^{T^{2*}} \left(p_i^t/p_i^0\right)\right]^{\frac{t-T^1}{T^2-T^1}}\right\}^{1/2},\end{aligned} \tag{31}$$

where $s_i^{T^{1*}} = (p_i^0/p_i^{T^1})s_i^{T^1} / \sum_i (p_i^0/p_i^{T^1})s_i^{T^1}$ and $s_i^{T^{2*}} = (p_i^0/p_i^{T^2})s_i^{T^2} / \sum_i (p_i^0/p_i^{T^2})s_i^{T^2}$ are price backdated expenditure shares. The first component, between square brackets, at the right-hand side of (31) is a Laspeyres price index, and the factor between braces approximates a Paasche price index. Expression (31) can easily be extended to four or more benchmark years.

Another possibility to approximate a direct superlative index would be to act as if (benchmark) year T^1 was just another intermediate year and ignore the observed expenditure shares. Replacing actually observed expenditure shares by estimated values is clearly not advisable. We will nevertheless try it out in Sect. 6 to get an impression of how well the various approximation methods perform in case of two very distant benchmark years.

The second scenario is to calculate a chained version of $\hat{P}^{0t}_{\mathrm{QF}}$ for $T^1 < t \leq T^2$, defined by $\hat{P}^{0t}_{\mathrm{QF,chain}} \equiv \hat{P}^{0T^1}_{\mathrm{QF}} \hat{P}^{T^1 t}_{\mathrm{QF}}$. Using (15) and the fact that $\hat{P}^{0T^1}_{\mathrm{QF}} = P^{0T^1}_{\mathrm{F}}$, we obtain

$$\hat{P}^{0t}_{\mathrm{QF,chain}} = P^{0T^1}_{\mathrm{F}} \left[\frac{\sum_i p_i^t q_i^{T^1}}{\sum_i p_i^{T^1} q_i^{T^1}}\right]^{1-\frac{t-T^1}{2(T^2-T^1)}} \left[\frac{\sum_i p_i^t q_i^{T^2}}{\sum_i p_i^{T^1} q_i^{T^2}}\right]^{\frac{t-T^1}{2(T^2-T^1)}}. \tag{32}$$

$\hat{P}^{0t}_{\mathrm{QF}}$ and $\hat{P}^{0t}_{\mathrm{QF,chain}}$ will usually differ, just like the unknown Fisher index and its chained counterpart will differ, though in practice the differences might be limited. Chaining has practical advantages. For example, there is no need to price backdate the expenditure shares relating to T^1 and T^2 to year 0. Further, statistical agencies must regularly revise commodity classification schemes. Changes in the number of commodity groups at the upper level of aggregation, or in their definitions, would make the computation of direct index numbers problematic.[10]

Chaining is also useful when estimating retrospectively Lloyd–Moulton price index numbers. The use of a direct index combined with a fixed value for the elasticity of substitution σ for $0 \leq t \leq T^1$ *and* $T^1 < t \leq T^2$ means that the index numbers in year T^1 or in year T^2 (or both) will differ from the true Fisher or Törnqvist index numbers. While not entirely consistent with the CES theory, it seems better to estimate separate values for σ for $0 \leq t \leq T^1$ and $T^1 < t \leq T^2$ so that $P^{0T^1}_{\mathrm{LM}}(\sigma) = P^{0T^1}_{\mathrm{F}}$ (or alternatively $P^{0T^1}_{\mathrm{LM}}(\sigma) = P^{0T^1}_{\mathrm{T}}$) and $P^{T^1T^2}_{\mathrm{LM}}(\sigma) = P^{T^1T^2}_{\mathrm{F}}$ (or $P^{T^1T^2}_{\mathrm{LM}}(\sigma) = P^{T^1T^2}_{\mathrm{T}}$).

6 Data and Empirical Evidence

6.1 Some Facts and Figures

The various methods will be illustrated on building blocks for the official Danish CPI. Our data set concerns 444 elementary aggregates. Monthly price index numbers are available from January 1996 to December 2006 (1996 = 100), and expenditure shares (CPI weights) for the weight-reference years 1994, 1996, 1999, and 2003. During this ten-year period Statistics Denmark made a number of changes in the set of elementary aggregates. To establish a coherent data set that allows us to calculate direct as well as chained index numbers, some elementary aggregates have been left out and some have been merged. In a few cases price changes have been imputed from those of similar elementary aggregates. These modifications, however, have a limited effect because the elementary aggregates concerned have low weights in the CPI.[11] Annual price index numbers are computed as arithmetic means of the twelve monthly index numbers. Price index numbers for 1994 and 1995 (1994 = 100) are not available. Using the expenditure shares for 1996, 1999, and 2003, we calculate "true" direct and chained superlative price index numbers for 1999 and 2003 and approximate superlative index numbers for the intermediate years 1997, 1998, 2000, 2001, and 2002.

Table 1 contains Laspeyres, Paasche, Geometric Laspeyres, Geometric Paasche, Fisher and Törnqvist price index numbers for 1999 and 2003. Let us focus first on the direct indexes for 1999 (1996 = 100) and 2003 (1999 = 100) shown in the first

---

[10] As will be explained in Sect. 6, our data have been adjusted to account for such changes.

[11] The resulting 444 elementary aggregates account for over 98% of the total CPI weight. Boldsen Hansen (2007) has shown that re-calculating the Danish (Young) CPI with this data set produces index numbers that differ only marginally from the officially published figures.

Table 1 Direct and chained price index numbers, 1999 and 2003

	Direct indexes			Chained indexes (1996 = 100)	
	1996 = 100		1999 = 100		
	1999	2003	2003	1999	2003
Laspeyres	106.69	117.90	110.74	106.69	118.15
Paasche	106.00	115.27	109.40	106.00	115.96
Fisher	106.34	116.58	110.07	106.34	117.05
Geometric Laspeyres	106.38	116.54	109.96	106.38	116.97
Geometric Paasche	106.38	117.15	110.16	106.38	117.20
Törnqvist	106.38	116.85	110.06	106.38	117.08

and third column. As expected, the Laspeyres index numbers are greater than the Paasche numbers. The Geometric Laspeyres is less than the ordinary Laspeyres – as it should be according to Jensen's Inequality – while the Geometric Paasche is greater than the ordinary Paasche. Notice that the Geometric Laspeyres and Paasche indexes in 1999 (1996 = 100) coincide while in 2003 (1999 = 100) the difference between them is very small. This suggests that the elasticity of substitution has a value of (almost) 1, which is extraordinarily high. In turn this suggests that, for this particular data set, the Geometric Laspeyres might be an acceptable approximation of a superlative price index for the intermediate years. We will come back to this issue in Sect. 6.2.

In accordance with our expectations the Fisher and Törnqvist index numbers are quite similar for 1999 (1996 = 100) and 2003 (1999 = 100). There is a surprisingly large difference, however, between the direct Fisher and Törnqvist price indexes going from 1996 to 2003 (116.58 and 116.85, respectively). The upper level substitution bias of the Laspeyres index, as measured by the difference with the Fisher, amounts to 0.11%-points on average per year during 1996–1999 and 0.17%-points during 1999–2003.[12]

The difference between the chained Laspeyres and Fisher price index numbers in 2003 (1996 = 100), shown in the last column of Table 1, is 0.15%-points on average per year. The difference between their direct counterparts in the second column is as large as 0.19%-points per year. What is surprising as well is that chain linking in 1999 raises the Laspeyres index in 2003 from 117.90 to 118.15, and also raises the Paasche, thereby raising the Fisher index from 116.58 to 117.05. The upward effect of chaining, though less strong, goes for the Törnqvist as well.

[12]These figures are in between the estimates of 0.2%-points per year on average by Shapiro and Wilcox (1997) for the US and 0.1 %-points per year by Balk (1990b) and De Haan (1999) for the Netherlands. The differences will be partly due to differences in the aggregation level. The Dutch figures are based on approximately 100 product categories, the US figures on more than 9,000 item-area strata.

A number of "unusual" data seem to have contributed to some of these unexpected findings. During 1999–2003 several services showed extreme price increases: the prices of financial services, car insurance and gardening rose by 33, 47 and 152%, respectively. At the same time their expenditure shares increased sharply, which is counterintuitive – one would expect consumers to substitute away from services that have become relatively much more expensive. We decided not to exclude these data as unusual things do happen now and then and reflect reality (assuming the expenditures were correctly measured).

6.2 Empirical Results

As mentioned in Sect. 5 we prefer chain linking in 1999 to calculating direct indexes. Table 2 lists approximate chained price index numbers for the intermediate years 1997–1998 and 2000–2002; the true numbers for the benchmark years 1999 and 2003, copied from Table 1, are also presented. The Fisher and Törnqvist index numbers in the third and sixth row are based on the "natural" approach of expressions (9) and (10) and differ only marginally from each other. Particularly during 2000–2002 the alternative "Quasi Fisher" index numbers based on expression (15) are slightly higher. In line with what was suggested in Sect. 3.2, the price index numbers estimated by the Time-series Generalized Fisher Ideal (TGFI) method (16) are lower.

Table 2 also contains two different versions of chained Lloyd–Moulton price index numbers. The value for the elasticity of substitution σ is estimated separately for each subperiod (1996–1999 and 1999–2003) such that the Lloyd–Moulton index equals either the Fisher or the Törnqvist in 1999 and 2003. The estimated elasticities differ appreciably; the values used for computing the Lloyd–Moulton index numbers in row nine are 1.11 (1996–1999) and 0.85 (1999–2003), whereas those used for computing the Lloyd–Moulton index numbers in row ten are 0.99 (1996–1999) and 0.87 (1999–2003). The estimates for the first sub-period are rather high

Table 2 Chained price index numbers (1996 = 100)

	1997	1998	1999	2000	2001	2002	2003
Laspeyres	102.11	104.03	106.69	109.88	112.59	115.62	118.15
Paasche	*102.03*	*103.74*	106.00	*108.86*	*111.20*	*113.81*	115.96
Fisher	*102.07*	*103.88*	106.34	*109.37*	*111.90*	*114.71*	117.05
Geometric Laspeyres	102.06	103.88	106.38	109.41	111.94	114.71	116.97
Geometric Paasche	*102.08*	*103.90*	106.38	*109.40*	*111.97*	*114.83*	117.20
Törnqvist	*102.07*	*103.89*	106.38	*109.41*	*111.96*	*114.77*	117.08
Quasi Fisher	*102.09*	*103.90*	106.34	*109.47*	*112.03*	*114.82*	117.05
TGFI	*102.04*	*103.83*	106.34	*109.29*	*111.83*	*114.68*	117.05
Lloyd–Moulton[a]	*102.06*	*103.86*	106.34	*109.40*	*111.95*	*114.75*	117.05
Lloyd–Moulton[b]	*102.06*	*103.88*	106.38	*109.43*	*111.99*	*114.79*	117.08

Approximations are shown in italics; [a]Fisher index as benchmark; [b]Törnqvist index as benchmark.

and might be related to anomalies in the data set used. Notwithstanding these estimates, the resulting Lloyd–Moulton index numbers for 1997 and 1998, as well as those for 2000, 2001 and 2002, are quite similar to the earlier approximations. This suggests that the "theoretically-oriented" CES-type method and our "statistically-oriented" methods based on taking linear combinations of the expenditures shares of the benchmark years 1996, 1999 and 2003, might both be considered by statistical agencies.

Table 3 shows the direct counterparts to the approximations presented in Table 2. The Quasi Fisher (alternative) approximations, shown in the last row, overestimate the corresponding natural approximations, shown in the third row, especially for 2000–2002. This may, at least partially, be caused by the fact that the Quasi Fisher approach essentially approximates a Törnqvist index – which in our case is greater than the Fisher – and then converts the result into a "Fisher-type" formula. Notice that the Geometric Laspeyres index numbers nearly coincide with the natural Fisher approximations.

Table 4 contains approximate direct index numbers (1996 = 100), computed as if expenditure data for 1999 were unavailable. Thus, the true 1999 index numbers in Table 3 are replaced by estimated values. Yet the index numbers in Table 4 calculated with the natural method are very similar to the corresponding numbers in Table 3, including the true 1999 numbers. This seems to indicate that this method works well even for rather distant benchmark years. As before, the Quasi Fisher index numbers are greater than those computed with the natural approach, the difference being 0.10%-points in 2000.

Table 4 confirms that the TGFI method understates our alternative, Quasi Fisher approximations. Moreover, in 1999 the difference with the true Fisher index (106.34) is as large as –0.31. The Lloyd–Moulton price index numbers have been calculated using $\sigma = 0.98$ (row nine) and $\sigma = 0.79$ (row ten) which make the numbers in 2003 equal to the true Fisher and Törnqvist index numbers. The Lloyd–Moulton estimates are slightly greater than our natural approximations, up to 0.12%-points for the Törnqvist in 2000. Also, the true Törnqvist index number in 1999 is 106.38 whereas the (Törnqvist-based) Lloyd–Moulton estimate is 106.45. Of course we should not draw the conclusion that the Lloyd–Moulton method in general overstates the true numbers.

Table 3 Direct price index numbers (1996 = 100)

	1997	1998	1999	2000	2001	2002	2003
Laspeyres	102.11	104.03	106.69	109.88	112.55	115.39	117.90
Paasche	*102.03*	*103.74*	106.00	*108.68*	*110.89*	*113.25*	115.27
Fisher	*102.07*	*103.88*	106.34	*109.28*	*111.72*	*114.31*	116.58
Geometric Laspeyres	102.06	103.88	106.38	109.29	111.72	114.29	116.54
Geometric Paasche	*102.08*	*103.90*	106.38	*109.39*	*111.95*	*114.80*	117.15
Törnqvist	*102.07*	*103.89*	106.38	*109.34*	*111.83*	*114.54*	116.85
Quasi Fisher	*102.09*	*103.90*	106.34	*109.39*	*111.84*	*114.45*	116.58

Approximations are shown in italics.

Table 4 Direct price index numbers (1996 = 100), excluding observed expenditure shares for 1999

	1997	1998	1999	2000	2001	2002	2003
Laspeyres	102.11	104.03	106.69	109.88	112.55	115.39	117.90
Paasche	*102.02*	*103.74*	*105.99*	*108.62*	*110.84*	*113.22*	115.27
Fisher	*102.06*	*103.88*	*106.34*	*109.25*	*111.69*	*114.30*	116.58
Geometric Laspeyres	102.06	103.88	106.38	109.29	111.72	114.29	116.54
Geometric Paasche	*102.07*	*103.90*	*106.37*	*109.32*	*111.88*	*114.75*	117.15
Törnqvist	*102.06*	*103.89*	*106.37*	*109.30*	*111.80*	*114.52*	116.85
Quasi Fisher	*102.08*	*103.92*	*106.40*	*109.35*	*111.78*	*114.39*	116.58
TGFI	*101.92*	*103.66*	*106.03*	*108.96*	*111.48*	*114.22*	116.58
Lloyd–Moulton[a]	*102.07*	*103.89*	*106.39*	*109.31*	*111.74*	*114.32*	116.58
Lloyd–Moulton[b]	*102.07*	*103.91*	*106.45*	*109.42*	*111.91*	*114.54*	116.85

Approximations are shown in italics; [a]Fisher index as benchmark; [b]Törnqvist index as benchmark.

The (direct) Laspeyres price index numbers, the natural Fisher approximations, the TGFI approximations and the Lloyd–Moulton (Fisher-based) estimates from Table 4 are depicted in Fig. 1. The figure nicely illustrates that, although there are

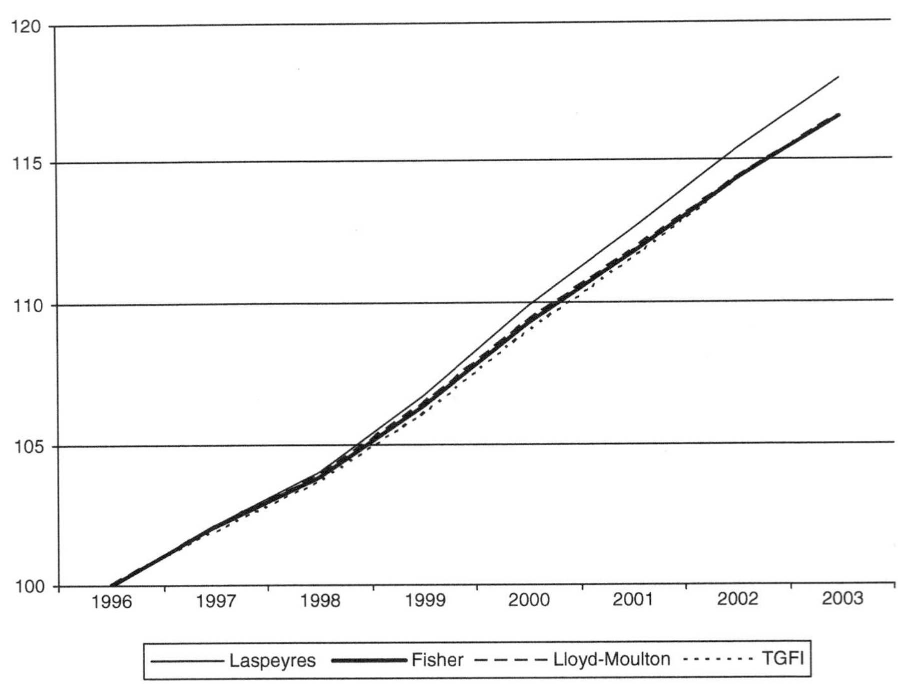

Fig. 1 Direct price index numbers (1996 = 100), excluding observed expenditure shares for 1999

differences between the natural and Lloyd–Moulton approximations, these differences are negligible compared to the differences with the Laspeyres index numbers. Figure 1 again makes clear that the TGFI method most likely produces (slightly) downward biased estimates of the Fisher price index, particularly for years in the middle of the period between two benchmark years.[13]

7 Conclusion

The results of the approximation methods discussed in this chapter are all numerically similar to those obtained with the Lloyd–Moulton approach. Ideally each method should be assessed on a data set that enables to calculate superlative price index numbers for intermediate years also. If both sorts of approaches work well then statistical agencies that wish to approximate retrospectively some superlative index can choose either. The Lloyd–Moulton price index has the advantage of being grounded in economic theory. Statistical agencies that are reluctant to rely on CES-assumptions or the like may find our pragmatic stance more attractive.

Suppose the Fisher index would be the preferred target. There is now a choice between the natural approach, for two benchmark years given by expression (9), and the Quasi Fisher alternative, given by expression (15). An advantage of the natural method is its greater flexibility. Data permitting, important expenditure shares can be estimated directly from available price and quantity data. [14] Linear combinations of benchmark year shares can then be used for the remaining shares.

References

Balk, B. M. (1990a). *Approximating a superlative price index when not all quantity data are available* (Research paper, BPA no. 6236-90-E7). Voorburg: Statistics Netherlands.

Balk, B. M. (1990b). On calculating cost-of-living index numbers for arbitrary income levels. *Econometrica, 58*, 75–92.

Balk, B. M. (2000). *On curing the CPI's substitution and new goods bias* (Research paper 0005). Voorburg: Department of Statistical Methods, Statistics Netherlands.

Balk, B. M., & Diewert, W. E. (2004). *The Lowe consumer price index and its substitution bias* (Discussion Paper No. 04-07). Vancouver: Department of Economics, University of British Columbia.

Boldsen Hansen, C. (2006, May 10–12). *Price-updating of weights in the CPI.* Paper presented at the Joint ECE/ILO meeting on Consumer Price Indices, Geneva.

[13]Balk (1990a) employed Triplett's (1989a) US data on office, computing and accounting machinery for 1972–1986. This data set had four benchmark (weight-reference) years, 1972, 1977, 1982, and 1986, so that there were three sub-periods: 1972–1977, 1977–1982, and 1982–1986. For the first sub-period the TGFI method actually yielded greater index numbers than the natural and Quasi Fisher methods, whereas for the second and third sub-periods the TGFI method produced smaller index numbers. Thus, though there are reasons to expect that the TGFI method in "normal circumstances" generates downward biased index numbers, this is not necessarily the case.

[14]This is what Nimmo et al. (2007) apparently did (see footnote 8).

Boldsen Hansen, C. (2007, October 9–12). *Recalculations of the Danish CPI 1996–2006.* Paper presented at the tenth meeting of the Ottawa Group, Ottawa.

Cage, R., Greenlees J., & Jackman P. (2003, May 27–29). *Introducing the chained consumer price index.* Paper presented at the seventh meeting of the Ottawa Group, Paris.

Diewert, W. E. (1976). Exact and superlative index numbers. *Journal of Econometrics, 4*, 115–145.

Diewert, W. E. (1978). Superlative index numbers and consistency in aggregation. *Econometrica, 46*, 883–900.

Diewert, W. E. (2000). *Notes on producing an annual superlative index using monthly price data* (Discussion Paper no. 00-08). Vancouver: Department of Economics, University of British Columbia.

Haan, J. de (1999). Demographic change and upper level bias in consumer price indexes. In M. Silver & D. Fenwick (Eds.), *Proceedings of the measurement of inflation conference.* Cardiff: Cardiff University, 31 August – 1 September 1999.

Haan, J. de (2005, December 12–13). *New and disappearing product varieties and the CES cost of living index.* Paper presented at the fifth Economic Measurement Group Workshop, Sydney.

Hill, R. J. (2006). Superlative index numbers: Not all of them are super. *Journal of Econometrics, 130*, 25–43.

International Labour Organization (ILO), IMF, OECD, Eurostat, United Nations, World Bank (2004). *Consumer price index manual: Theory and practice.* Geneva: ILO Publications.

Ivancic, L. (2007). *Scanner data and the construction of price indices*, PhD thesis, Sydney: University of New South Wales.

Lloyd, P. J. (1975). Substitution effects and biases in nontrue price indices. *The American Economic Review, 65*, 301–313.

Melser, D. (2006). Accounting for the effects of new and disappearing goods using scanner data. *Review of Income and Wealth, 52*, 547–568.

Moulton, B. R. (1996). *Constant elasticity cost-of-living index in share-relative form.* Mimeo, Washington, DC: Bureau of Labor Statistics.

Nimmo, B., Hayes, S., & Pike, C. (2007, October 9–12). *Consumers price index: Retrospective superlative index and impact of alternative housing weights.* Paper presented at the tenth meeting of the Ottawa Group, Ottawa. Retrieved from www.stats.govt.nz/developments/price-index-developments.

Shapiro, M. D., & Wilcox, D. W. (1997). Alternative strategies for aggregating prices in the CPI. *Federal Reserve Bank of St. Louis Review, 79*(3), 113–125.

Triplett, J. E. (1989a). *Superlative and quasi-superlative indexes of price and output for investment goods: Office, computing and accounting machinery* (Discussion Paper 40). Washington, DC: Bureau of Economic Analysis.

Triplett, J. E. (1989b). Price and technological change in a capital good: A survey of research on computers. In D. W. Jorgenson & R. Landau (Eds.), *Technology and capital formation.* Cambridge/London: MIT Press.

Harmonized Cross Region and Cross Country CPI Time-Space Integration in the Euro-Zone

Three Characters in Search of an Author

Guido Ferrari, Tiziana Laureti, and José Mondéjar Jiménez

1 Introduction

The European Union (EU) system of compilation and dissemination of the time Consumer Price Indexes (TCPIs) and of the space Consumer Price Indexes (SCPIs) (or Purchasing Power Parities (PPPs)) is not integrated. Obviously, the system inside the Euro-zone countries is not integrated either.

As a result, and just to quote the most evident discrepancies, the two sets of indexes are produced:

(i) on the basis of not fully coordinated surveys and elaboration methodologies;
(ii) on the basis of different baskets for time and space;
(iii) using different weighting structures;
(iv) using different methods for the treatment of the quality modifications, of some kinds of non market services, and of seasonality; and
(v) disseminated in different times.

Obviously, although some of these inconsistencies are difficult to conciliate and some are only partly reconcilable due to national specificities, it would be highly advisable to integrate the above system, in order to coordinate it and put it in an overall consistent frame.

To this aim, a very relevant point worth working on is the question of baskets of both the TCPIs and the PPPs and this is what we intend to do with the reflections that will follow, not disjoined from the considerations concerning the surveys, although to the extent allowed by this basket construction oriented analysis conducted in a Euro-zone harmonized framework.

G. Ferrari (✉)
Dipartimento di Statistica, Università di Firenze, Firenze, Italy and Renmin University of China, PRC
e-mail: ferrari@ds.unifi.it

L. Biggeri, G. Ferrari (eds.), *Price Indexes in Time and Space*, Contributions to Statistics, DOI 10.1007/978-3-7908-2140-6_3, © Springer-Verlag Berlin Heidelberg 2010

More precisely, this paper's objective is to discuss a harmonized cross region and cross country TCPIs and PPPs integrated approach in the Euro-zone, or European Monetary Union (EMU).[1] Countries and make some remarks and proposals for its implementation.

The main stream of the paper will insert in the discussion carried out in Ferrari, Laureti, and Mostacci (2005) by enlarging and developing it, particularly at a regional level and by introducing new points and perspectives.

This will be done, first by analyzing the debate that is being formed on CPIs time-space integration, retracing the (relatively) brief history of the matter, focussing on some ideas and pursuing their consolidation, developing them and putting on the table food for discussion.

We would like to underline that in order to get this harmonized integration one should be willing to accept to pay some costs in terms of precise identification of goods and services and of comparability. Thus, the problem might be shifted to the question whether these prices are worth being paid or not.

Before closing this Introduction, let's stress an as obvious as important point: to refer to a subset of countries such as those of the Euro-zone does not constrain the analysis, as evidently it can easily be extended to the rest of EU as well as to he rest of the world.

In order to proceed as clearly as possible, we first analyze the current state of the art of the cross country TCPIs-PPPs integration, which, as said above, has not been adequately handled yet, and discuss the achievements got so far (Paragraph 2).

Then, a discussion on the TCPIs and the PPPs currently being produced will be undertaken in Paragraph 3, whereas a critical investigation about the duality of CPIs is performed in Paragraph 4.

All that in order to come to proposals for the cross-country integration in a harmonized frame that will be made in Paragraph 5. Unfortunately, data is currently not available to seek empirical confirmation of the ideas we are going to present and defend. Therefore, we are bound to theoretical/methodological assertions only and our endeavour will be to try to set up a convincing scheme of reasoning strongly supported by objective and generally accepted basis.

Some remarks in Paragraph 6 will conclude the paper.

2 The State of the Art

The story of the time-space integration of the CPIs, although still at its early steps, seems to us to be intricate, contradictory and, in some respects, nearly

---

[1]On 17th May 2006 the European Commission has expressed favourable opinion to the entering of Slovenia into Euro-zone. Afterwards, this opinion has been ratified by the Ministries of Finance of the 12 countries. Thus, since 1st January 2007, Slovenia has thus substituted the Taler with the Euro, by raising to 13 the number of European countries that adopt the so called "unique currency". Slovenia adds itself to the list of the "twelve" composed by: Austria, Belgium, Finland, France, Germany, Greece, Ireland, Italy, Luxembourg, Netherlands and Spain.

incomprehensible if one looks at it in the light of the current acquirements, but better understandable if regarded in its time path.

To begin with the discussion, we stress that in time and space, CPIs elaboration reveals, *mutatis mutandis*, the same conceptual, methodological and practical problems and therefore, that it should be treated in a unified way.

There is no reason to handle the two contexts as if they were disjoined fields, except the reasons of the different roads followed by them, mostly due to different concrete needs that in turn have stimulated the differentiated development and consolidation of principles, theories and strategies.

This has led to a different development of time and space CPIs methodology and practice as well, so that, in many respects, until recently and also presently, they have mostly been considered as nearly specific, different fields and their close interrelation neglected both by scholars and organisms.

Almost all theorists and researchers – with the exceptions, as far as we know, represented by Rao (2001) and Ferrari et al. (2005) – have approached them accordingly and CPIs producers, namely international institutions and National Statistical Offices (NSOs) have elaborated them without taking account of the duality of the two dimensions. Consequently, even the level of development and achievements of the two kinds of CPIs is currently largely different.

In fact, the two indexes have received an unequal amount of attention, both in theoretical analyses and applications: Therefore, the levels of statistical achievements as regards survey design and sampling procedures both for outlet identification and for price basket selection differ. Concretely, TCPIs and PPPs are elaborated on the basis of different baskets and surveys and disseminated, by the official producers, on a monthly basis the first ones, and annually, the second ones.

As argued above, it is very likely that this is due, at least partly, to the above cited different story of the TCPIs and of the PPPs.

The former are older than the latter, some of them invented just at the dawn of the XXth, pushed by the need of measuring the price differential in time, whatever the goal might be, whereas the latter are more recent, originating from increasing requirements of price level comparisons in a framework of Gross Domestic Product (GDP) space comparisons.

TCPIs have developed individually, as the time measurement of the price level variation has involved the interested area (country or region) only, isolated from other areas. Only in recent years, with the European harmonization program, they had to take account of the overall EU spatial frame and its constraints as well.

PPPs, vice-versa, have been developed for groups of countries, inside the International Comparisons Programme (ICP), by increasingly adding a number of countries, with all the involved problems of statistical conciliation and actual coordination.

This non-integrated treatment has no advantages, whereas it presents several disadvantages.

Among the most relevant drawbacks, the first one, eminently ideal/theoretical, is represented by the fact that the two dimensions, the time and space ones, are regarded and approached as if they were two different fields, which drives to the loss of the awareness of the uniqueness of the problem and of the need of developing a dual approach to the matter.

This leads, as a second disadvantage, to the useless and sometimes detrimental repetitiveness of proposals of methods and procedures that, instead, should be the same and to the non-perception of the uniqueness of the problem, and to a considerable confusion as regards possible solutions.

On the practical side, there is a useless dispersion of energies, duplication of efforts, overlapping of suggestions and, above all, diversified production and dissemination of indexes and information which implies, on one hand, the increasing risk of omissions and errors and, on the other hand, high costs in terms of both time and money.

Again, many inconsistencies are possible, if one thinks that in Euro-zone countries, as well as in the other EU countries, monthly national TCPIs, and monthly harmonized TCPIs, the Harmonized Indexes of Consumer Prices (HTICPs), based on different baskets, are produced and disseminated and annual PPPs are produced and disseminated by Eurostat based on a price survey that is periodically carried out in each capital city (thus with the PPPs that refer to that city and not to the whole country), and on a basket of goods and services that in turn is different from the others. The weighting structures, although based on the COICOP classification, differ across countries.

The two sets of price indexes are elaborated independently by NSOs and Eurostat and so far no attempt has been undertaken in order to get an integrated treatment.

If one refers to sub national areas, the situation is by far different.

In time domain, it must be stressed that the common perception of a need of measuring the time inflation at regional level already existed since long time all over Europe.

This perception was first mostly felt by official statistical information producers such as the NSOs, which, well aware of the utility of providing the national economic and social community with a regional price index, started producing and disseminating regional TCPIs, obtained as an intermediate step in the process of national TCPIs aggregation.

Nevertheless, it should be said that there was no clear project behind it, and a demand from economic and social users did not clearly emerge, at least initially. It was more an offer due to enlightened statisticians' hunch than a precise and convinced request of households and entrepreneurs.

As a matter of fact, when thinking of time inflation, users are more concerned with the national level and are content to be provided with the national average inflation, and only seldom they need and require regional specific measures.

After all, when these subjects want to compare their individual or household purchasing power changes, it is enough for them to use the national TCPI. And also, the regional political authorities and policy makers, when comparing in time the regional GDP, can use with no question the national TCPI to deflate the current price figures.

In space domain, things are somewhat different. The need for making cross region inflation comparisons and being provided with a cross regional inflation measurement instrument is perceived more clearly than that of a regional time inflation measurement by citizens, entrepreneurs and local governments.

The increasing relevance gained by regions in the EU bureaucratic and administrative organization (so that more and more officials, documents, reports refer to the "Europe of regions") and the key political and economic role taken by them has contributed to further stimulate and strengthen this need.

Citizens have always perceived the importance of the cross region and cross city price level differences, as they have concretely experienced them: for example, between Florence and Rome there are price differences for the same good, sometimes even somewhat high, and a person has daily opportunities of making comparisons just by drinking a coffee or a beer, having a meal in a restaurant or simply watching the shop window prices.

The same, and even in a stronger way, occurs for the entrepreneurs who feel how necessary is to take account of regional price differences in order to operate at the best and make the best choices for competitiveness.

Also, this need is strongly felt by regional institutions and governments for which the knowledge of the regional price level differentials is crucial to design and implement regional policy and take economic and social decisions, as well as to make strategies and plans.

The above need for cross region price level comparisons has been fully highlighted and brought up-to-date by the introduction of the Euro as the common currency in the group of EU countries which gave life to the first wave.

Indeed, the movements of people across the Euro-zone countries for reasons of production, trade, tourism, already previously existing although with perhaps a lower relevance, drive people to look more carefully at the prices here and there and as they are now expressed in the same currency, to nearly spontaneously make price comparisons and to ask themselves why in one country the same good or service has a different cost than in another country, above all when this difference is macroscopic.

As a result, price comparisons among regions belonging to different countries as well as the above discussed within countries cross region comparisons (in the second case even strengthened by the use of the same language which makes comparisons more immediate and effective) become even more perceived and a request for regional PPPs does increasingly materialize.

Obvious reference to cross region comparisons of individual and household income and welfare level is implied, which is a crucial point in to-days life and has a key impact on people communities relations and on governments' economic, social and political decisions.

In addition, cross regional GDP comparisons are claimed in order to supplement the above decisions and eventually integrate with additional information the existing GDP based EU system of economic assistance to the EU less developed regions and better ensure its equity, consistency and sustainability.

Although all the subjects involved in regions' economic, social and political life should be provided with a tool capable to measure the above price level differences, strangely enough this need has not come up so far and no regional PPPs compilation has been undertaken by any NSO of the Euro-zone.[2]

This increasingly strengthening need for cross region price level measurement and therefore, for suitable tools like regional PPPs, frames in the existing and consolidated production of regional TCPIs and therefore, when implementing a system of PPPs one should take account of them.

To resume, at the Euro-zone country level the situation is such that two kinds of TCPIs (the national and the harmonized ones) and one set of PPPs are elaborated and disseminated, based on different baskets and two different surveys. At the Euro-zone regional level, only the TCPIs are produced and disseminated, the PPPs being inexistent.

Consequently, it is our opinion that this is the time to think of a harmonized cross region and cross country integrated approach and that a suitable and feasible methodology should be implemented to this aim.

As anticipated by the paraphrase of Pirandello's masterpiece in the title, there are three characters in search of an author.

The discussion of this methodology and of its possible implementation closely involves the discussion of the survey design, both for outlets identification and for goods and services prices collection, as the two stages are closely connected.

In what follows, we will discuss some ideas in this direction.

---

[2]Notwithstanding the question of elaborating regional PPPs did not receive yet the attention it deserves by researchers and scholars and, above all, by Euro-zone statistical information institutional producers, such as NSOs and Eurostat (and not even by the other world institutional statistical offices such as the United Nations Statistical Division, the OECD, the World Bank, the World Bank), some countries have started working in this direction. As an example, the UK Office for National Statistics (Wingfield, Fenwick, and Smith, 2005) made in 2000 the first exercise of estimating regional price level comparisons, then updated in 2003. In 2004 new estimates have been produced and disseminated. The basic approach to calculating regional price level differences was to measure the cost of purchasing a common basket of goods and services (the Retail Price Index (RPI) basket) in each region and express that cost relative to buying the same basket nationally. Despite this "star" approach, and although the estimates are conducted on a time-to-time basis (and not on a regular monthly one as envisaged in this paper), this work represents an important step forward the time-space integration.

The second example is provided by the Australian Bureau of Statistics in a paper by Waschka et alii (2003), with the development of experimental spatial price indexes to measure the price differences between the eight Australian capital cities. Here too, the TCPI basket was used, and the GEKS formula was applied, as it does not require prices of specific commodities to be available in every city, since the price comparisons can be made as long as prices are available in at least two cities.

The third example is given by Aten (2005) who used the CPD method to estimate differences in price levels across 38 geographic areas in the US, based on prices collected for 2003 CPI.

Later on (Aten, 2006), she proposed an experimental index of the price level differences in US selected areas for 2003 and 2004, by refining the analysis previously conducted by Koroski, Cardiff, and Moulton (1994), and Koroski, Moulton, and Zieschang (1999). In the Euro-zone, the Italian NSO, Istat, has recently started a programme aiming at elaborating a system of PPPs at a regional level (Biggeri, De Carli, and Laureti, 2008).

3 The TCPIs and the PPPs Elaborated by the EU Individual Countries and by Eurostat. Lost in the Intricacy of the Jungle

As stressed above, there is no regional PPPs compilation activity, neither at the Euro-zone NSOs level nor at the Eurostat one, except some experiments, such as the Istat program mentioned in footnote (2). Instead, regional TCPIs are produced and disseminated.

As regards the country level, yearly PPPs are elaborated by Eurostat in the framework of the International Comparison Program (ICP) and of the Eurostat-OECD European Comparison Program (ECP) for both EU and Euro-zone countries.

Eurostat started the program of elaboration in the framework of the ICP in 1970, and thereafter in the framework of the ECP, and PPPs were produced on a quinquennial basis. Since 1991 the reference period of PPPs elaboration has become the year. Currently, Eurostat elaborates yearly PPPs for 30 countries plus Switzerland. It is worth stressing that the basket of goods and services, unlike that of HTICPs, is decided by Eurostat (Eurostat, 2004a).

As far as the TCPIs are concerned, each NSO belonging to EU and Euro-zone elaborates monthly TCPIs for the respective country.[3]

The requirement for comparability claimed by the Maastricht Treaty spurred Eurostat into action in 1993 to ask Member States to provide TCPIs which were comparable with one another. The new indexes should have been comparable with one another within fairly tight limits: comparability was generally defined as a tolerance of $\pm$ 0.1% points on the overall indexes.

Common rules in methodology were claimed, but in a frame of flexibility which resulted also in avoiding to impose a unique basket (the basic point for strict comparability) and each country was left free to have its own one.[4]

The idea underlying this decision was that each country in Europe is different, their consumers spend their money on different goods and services in different proportions.

To fix a common basket would have meant to require Member States to compile an index unrelated to the actual consumption pattern and hence to the economic situation in their country. Italians, Germans, Spaniards eat and drink differently. There is no need to invent an average European (Astin, 1999).

This gave life to the HTICP, which flanked the existing TCPIs.

Until 1992, the concepts of "compensation", "cost of living", "national accounts deflator" underlined a number of TCPIs.

---

[3]Before the Maastricht Treaty in 1992, these indexes had developed according to the perceived needs of each country, often with the objective of having an index which could be used to uprate the wages of the workers in order to preserve their purchasing power (a "compensation" index). The result was that a single comparison of national CPIs did not necessarily give an accurate comparison of relative rates of consumer price inflation (Astin, 1999).

[4]Eurostat view was that the Member States should be allowed to decide on their own specific procedures, provided that comparability was not threatened (Astin, 1999).

Eurostat decided that HTICP should be a "pure" inflation index. For it to be "pure", Eurostat established that it would be concerned only with actual monetary transactions. So, for example, in the field of housing the imputed rent method to measure the price of owner-occupied housing would not be used.

The HTICP basket is not uniform across countries. The differences between the HTICP and the national TCPIs concern subsidies, health and education treatment, as well as the treatment of the own property houses (the HTICP takes account of the net price paid by consumers after reimbursement whereas many TCPIs take account of the gross price).

Also, the cost of borrowing money, which is neither a good nor a service, would not be included in HTICP. Thus, interest payments were excluded.

Laspeyres formula was preserved and the only question was to decide a maximum period between re-basing of the basket.

As regards the geographic coverage, the domestic concept was chosen. Quality adjustments have been indicated as the most important technical issue, that most likely has potentially the biggest single effect on the accuracy of the index and may hardly affect comparability.

As a result, each country provides to Eurostat the monthly HTICPs.

Eurostat compiles the HTICP averages for the EU Member States. Currently, it monthly produces, as a weighted average of the individual HTICPs: (i) a HTICP (called Monetary Union Index of Consumer Prices MUICP)) for the Euro-zone (the twelve, EU12, and the thirteen countries (EU13, including Slovenia)) and (ii) a HTICP (called European Index of Consumer Prices (EICP)) including: (ii_1) a HTICP for the 15 countries (the twelve founders countries plus Denmark, Great Britain and Sweden (EU15); (ii_2) a HTICP for the 25 countries[5] (EU25); (ii_3) a HTICP for the EU25 plus Island and Norway (EU27) (Eurostat, 2004a).[6]

The weights currently being used are derived from previous year national accounts, price updated to December of the year concerned.

To conclude and summarize: there are three kinds of price indexes for every country: two time indexes, a TCPI (in Italy just two), a HTICP and a space one, the PPP. The first two are under country's control, the third one is not.

4 The Dual Nature of CPIs: A Critical Appraisal of the National TCPIs/HTICPs and PPPs Elaborated by the NSOs of the EU Countries and by Eurostat

As said above, the implementation of national TCPIs and HTICPs occurs on different basis than that of PPPs. Indeed:

---

[5]The group of the 15 countries plus the 10 countries (Cyprus, Estonia, Latvia, Lithuania, Malta, Poland, Czech Republic, Slovakia, Slovenia and Hungary) that entered the EU starting from the 1st May 2004.

[6]Since 1st January 2007 two additional countries, Bulgaria and Romania have entered the EU. These countries too, as well as Turkey, a candidate country, are included in the average HTICP. Therefore, the total number of countries involved in average HTICP elaboration accounts for 30.

(i) the surveys underlying the collection of the information on prices are different;
(ii) the baskets of goods and services are different as well;
(iii) the timing of the updating is different: monthly in national TCPIs and HTICPs case, yearly in PPPs case[7];
(iv) the geographical coverage is different: referred to the whole country, in TCPIs case, but constrained to the capital cities only in PPPs case;
(v) the motivations which lead to the implementation of the two kinds of indexes are different.

The latter point is a very relevant one. On one hand, the national TCPIs are currently used to measure inflation, although, as said above, many of them were used in the earlier years of the twentieth century with the purpose of representing compensation factors, or cost of living measures or deflators, and the HTICPs have been introduced to allow the European Central Bank (ECB) to monitor price increasing[8] in Euro-zone countries and therefore to check whether the yearly inflation is lower than 2%, the level assumed as the one which ensures the medium term objective of price stability, as claimed by the Maastricht Treaty.

On the other hand, the PPPs are elaborated for the above 31 European countries cared of by Eurostat inside the broader Eurostat-OECD program, to the purpose both of getting a set of space price deflators which allows, for example, to allocate a relevant part of the EU structural funds to those countries and regions whose PPPs deflated GDP is lower than a given percentage of the community average and of comparing the consumer price level of the different European countries.

The dual nature of national TCPIs and PPPs clearly emerges from the above remarks. In both dimensions, they represent the parity which compensates price level differences: in time, the index represents the parity which compensates the different price level between two times, t and 0 in a country, whereas in space the index represents the parity which compensates the different price level between two countries, A and B, at time t.

Some, in theoretical/applied research, have started to proceed in the direction of taking account of this duality and of elaborating scenarios and schemes able to represent it simultaneously.

---

[7] While the updating is yearly, the survey is conducted every three years. In fact, Eurostat has the task of producing yearly PPPs (with the further obligation set up by the European Commission of publishing final results within 36 months from the reference year), whereas the NSO have given the task to collect and transmit to Eurostat the prices every three years. In order of not overcharging the prices collection work, the goods and services list has been subdivided in six parts and the ONSs conduct six differentiate surveys, two each year (one in the first semester and one in the second semester), over three years. As a consequence, the surveyed prices refer for one third to year t, for one third to year t–1 and for another one third to year t–2. This implies that the NSOs must provide the so-called "time rectification factors" with which to "centre" the prices collected in the three year period to a reference year t.

[8] Actually, in the last decades all countries exhibited consumer price increasing, except Japan, the only big western economy which has experienced a period of consumer prices decreasing.

Rao (2001), foretells the time-space integration of CPIs and supports the "best possible integration" between the basket of goods and services used in CPIs framework and the basket utilized for the compilation of the PPPs. Since his perspective is from ICP, he stresses that the selection of the product list is probably the most difficult and most contentious issue in constructing PPPs and that to keep the quality of the products constant to the maximum extent possible, a very detailed set of product specifications is employed and nearly perfect product identity is ensured. This is done in order the PPPs to measure price level differences rather than differences in product qualities.

Identity, and the closely related concept of comparability, of goods and services conflicts with representativeness; in fact, items that do exist and are identical in two or more countries may not be representative in some of them.

As a consequence, if the best integration were pursued, a compromise between representativeness (also called characteristicity and marked with an asterisk) and identity should be found.

Ferrari et al. (2005), after stressing the drawbacks of the current systems for estimating both TCIP, HTICP and PPPs, and the very low number of common products specifications, claim for an integrated view that might lead to a unique operational and organizational system.

To this aim, they focus both on the modalities of the basket selection and on the sample of outlets and suggest a three step procedure for the integration of the product list, with the creation of a single list for the then 12 countries of the Eurozone made possible by the adjustment of definitions and selection criteria by the 12 NSOs, to obtain a list of products wider than the one currently needed for time comparisons but smaller than that currently utilized for space comparisons.

Finally, in order to identify the specifications to be surveyed in both time and space, they suggest to give up, in some cases, the perfect identity in space by resorting to methods of evaluation of quality differences, according to what is suggested by Rao (2001).

As regards the outlets selection criteria, uniqueness of ways of selection and keeping of the same survey units is suggested.

Although no attempt has been undertaken to try to take account of the above duality by any NSO or international institution in their activity of time and space indexes elaboration, in recent times the sensitivity towards the theme of the time-space integration has increased.

In fact, in the framework of the international institutions which work to PPPs elaboration, the subject is discussed, even though concrete attempts towards some form of integration have not been undertaken so far.

For example, the International Labour Office (ILO) in 2004 stressed the global advantages that would derive from the integration in ICP framework.

In particular, and by referring to developed countries only, where there are NSOs which have reached a good technical and methodological level in the compilation of their CPIs, the integration would be beneficial both to the analysis of the quality modifications and to the treatment of the new products.

As far as these two problems are concerned, which occur, the first one, in a time framework and, the second one, in a spatial framework, and that lead to quality adjustments that do not prevent the time and space comparability between goods and services having partly different characteristics, it is stressed that the integration would imply synergies which probably could drive the efforts in the direction of the development of more suitable statistical methods for the treatment of quality changes.

In 2006, the World Bank (WB) dealt with the integration, both from the point of view of the price information collection and from the view of the processing of this information as well. It was admitted that, above all in developing countries, the price collection excludes some kinds of households and some kinds of goods and services in time indexes elaboration, whereas this is not true in the price survey concerning WB PPPs elaboration, and that the price collection for the time indexes is generally undertaken monthly, with the time lag between information collection and time indexes publication reduced to a few days, whereas the price collection for PPPs is fractioned in time and the publication of PPPs is much delayed.

Notwithstanding the above differences in terms of coverage (of households and of goods and services), of periodicity of the survey and of timeliness of the procedure of information elaboration, an as close TCPIs and PPPs integration as possible is claimed, in order to achieve significant returns to scale.

Parallel to what is done by ILO, an as extended as possible utilization of price data collected for the TCPIs is recommended.[9]

Both ILO and WB positions seem to focus on the choice of whether to prefer representativeness or identity.

Eurostat appears not to devote the same interest to the topic of TCPIs-PPPs integration. No mention to the question is made in Eurostat (2001), whereas in Eurostat (2004b) it seems possible to perceive a certain coldness, since, rather than to catch the advantages, only the difficulties of integration are stressed.

Apparently, above all in EU framework, a new mentality towards the theme of the time-space integration hardly seems to appear, capable to catch all the synergies that spring from the interaction between TCPIs and PPPs survey and compilation.

5 Cross Region and Cross Country Integration in an Euro-Zone Harmonized Frame: Some Comments and Proposals

In order to proceed as clearly and smoothly as possible in the discussion of integration, let's begin with the national TCPIs and the HTICPs and analyze their role and characteristics.

---

[9]To this regard, the ICP Tool-Pack software implemented by the WB in TCPI 2003–2006 framework allows to import the information on prices and on products specifications collected through the surveys on national TCPIs that are believed useful also in PPPs implementation.

Although, according to Astin (1999), Eurostat decided initially that it would not be prudent to proceed on the assumption that the HTICPs would replace the national TCPIs, as they were designed primarily to facilitate international comparisons of consumer price inflation across the EU, an objective that clearly is not an objective of national CPIs, our view is that it is now the time for the HTICPs to fully replace the CPIs so that the countries produce the former only.[10]

This evidently was the view that motivated Luxembourg to adopt from the beginning the HTICP as its national CPI, an evidence that confirms that the replacement is not shocking and does not produce troubles at all.

There are many reasons that support our suggestion.

First, it should be stressed that many of the technical aspects of HTICPs construction are also used in national CPIs; for example, they use the same set of sampling rules and the same samples of outlets.

On the other hand, Eurostat itself was inclined to expect a gradual convergence over time between national CPIs and the HTICPs.

With two sets of time indexes, there are, (i) higher costs, (ii) higher confusion and possibility of errors in the compilation of them and, above all, (iii) difficulties in explaining to the users the reason why to have two indexes that are basically equal. Even though one aims at measuring a cross country comparable inflation and one intends to measure the within country inflation, both measure the same phenomenon and in the same way (same formula, same methodology, nearly same basket of goods and services, same sample outlets). The users might feel confused and, above all, might become suspicious towards the activity of NSOs and the reliability of official statistics.

One can therefore claim: let the HTICPs measure the comparable inflation; nothing prevents them to measure the inflation of the concerned country as well.

Moreover, the gain in NSOs bureaucratic procedures simplification and in overall outlet sample survey and price collection work would be great indeed.

The regional TCPIs would be replaced then by the regional HTICPs, with no complication at all, neither theoretical nor practical.

Incidentally, in the above discussion, the question of the objectives of the indexes elaboration, although not in a clearly stated way, already emerged and permeated all the questions we touched.

Indeed, this is the crucial point that should always be taken into account when discussing integration.

Moreover, a cost-benefit mentality should lie beneath any views and proposals.

Settled on this premise, i.e., having established a well behaved set of regional HTICPs – which is easy as one can fully inherit the methodology already used for regional TCPIs – let's move to discuss a system of time integrated cross region PPPs.

---

[10] A discussion of the properties that should be possessed by a harmonized price index is given in Diewert (2002).

The starting point, and in the meantime the supporting basis, is represented by the sample of outlets and by the products basket used for the regional HICPs compilation, which of course are the sample of outlets and the products basket used for the national HTICP.

For the sake of simplicity, let's take Italian regions A and B. Both of them have elaborated their own specific outlet sample survey and arranged their own products basket, although in the framework of Istat very precise general instructions and rules, which are based on the "representative positions" of each product, in turn detailed through descriptions (one or more each) or specifications. For example, the representative position "rise" may be specified through parboiled rise, brand x, 1 kg, in package, and so on. This description corresponds to the "most sold" representative position, established on the basis of market evidence, used for one year and subject to change if necessary.

The price of each specification (description) is collected in different outlets in order to get "references".

The price ratios (base December previous year) of these references are averaged (geometric mean) to obtain the price ratios of the specifications which may coincide with the representative position or not. If not, an average (geometric mean) of the price ratios of the specifications is made in turn, to get the price ratio of the representative position.

The products references surveyed in region A may not coincide with those surveyed in region B and the products specifications may not have the same content in both regions.

Some of them, likely very few, will differ from each other. But, since they are regions of the same country, where consumers' habits, tastes, kinds of products available in the market, kind of sales structure are very similar, it is likely that these differences – and therefore the phenomenon of "non equi-characteristicity" – are very low.[11]

To compile PPPs between A and B, there is no strict need to compare the above specifications. It is enough to compare the representative positions, which mediate them. In this way, quality differences in compared representative positions in cases where the specifications content is not the same may be implied.

By and large, a flexible mentality should be adopted, similar to that used by Eurostat in defining comparability limits for HTICPs.

The same attitude that, on the other hand, just in the field of TCPIs compilation and use, leads to employ the Laspeyres index as if it was transitive, while it is not, being "quasi-transitive" only.

Eventually, in specific situations, some methods for quality modifications treatment might be used with nearly not significant cost and time increasing.

---

[11] "Characteristicity" is a concept first introduced by Drechsler in 1973 and used by United Nations Statistical Office (UNSO) and Eurostat, though the term has acquired quite a different meaning with time, signifying that the item is typical of that country; in other words, that it is consumed in a large quantity.

Thus, one basket only, the one used for HTICP compilation and one survey only would be necessary for regional HTICPs and for cross region PPPs.

Here, the question which should drive the NSOs would be the costs-benefits one.

Which are the costs? It seems to us that, apart from an obvious additional work that would be requested to NSOs, there is no additional fees.

Which might be the benefits? Basically, to get monthly PPPs fully integrated with the regional HTICPs. Taking into consideration the above stressed demand for cross region PPPs, this would be a remarkable result.

As regards the cross country national baskets integration, things may appear more difficult, if not because of the fact that the national HICPs baskets are different from each other.

Nevertheless, the situation may not be as complicated as it appears at first glance.

Indeed, in ECP framework, each country provides Eurostat with the information concerning the prices of the items which are the specifications of the "basic headings". It may happen that an item whose price is surveyed in one country does not have a corresponding item – and therefore price – in another country, either because it is not relevant in the related basket of consumer goods, or because it has been for some reason impossible to survey it. No weight is attached to any price; but the countries indicate whether items are important in their consumption by an asterisk on the price: the characteristicity of the specification. In the calculation of the basic headings ratios, ratios of prices in which at least one asterisk appears are considered only (Ferrari & Riani, 1998).

This frame is equivalent to the time one, and the above methodology can be used in time domain.

In fact, in time domain the representative position is equivalent to the basic heading in space domain. The price ratio for the representative position is formed in the same way as that for the basic heading: by averaging the price ratios of the most sold specifications, exactly the same as for basic heading, whose price ratio is formed by averaging the price ratios of the most sold items.

To refer to Italy, the price ratio of the representative position "ham", for example, is formed by the average of the price ratios of the references "Parma ham, brand x, quantity y, package w, and so on" and "Steamed ham, brand x; and so on".

In Spain, the representative position (or whatever named, as long as according to CICOP and at the same digit level) "ham" will be formed, for example, by "Jamon ibérico, brand x, and so on" and "Steamed ham, and so on".

Exactly the same information, of course, in a broad sense, is transmitted by the two NSOs to Eurostat as regards the basic headings in the framework of the ECP.

Therefore, in the Euro-zone cross country time-space basket integration it seems possible to identify a common basket derived from the best intersection of the basket each country uses for the national HICP.

Of course, some additional information and goods and services might be requested and specifications, warnings and adjustments might become necessary. However, in principle no substantial obstacles would seem to prevent to explore this road which would allow also a useful convergence of quality modification treatment and seasonality adjustment methods and procedures.

This would also imply the abandoning of the double list of products, the one for HTICP independently defined by each NSO which emphasizes representativeness and the one, the same for all countries, decided by Eurostat which emphasizes comparability (and therefore, identity) and to uses a list derived from, and closely linked to the list used by each country for HTICP purposes only.

Some further and concluding remarks are worth stressing to support our proposal:

(i) the TCPIs and the PPPs are published and disseminated with one decimal figure only;
(ii) no statistical significance statement is attached to them.

Then, why be so choosy as to want such a large number of goods and services (more than 800 in the Italian case), to want the extremely exact specification of all characteristics, to want the absolute time and space comparability, without not even taking into account the recommendations of Eurostat itself which suggests a tolerance of $\pm$ 0.1% points on the overall indexes?

It would perhaps be better to select a reduced number of goods and services, and concentrate instead on important points such as the outlets choice (currently not totally controlled and updated in many countries), the surveyors training, an adequate data elaboration and exploitation, a good weighting system (a question currently still very disputable).

Finally, it should be stressed that Istat as well, in the framework of the implementation of the regional PPPs for Italy, is conducting experiments to the purpose of analysing the possibility of introducing a unique basket for time and space regional price indexes.

6 Conclusion

In to-days global world it is necessary to look for the simplification and the weakening of the data survey statistical procedures and information production and dissemination.

This is an objective at which primarily one should aim, even at the cost of some approximation and loss. Of course, as long as these are sustainable, both from a theoretical and an operational point of view. Objectives, costs and time would be the reference points that should always guide our way.

The statistical information apparatus has reached to-day such a huge relevance and weight to make the information production costs and the time of its dissemination crucial variables.

This is what should be taken into account and what actually is needed in the field of the elaboration of the TCPIs and of the PPPs as well.

It is a largely shared opinion that in this field there are too many useless complications, duplications, missed synergies, needless efforts and objectives dispersion, lack of coordination, which make the work unduly complex, costly and confused and lastly not as efficient and effective as it might and should be. *A fortiori*, currently, since indexes that measure the regional price level differences are increasingly requested.

One of the most crucial points which reflects this situation is represented by the lack of integration of the baskets for the price collection for the compilation of TCPIs and HICPs on one hand and PPPs on the other hand.

In this field, the existing situation in Euro-zone framework, but also in the rest of Europe and in the world, appears to be unsatisfactory.

It seems to us that the times are mature for a deep re-thinking of the matter and for a revision of the procedures so far used and perhaps, in some way, of the "mentality" underlying the approach to it.

For this reason, in this paper we have risked a global reflection with the aim of re-discussing all the aspects, even those that might seem more consolidated, in an endeavour of disentangling a considerably embrangled hank and provide some answers – how convincing will be judged by the reader – to questions that have become no longer deferrable.

Some conclusions we have reached may seem, on one hand, too rough and simplified and, on the other hand, too hazarded, but they have been raised both as we are convinced they are supportable, and to "cast a stone" in a pool that seemed to us too quiescent and little willing to re-discuss principles and ideas and perhaps too devoted to the management of the everyday and of its slow modification, obviously also due to the obligation of the quick publication of the information.

All done in the direction of stimulating the discussion and the reflection, in order to re-start analysis and activity, following the "leitmotiv" of the historical path that has led to the present situation and has formed the methodologies and the practices currently used.

We think that one can agree on the need of re-thinking the whole approach, if only to give a fully convincing answer to those who are the final receivers of the information on time and space inflation, that is the users.

References

Astin, J. (1999). The European Union Harmonized indices of consumer prices (HTICP). *Statistical Journal of the United Nations ECE, 16*, 123–135.

Aten, B. H. (2005, November). Report on Interarea Price Levels (BEA Working Paper No. 11).

Aten, B. H. (2006, September). Interarea Price Levels: an Experimental Methodology. Monthly Labor Review.

Biggeri, L., De Carli, R., & Laureti, T. (2008). The Interpretation of the PPPs: a Method for Measuring the Factors that Affect the Comparisons and the Integration with CPI Work at Regional Level. Paper presented at the joint UNECE/ILO Meeting on Consumer Price Indexes. Geneva.

Diewert, E. (2002, March). *Harmonized indexes of consumer prices: Their conceptual foundations* (Working Paper No. 130). European Central Bank.

Eurostat. (2001). *Compendium of HTICP – Reference documents*. Luxembourg.

Eurostat. (2004a, March). *Harmonized indices of consumer prices (HTICPs) – A short guide for users*. Theme 2, Economy and Finance. Luxembourg.

Eurostat. (2004b). *Eurostat-OECD methodological manual on purchasing power parities.* Luxembourg.

Ferrari, G., Laureti, T., & Mostacci, F. (2005). Time-space harmonization of consumer price indexes in Euro-zone countries. *International Advances in Economic Research, 11*(4), 359–378.

Ferrari, G., & Riani, M. (1998). On purchasing power parities calculation at the basic heading level. *Statistica,* anno *LVIII*(1), 91–108.

International Labour Office (2004). *Consumer price index manual: Theory and practice*, Geneva: International Labour Office.

Kokoski, M., Cardiff, P., & Moulton, B. (1994, July). Interarea Price Indices for Consumer Goods and Services: An Hedonic Approach Using CPI Data (BEA Working Paper No. 256).

Kokoski, M., Moulton, B., & Zieschang, K. (1999). Interarea price comparisons for heterogeneous goods and several levels of commodity aggregation. In A. Heston & R. Lipsey (Eds.) *International and interarea comparisons of income, output and prices*. University of Chicago Press.

Rao, D. S. Prasada (2001). *Integration of CPI and PPP: Methodological issues, feasibility and recommendations*, Joint World Bank – OECD Seminar on PPP (pp. 1–26). Washington, DC.

United Nations Statistical Commission and Economic Commission for Europe. (1996). *International Comparison of Gross Domestic Product in Europe. Results of the European Comparison Programme*, United Nations.

Waschka, A., Milne, W., Khoo, J., Quirey, T., & Zhao, S. (2003). Comparing Living Costs in Australian Capital Cities. 32nd Conference of Economists. Australia.

Wingfield, D., Fenwick, D., & Smith, K. (2005, February). Relative Regional Consumer Price Levels in 2004. UK Office for National Statistics. Economic Trend 615.

World Bank. (2006). *ICP 2003–2006 handbook.* New York: World Bank.

Part II
Consumer Price Indexes in Space

Modelling Spatially Correlated Error Structures in the Time-Space Extrapolation of Purchasing Power Parities

Alicia N. Rambaldi, D.S. Prasada Rao and K. Renuka Ganegodage

1 Introduction

There is considerable literature focusing on the problem of explaining the exchange rate deviation index, more commonly referred to as the national price level. If ER_{it} denotes the exchange rate of the currency of country i at time t, then the national price level for country i (or exchange rate deviation index) is the ratio of PPP_{it}, the *Purchasing Power Parity* of country i at time t to ER_{it}. The notation R_{it} is used to denote the *Price Level* of country i at time t.

$$R_{it} = \frac{PPP_{it}}{ER_{it}} \tag{1}$$

For example, if the *PPP* and *ER* for Japan, with respect to one US dollar, are 130 and 110 yen respectively, then the price level in Japan is 1.18 indicating that prices in Japan are roughly twenty per cent higher than those in the United States. A value of this ratio greater than one implies national price levels in excess of international levels and vice versa.

PPP-converted real per capita incomes are used in influential publications like the World Development Indicators of the World Bank (World Bank, 2006 and other years) and the Human Development Report (ADB, 2005) which publishes values of the Human Development Index (HDI) for all countries in the world. The *PPPs* are also used in a variety of areas including: the study of global and regional inequality Milanovic, (2002); measurement of regional and global poverty using international poverty lines like \$1/day and \$2/day (regularly published in the World Development Indicators, World Bank); the study of convergence and issues surrounding carbon emissions and climate change (McKibbin & Stegman, 2005; Castles & Henderson, 2003); and in the study of catch-up and convergence in real incomes (Barro and SalaiMartin, 2004; Durlauf, Johnson, & Temple 2005; SalaiMartin, 2002).

A.N. Rambaldi (✉)
School of Economics, The University of Queensland, St Lucia 4072, Australia

L. Biggeri, G. Ferrari (eds.), *Price Indexes in Time and Space*, Contributions to Statistics, DOI 10.1007/978-3-7908-2140-6_4,

The only source for PPPs for the economy as a whole is the International Comparison Program (ICP). The *PPP* data are compiled under the ICP which began as a major research project by Kravis and his associates at the University of Pennsylvania in 1968 and in more recent years has been conducted under the auspices of the UN Statistical Commission. Due to the complex nature of the project and the underlying resource requirements, it has been conducted roughly every five years since 1970. The latest round of the ICP for the 2005 benchmark year was released in early 2008. The final results are available on the World Bank website: http://siteresources.worldbank.org/ICPINT/Resources/ICP_final-results.pdf. In more recent years, beginning from early 1990's, the OECD and EUROSTAT have been compiling PPPs roughly every three years. The country coverage of the ICP in the past benchmarks has been limited with 64 countries participating in the 1996 benchmark comparisons. However this coverage has increased dramatically to 146 for the 2005 benchmark year. Details of the history of the ICP and its coverage are well documented in the recent report of the Asian Development Bank (http://adb.org/Documents/Reports/ICP-Purchasing-Power-Expenditures/default.asp).

International organizations such as the World Bank and the United Nations, as well as economists and researchers, seek *PPP* data for countries not covered by the ICP and also for non-benchmark years. For most analytical and policy purposes, there is a need for PPPs covering all the countries and a three to four-decade period.[1] The Penn World Tables (PWT) has been the main source of such data. Summers and Heston are pioneers in this field. Summers and Heston (1991) provides a clear description of the construction of the earlier versions of the PWT. The most recent version, PWT 6.2, available on http://pwt.econ.upenn.edu, covers 188 countries and a period in excess of five decades starting from 1950. In addition to the PWT, there is the real gross domestic product (GDP) series constructed by Angus Maddison (Maddison, 1995, 2007). The Maddison series is available on Groningen Growth and Development Centre website: www.ggdc.net/dseries/totecon.html. The series constructed by the World Bank are available in various issues of World Development Indicators publication. The Maddison series make use of a single benchmark and national growth rates to construct panel data of real GDP and no estimates are available for non-benchmark countries. The World Bank series are based on the methodology described in Ahmad (1996) and the series makes use of a single benchmark year for which extrapolations to non-benchmark countries are derived using a regression-based approach. The benchmark and non-benchmark PPPs are extrapolated using national growth rates[2] in national prices.

Recent work by Rao and colleagues at the University of Queensland has proposed a new method to construct a tableau of PPPs similar to the PWT (see Rao,

[1]For example, the Human Development Index is computed and published on an annual basis. Similarly, the World Development Indicators publication provides PPP converted real per capita incomes for all the countries in the world for every year.

[2]We define "national growth rates" in the next section.

Rambaldi, & Doran 2008, 2009 for details), which we will refer to as the RRD method. The method proposed has several desirable properties including providing PPPs with standard errors from an econometric method that is invariant to the reference country and makes use of all available ICP benchmark information to date. The new method, which is described briefly in Sect. 3, like PWT and the World Bank methods, includes as a component a *model of the price level* whose main function is to assist with the prediction of PPPs for non-participating countries. One of the differences between the approach of RRD and other alternatives is that the model for price levels is specified with a spatially correlated error. In this paper we evaluate the sensitivity of the final results to the use of a spatial error, as well as alternative spatial specifications in this component of the method. Given the objectives of the paper, we present and discuss the modelling of national price levels before moving to the description of the RRD method.

2 The National Price Levels Model

Most of the explanations of price levels are based on productivity differences in traded and non-traded goods across developed and developing countries. Much of the early literature explaining national price levels (Kravis and Lipsey, 1983, 1986) has relied on the structural characteristics of countries such as the level of economic development, resource endowments, foreign trade ratios, education levels. More recent literature has focused on measures like openness of the economy, size of the service sector reflecting the size of the non-tradable sector and on the nature and extent of any barriers to free trade (Ahmad, 1996; Bergstrand, 1991, 1996; Clague, 1988).

It has been found that for most developed countries the price levels are around unity and for most developing countries these ratios are usually well below unity. In general it is possible to identify a vector of regressor variables and postulate a regression relationship:

$$r_{it} = \beta_{0t} + \boldsymbol{x}'_{it}\boldsymbol{\beta}_s + u_{it} \qquad (2)$$

where,

$r_{it} = \ln(PPP_{it}/ER_{it})$
$\boldsymbol{x}'_{it}$ is a set of conditioning variables
β_{0t} intercept parameter
$\boldsymbol{\beta}_s$ a vector of slope parameters
u_{it} a random disturbance with specific distributional characteristics.

Provided estimates of β_{0t} and $\boldsymbol{\beta}_s$ are available, model (2) can provide a prediction of the $ln(PPP_{it})$ consistent with price level theory.

$$\hat{p}_{it} = \hat{\beta}_{0t} + \boldsymbol{x}'_{it}\hat{\boldsymbol{\beta}}_s + ln(ER_{it}) \qquad (3)$$

where, $\hat{p}_{it}$ is a predicted value of $ln(PPP_{it})$. We return to the estimation of β_{0t} and $\boldsymbol{\beta}_s$ in Section 3.

2.1 The Spatial Error Structure

In RRD the errors u_{it} in the regression relationship (2) are assumed to be spatially correlated. The error structure is of the form

$$\mathbf{u}_t = \phi \mathbf{W}_t \mathbf{u}_t + \mathbf{e}_t \tag{4}$$

where $\phi < 1$ and $\mathbf{W}_t(N \times N)$ is a spatial weights matrix. That is, the diagonal elements, w_{ii}, are zero and the off diagonal elements, w_{ij}, measure the "distance" between observations. We assume that $\mathbf{e}_t$ is a vector of random variables with zero mean vector and covariance matrix proportional to I. The term "spatial distance" in the present context refers to economic distance rather than the traditional geographical distance, and we discuss shortly the alternative measures considered in this paper.

It is customary for the rows of $\mathbf{W}_t$ to add up to one (in which case $\boldsymbol{W}_t$ is (right or) row stochastic) and this implies $\phi < 1$ (see Ord, 1965, Kräemer, 2005).[3] It follows that $\mathrm{E}(\mathbf{u}_t\mathbf{u}_t')$ is proportional to $\boldsymbol{\Omega}$, where $\boldsymbol{\Omega} = (\mathbf{I} - \phi\mathbf{W}_t)^{-1}\,(\mathbf{I}{-}\phi\mathbf{W}_t)^{-1\,\prime}$.

It is easily seen that if ϕ is zero in (4), the error term $\mathbf{u}_t = \boldsymbol{e}_t$, and $\boldsymbol{e}_t$ is assumed not to suffer from auto correlation or heteroskedasticity. That is, when $\phi = 0$ the model in (2) is a multiple regression with spherical errors. Through the size of the parameter ϕ and the specification of $\mathbf{W}_t$ it is possible to study the influence asserted by the spatial specification on the final predictions, p_{it}^*. Section 4 specifically discusses the alternatives used in this paper to evaluate the role of the spatial specification.

The remaining of the paper is structured as follows. Section 3 briefly presents the RRD method to show where the spatial error appears in the method. Section 4 describes a series of alternative specifications of the spatial structure that are used in the empirical estimations as well as the data used in the estimations. Section 5 presents and discusses the empirical findings and Sect. 6 concludes.

3 The RRD Method

RRD is a smoothing method grounded in an econometric model designed to make use of all the information available for the purpose of constructing a panel of *PPPs*. Sources of data available from national and international sources are combined in an econometric model used by the smoothing algorithm. The description is brief and the reader is referred to Rao et al. (2008) for a complete version.

[3]This follows because ϕ is strictly bounded by the inverse of the eigenvalues of $\boldsymbol{W}_t$.

3.1 The Econometric Formulation of RRD

In this section we describe the econometric model underlying RRD. The description is brief and intended to show where the price level model fits within the smoothing method.

The variable of interest is denoted by $p_{it} = ln(PPP_{it})$ for country $i = 1,\ldots,N$ and time $t = 1,\ldots,T$ where PPP_{it} represents the purchasing power parity of the currency of country i with respect to a reference country currency. Although it is directly unobservable, several noisy sources of information can be combined to obtain an optimal prediction, p^*_{it}. The econometric model that encompasses these sources of information is written in state-space form and thus given by two sets of equations, the observation equations and the transition equations.

3.1.1 Observation Equations

The observation equations map the observations (ICP and predictions) and explanations (conditioning structural variables) to the unknown vector of variables of interest known as the "state vector". In RRD they take the following form:

$$\mathbf{y}_t = \mathbf{Z}_t\boldsymbol{p}_t + \mathbf{B}_t\mathbf{X}_t\boldsymbol{\theta} + \boldsymbol{\zeta}_t \tag{5}$$

where,

$\mathbf{y}_t = \begin{bmatrix} 0 \\ \mathbf{S}_{np}\hat{\boldsymbol{p}}_t \\ \tilde{\boldsymbol{p}}_t \end{bmatrix}$ is a vector of "observations" of the state vector. The observations are the reference country constraint with a value of zero, predictions of the state vector from (3), $\hat{\boldsymbol{p}}_t$ and benchmark observations (in benchmark years), $\tilde{\boldsymbol{p}}_t$.

$\mathbf{S}_{np}$ is a known selection matrix for non-participating countries at time t.

$\mathbf{Z}_t,\mathbf{B}_t$ are known selection matrices that map the unobservable state vector and conditioning variables to the observations in $\mathbf{y}_t$.

$\mathbf{X}_t$ is a matrix of observable socio-economic variables.

$\boldsymbol{p}_t$ is the unobserved state vector

$\boldsymbol{\theta}$ is a vector of parameters to be estimated and is a known form of the vector $\hat{\boldsymbol{\beta}}$ in (3)

$\boldsymbol{\zeta}_t$ is an error with $\mathrm{E}(\boldsymbol{\zeta}_t) = \mathbf{0}$ and $E\left(\boldsymbol{\zeta}_t\boldsymbol{\zeta}'_t\right) \equiv \mathbf{H}_t$

The observation vector $\mathbf{y}_t$ contains two sources of noisy observations of the (unobservable) state vector, $\boldsymbol{p}_t, \tilde{\boldsymbol{p}}_t$ which are the ICP collected observations of *PPPs* for participating countries in benchmark years, and $\hat{\boldsymbol{p}}_t$ from the predictions of the model in (3) where the spatial error structure is located (see Sect. 2.1).

Observations from the ICP

Data in the form of *PPPs* from the ICP benchmarks are a crucial component of the model. These are *PPPs* compiled by the global office of the ICP or regional offices of the ICP by conducting extensive price surveys in the participating countries, i.e., from the first benchmark comparison in 1970 till to date inclusive of the recently completed 2005 benchmark. The data on PPPs compiled over the past benchmarks may be best viewed as an incomplete panel due to the differing degrees of participation of countries in different benchmarks and due to the fact that the benchmark comparisons are conducted roughly once in five years. Due to the complexity in the design and collection of the ICP benchmark data (see Chap. 4, 5 and 6 of the ICP Handbook which can be found on the World Bank ICP website:www.worldbank.org/data/ICP), the observed *PPPs* are likely to be contaminated with some measurement error. As the surveys for these benchmark exercises are conducted by national statistical offices, the availability of resources to national statistical offices is likely to be positively related to the level of resources (technical and human) available in individual countries and it is likely to be reflected in the quality of the data collection by individual countries. Thus, ICP benchmark observations are assumed to be measured with error, giving rise to the bottom partition in (5) and (8) (as shown shortly),

$$\tilde{p}_{it} = p_{it} + \xi_{it} \tag{6}$$

where,

> $\tilde{p}_{it}$ is the ICP benchmark observation for participating country i at time t; and
> ξ_{it} is a random error accounting for measurement error.
> $E(\xi_{it}) = 0, E(\xi_{it}^2) = \sigma_\xi^2 V_{it}$

The measurement error variance-covariance is of the form[4]

$$\boldsymbol{V}_t = \begin{bmatrix} 0 & \boldsymbol{0} \\ \boldsymbol{0} & \sigma_{1t}^2 \boldsymbol{jj}' + diag(\sigma_{2t}^2, \ldots, \sigma_{Nt}^2) \end{bmatrix} \tag{7}$$

where, σ_{it}^2 is the variance of country i at time t, which is measured as the inverse of the a country's degree of development,[5] and σ_{1t}^2 is the variance of the reference country. This form of the covariance is sufficient for the invariance of the method to the choice of reference country (see Rao et al. (2008) for details and proof).

Predictions from the Price Level Model

The conditioning variables, $\boldsymbol{X}_t$, and predictions from the price level model, $\hat{\boldsymbol{p}}_t$, are mapped through the matrices $\boldsymbol{Z}_t$ and $\boldsymbol{B}_t$ to the state vector, $\boldsymbol{p}_t$. This is achieved by re-writing Eq. (3) (details are shown in Rao et al., 2008).

---

[4]The form of this covariance follows from the definition of PPP. The interested reader is referred to Rao et al. (2009) for details.

[5]In the empirical implementation we model σ_{it}^2 as inversely related to GDP_{it} per capita measured in $US (exchange rates adjusted). It is well known that exchange rates adjustments accentuate the difference between developed and developing countries and thus provide a suitable measurement of the desired effect.

The Reference Country Constraint

The first row in the observation Eq. (5) and subsequent definitions correspond to the reference country (this is without loss of generality). The form of the observation and its variance emanate from the basic concept of PPPs. The *PPP* for a particular country's currency is always defined or measured relative to the currency of a selected reference country. For example, the *PPP* for the currency of a country, say India, with respect to the currency of a reference country, say the United States, is defined as the number of currency units of Indian rupees required to purchase the amount of goods and services purchased with one US dollar. Hence, PPPs are determined only when the currency of a country is chosen as the base or reference currency. Therefore, by definition the *PPP* of the *reference currency* is always equal to unity in all periods. So, if country *1* is chosen as the base currency, the $p_{1t} = 0$ for all t with variance zero is set as a constraint which insures that the time series $\{\widehat{PPP}_1\}$ predicted by RRD is equal to 1 with standard error zero in all periods.

The Covariance of the Observation Equations' Error

Equation (5) takes a different form in benchmark and non-benchmark years as in non-benchmark years only observations $\hat{\boldsymbol{p}}_t$ are available. Accordingly, so does the covariance of $\boldsymbol{\zeta}_t$ in (5), $\boldsymbol{H}_t$. The covariance of the spatial error, which will be presented in detail below, is in the matrix $\boldsymbol{H}_t$, which is partitioned to account for the reference country's constraint[6] and the variance-covariance structure associated with the $\hat{\boldsymbol{p}}_t$ and $\tilde{\boldsymbol{p}}_t$. In a benchmark year, its form is:

$$\mathbf{H}_t = \begin{bmatrix} 0 & \mathbf{0} & 0 \\ 0 & \sigma_u^2 \mathbf{S}_{np} \boldsymbol{\Omega} \boldsymbol{S}'_{np} & 0 \\ 0 & 0 & \sigma_\xi^2 \boldsymbol{S}_p \boldsymbol{V}_t \boldsymbol{S}'_p \end{bmatrix} \tag{8}$$

$\boldsymbol{S}_{np}$ and $\boldsymbol{S}_p$ are known selection matrices for non-participating and participating countries at time t, respectively.

In non-benchmark years the bottom partition is not present as that is the covariance of the measurement error in the ICP benchmark observations.

3.1.2 Transition Equations

The transition equations model the movement of the state vector across time. In the RRD case the transition equations model the movement of *PPPs* over time.

A major consideration in the construction of a panel of PPPs, and real incomes, is that the growth rates in real income obtained using PPPs should be the same or close to the national growth rates in prices observed and reported by the respective

---

[6]The *PPPs* between currencies of two countries are invariant to the choice of the base country, which in turn requires the predictions of the reference country to be zero with variance zero in all time periods, $p_{US,t} = 0$. See Rao et al. (2008) for a proof that the method is invariant to the choice of the reference currency.

national statistical offices. This is considered an important property to be satisfied by the extrapolated *PPPs*. The currently available series from PWT, the World Bank's World Development Indicators and the Maddison series all adhere to this important principle. Thus, the national growth rates and the price movements implicit in such growth rates which are referred to as GDP deflators offer crucial information for updating PPPs.

The temporal movements in PPPs are, therefore, governed by a simple relationship presented in Eq. (9). It is easy to see that equation (9) is a simple identity if PPPs were the price of a single commodity. However in the case of PPPs at GDP level, GDP is treated as a composite commodity.

$$PPP_{i,t} = PPP_{i,t-1} \times \frac{GDPDef_{i,[t-1,t]}}{GDPDef_{US,[t-1,t]}} \tag{9}$$

where $GDPDef_{i,[t-1,t]}$ denotes the GDP deflator showing price movements from period $t-1$ to t in country i, and the US is the reference country.

Equation (9) simply provides a mechanism for updating PPPs using movements in the GDP deflator of the country concerned and therefore, provides a definition for the growth rate of PPP_{it}.

Thus, implicit GDP deflators provide a measure of movements in prices in different countries over time. These deflators provide critical information on country-specific temporal movements in prices. The main source of data on deflators is the national accounts published by countries, generally on an annual basis. In the econometric specification, Eq. (9) is modified to include a random disturbance term that makes it possible for the extrapolated *PPPs* to deviate from the *PPPs* implied by the growth rates in GDP deflators.

Therefore, the 'transition equations' of the state-space representation follow from Eq. (9) and are given by:

$$\boldsymbol{p}_t = \boldsymbol{p}_{t-1} + \mathbf{c}_t + \boldsymbol{\eta}_t$$

where,
$\boldsymbol{c}_t$ is the observed growth rate of $\boldsymbol{p}_t$ with elements $c_{it} = ln\left(\frac{GDPDef_{i,[t-1,t]}}{GDPDef_{US,[t-1,t]}}\right)$, $\boldsymbol{\eta}_t$ is a measurement error with $E(\boldsymbol{\eta}_t) = 0$ and $E\left(\boldsymbol{\eta}_t\boldsymbol{\eta}_t'\right) \equiv \mathbf{Q}_t = \sigma_\eta^2\mathbf{V}_t$

As GDP data are collected by national statistical offices, the argument made previously to justify the structure of the measurement error holds in this case also.

Equation (10) simply updates *PPPs* from period $t-1$ using the observed price changes over the period represented by $\boldsymbol{c}_t$. We note that Eq. (10) offers flexibility in imposing the national growth rates. If η_{it} is set to zero or its variance set close to zero then the temporal movements in *PPPs* track the relative movements in national GDP deflators, and hence the implied growth rates in income are also maintained. If there is no restriction that national movements are to be tracked by the extrapolated *PPPs* then η_{it} can be determined by the data rather than any a priori restriction imposed on its variance.

3.2 Estimation and Prediction of PPPs

The unknown parameters in Eq. (5) and (10) are estimated by maximum likelihood (the estimation method is described in detail in Rao et al., 2008). The likelihood function is computed by running the Kalman filter through the state-space equations. Upon convergence, and given these estimates and the initial distribution of the state vector, $\mathbf{p}_0$, the Kalman filter computes the conditional mean (based on the information available at time t), $\check{\boldsymbol{p}}_t$, and corresponding covariance matrix, $\boldsymbol{\Psi}_t$, of the distribution of $\mathbf{p}_t$. Further, $\check{\boldsymbol{p}}_t$ is a minimum mean square estimator (MMSE) of the state vector, $\boldsymbol{p}_t$, under Gaussian assumptions. When Gaussian assumptions are dropped, the Kalman filter is still the optimal estimator in the sense that it minimizes the mean square error within the class of all linear estimators (see Harvey, 1990 pp. 100–12), Durbin and Koopman (2001) Sect. 4.2 and 4.3). A fixed interval Kalman smoother is used to smooth the Kalman filter predictions and generate *PPPs* for all the countries and years in the data set. The interested reader is referred to Rao et al. (2008) for full description of the maximum likelihood estimation, the Kalman filter and smoother used in RRD.

Since the state vector $\boldsymbol{p}_t$ is in fact $ln(PPP_{it})$ and our interest is in *PPPs*, the following transformation is used to derive the predicted *PPPs*.

$$\widehat{PPP}_{it} = e^{p^*_{it}} \tag{11}$$

where, p^*_{it} is the corresponding Kalman smoothed element. The estimator $\widehat{PPP}_{it}$ is a biased estimator of PPP_{it}[7], which is a positive random variable.

The standard errors for the predicted *PPPs* are computed as follows[8]:

$$se(\widehat{PPP}_{it}) = \sqrt{e^{2p^*_{it}} e^{\psi^*_{ii,t}} (e^{\psi^*_{ii,t}} - 1)} \tag{12}$$

where,
$\psi^*_{ii,t}$ is the *ith* diagonal element of the estimated smoothed covariance of the state vector, $\boldsymbol{\Psi}^*_t$. Given that the distribution of $\widehat{PPP}_{it}$ is skewed, the standard error in eq. (12) should be used carefully when confidence intervals are constructed. The simplest approach would be to construct confidence intervals for $\ln(PPP_{it})$ and then take exponential of the resulting interval.

4 Spatial Specifications Considered in the Analysis

As briefly mentioned in the introduction, the latest round of the ICP was conducted in 2005 and the *PPPs* for the 146 participating countries released in early

[7] However, if the distribution of the disturbances in the state-space from are symmetric, it is median-unbiased. We thank an anonymous referee for making this point.

[8] The standard errors are computed under the assumption of the lognormality of the predictions.

2008. This round was the first global round of the ICP since 1996 when only 64 countries participated and there was no systematic linking of the results across countries. Thus, the 2005 round is a milestone both because of the number of countries that participated as well as the careful methodology applied to the data collected in order to produce the benchmark *PPPs* (see http://adb.org/Documents/Reports/ICP-Purchasing-Power-Expenditures/default.asp for a detailed report). It is then of interest to assess whether a spatial specification became less relevant once the comprehensive 2005 benchmark data were available. Specifically, we explore two hypotheses:

a) A spatially correlated error made a significant difference to the predictions of *PPPs* for non-participating countries when the ICP 2005 round was unknown
b) The use of a spatially correlated error does not significantly add to the predictions of *PPPs* when the 2005 ICP benchmarks are used.

To assess these hypothesis, we use RRD to construct tables of *PPPs* for the period 1970–2005 under the following alternatives:

No05 Assuming the year 2005 was not a benchmark year (this table is comparable to PWT 6.2) to produce tables from the model with and without a spatial error specification. Estimates based on assuming spherical errors in (4), $\mathbf{u}_t = \boldsymbol{e}_t$, are obtained by setting the spatial parameter ϕ to zero and will be denoted by the post-fix "No05_NoSpt" indicating the 2005 ICP data have been ignored and the errors are not spatial. Estimates obtained allowing for spatial errors (without constraining ϕ to zero) will be denoted with the post-fix "No05."

Y05 Including the benchmark ICP information for the year 2005 to produce tables from the model with and without a spatial error specification. Similar to above this is achieved by restricting the spatial parameter to zero to obtain a table without spatial errors. The estimates will be denoted by the post-fix "NoSpt" if they are obtained without the spatial error.

The definition of the weights matrix, $\boldsymbol{W}_t$ (see eq. (4)) could potentially have an influence in the results. It is important to reiterate that the objective of the spatial structure is to capture the "economic distance" between pairs of countries in an effort to improve the predictions of *PPPs* from the national price level's regression for those countries that did not participate in the ICP benchmark exercises. We explore three alternative specifications within each of the two above stated scenarios.

EW *Equal Weights*: The spatial weights matrix is constructed by identifying the five nearest trade "neighbours" for each country in the study. We define trade neighbours through bilateral trade flows. For each country, i (the ith row of $\boldsymbol{W}_t$), five columns have a value of 0.25 corresponding to its five major trading partners.

TW *Trade Weights*: Each row of $\boldsymbol{W}_t$, i, has columns with values that are directly proportional to the volume of bilateral trade between country i and j ($j \neq i$). Thus, any pair of countries in the sample that trades will have a non-zero weight and the sum of the weights for each row is one.

PCW *Economic Distance Factor Weights*: The spatial weights matrix, $\boldsymbol{W}_t$, is derived from a measure of economic distance constructed by the authors. The measure is constructed by extracting a common factor (through principal components analysis) for each country that combines trade closeness, geographical proximity, and cultural closeness. We present a brief description of its construction next and a more detailed description in Appendix 1.

4.1 Variables Included in the Measure of Economic Distance

- *Trade closeness* is measured as the percentage of bilateral trade between each country and all others in the sample (compiled using data from Rose, 2004 and IMF Trade Directions).
- *Geographical proximity* is measured by a series of dummies for *border* (both land and sea proximity), and *regional membership* (such us Asia pacific region, Europe, south America, north and central America, sub Saharan Africa, middle east). The data were constructed using Atlas, CIA factbook and individual country references.
- *Cultural and colonial closeness* dummies are used for *common language* and *common colonial history*. The data were constructed from the CIA factbook and individual country references.

4.2 Construction of the Distance Score

The objective is to measure "an economic distance" between pairs of countries. The steps involved in the construction of the measure can be summarised as follows:

1) A separate principal components (PC) model is estimated for each country to measure the distance between the respective country and each of the other countries in the sample. Therefore, for each time period 141 models are estimated. The analysis was conducted for the years 1970, 1975, 1980, 1985, 1990, 1995, 2000 and 2005 to account for the changing patterns in bilateral trade over time.
2) After the PC are extracted for a particular country and time period only one PC is retained since the number of variables is small. This is the *common factor* for each country and time period. That is, it is a linear combination of the variables.
3) A factor score is computed using the estimated factor loadings and the data (see Appendix 1 for details). These scores are not bounded; therefore, they are rescaled to prepare a *proximity matrix* using the formula:

$$S_{ij} = \left[\frac{f_{ij} - f_{\min}}{f_{\max} - f_{\min}}\right] \tag{13}$$

where, S_{ij} is the proximity score, $f_{\min}$, $f_{\max}$ and f_{ij} are respectively the minimum value, maximum value and factor score of country i in relation to j. These rescaled factor scores are in the range of 0–1, and if country g and j are the same (e.g. $i = j = 1$), the rescaled value is zero. The distance or proximity score is assumed to be constant within the five yearly intervals (e.g. from 1970 to 1974, 1975 to 1979, and so on).

4.3 Construction of the Weights Matrix

The proximity matrix is transformed into a row stochastic matrix $\boldsymbol{W}_t$ (i.e. rows add up to one) by simply dividing each proximity score within a row (which represents a country) by the sum of that row, and thus creating weights.

All three spatial specifications are estimated for both the No05 and Y05 cases. The next section presents the data used in the estimations.

5 Data and Empirical Results

In this section we present the empirical comparisons of the alternative scenarios described above. Section 5.1 describes the dataset used and Sect. 5.2 presents and discusses the empirical results

5.1 Data Compilation and Data Construction

The data set used in the study covers 141 countries over the years 1970–2005. It is worth noting that only 110 of the 141 countries in the sample participated in the 2005 round of the ICP. The dimensions of the data set were largely determined by data availability. That is, a number of countries were excluded because of missing data. The reader is referred to the data appendices in Rao et al. (2008) for a detailed description.

5.1.1 PPP Data

The state variable in the state space model is $ln(PPP_{it})$, and observed values (which define the dependent variable in the measurement equation) are obtained from all the benchmarks conducted since 1975. Thus *PPP* data are drawn from the early benchmarks of 1975, 1980 and 1985 as well as from more recent benchmark information for the years 1990, 1993, 1996, 1999, 2002 and 2005. Several features of the *PPP* data are noteworthy. The first benchmark covered 13 countries. The 1980, 1985 and the recent 2005, benchmarks represent truly global comparisons with *PPPs* computed using data for all the participating countries. For the years beginning from 1990 to 2002, data are essentially from the OECD and EU comparisons with the exception of 1996.[9] The 1996 benchmark year again is a global comparison with *PPPs* for countries from all the regions of the world. However, the 1996 benchmark may be considered weaker than the 1980, 1985 and 2005 benchmark comparisons as no systematic linking of regional *PPPs* was undertaken. In terms of reliability, one would consider the 1996 benchmark *PPPs* to be less reliable. Another related point of interest is the fact that *PPPs* for all the benchmarks prior to 1990 were based on the Geary-Khamis method and *PPPs* for the more recent years are all based on the EKS method of aggregation.[10] In the current empirical analysis, we have not made any adjustments to the *PPP* data but making the series comparable through the use of the same aggregation methodology is part of our ongoing research program.

5.1.2 Socio-Economic Variables Included in the Price Level Regression

The variables used come under two categories. We use a set of variables that are essentially dummy variables designed to capture country-specific episodes that may influence the exchange rates or *PPPs* or both as well as benchmark dummies (these are sufficient to insure invariance of the method to the choice of reference country). The second set of variables are more of a structural nature commonly discussed in the works of Kravis and Lipsey (1983), Kravis and Lipsey (1986), Clague (1988), Bergstrand (1991, 1996) and Ahmad (1996).[11] It is important to recall at this point that the role of the regression component of the model is to provide a prediction of the ln(*PPP*) and thus the emphasis is not on the marginal effect of individual variables but on its overall prediction ability.

5.1.3 Covariance Variables

Measuring Spatial Correlation

The alternative spatial weights matrices, W_t, used in modeling spatial autocorrelation were discussed in Sect. 4 above, and Appendix 1 presents a detailed description of the method and summary statistics of the data used to generate each alternative for a subset of countries. Due to space constrains not all countries are shown. However, the data are available upon request from the authors.

---

[9]We are indebted to Ms Francette Koechlin (OECD) for providing ICP benchmark data for these years. PPPs for those countries which joined in the Euro zone, the pre-Euro domestic currencies were converted using the 1999 Irrevocable Conversion Rates (Source:http://www.ecb.int/press/date/1998/html/pr981231_2.en.html). The irrevocable conversion rate of the drachma vis a vis the euro was set at GRD 340.750 Source: http://www.bankofgreece.gr/en/euro.

[10]This was brought to our attention by Steve Dowrick who attended a seminar on the topic presented at the Australian National University in October 2007.

[11]We are conscious of the fact that serious multicollinearity issues may be present here as the variables are potentially correlated. As the main purpose of inclusion of these variables is to improve the quality of the predictions, we decided to leave the variables in the model with the view that the model results in better predictions.

Accuracy of Benchmarks and National Accounts' Growth Rates
The specification includes the modeling of the accuracy of benchmark *PPPs* and national growth rates. We assume that the measurement errors in both cases have variances that are inversely proportional to the per capita GDP expressed in US dollars. This means that countries with higher per capita incomes are expected to have more reliable data, as reflected by lower variances associated with them.[12]

5.2 Empirical Evidence

5.2.1 Parameter Estimates and Tests for Spatial Correlation

Tables 1, 2 and 3 summarise the results obtained from the estimation of the models and testing for spatial autocorrelation. Table 1 presents the estimates of the Y05 model, that is when the 2005 ICP data are included (2005 is a benchmark year) under the three spatial weights matrices as well as when the model assumes no spatially correlated errors. Panel 1 presents the least squares estimates of the price level model while the other panels present the maximum likelihood estimates of all the parameters in the state-space model. Table 2 shows the same information as Table 1 for the No05 model, that is, the 2005 ICP data are not included since 2005 is treated as a non-benchmark year. Table 3 presents the computed LM Statistics for the null hypothesis of no spatial errors (with p-values) computed for two global ICP benchmark years, 1985 and 2005.

The results can be summarised as follows,

- The goodness of fit of the price level regression is high. The R^2 of the pooled regression including the 2005 data is 0.737 (see Panel 1, Table 1). The sample contains 449 benchmark points from 1975 to 2005. The R^2 of the pooled regression for the model not including the 2005 data is 0.753 (see Panel 1, Table 2), and the sample size is 339. The same set of conditioning variables are used in both cases. These results indicate that this important component of the state-space model is strong and able to provide reasonable predictions for non-participating countries and non-benchmark years.
- The estimates of some of the regression slopes change when these parameters are estimated jointly with the covariance parameters by maximum likelihood in the state-space model, although some do not change substantially. For example the coefficient for PHONES, RURPOP and LIFE remain almost unchanged.

[12]We make use of exchange rate converted per capita incomes to overcome the problem of possible endogeneity arising out of the use of *PPP* converted exchange rates. These data are drawn from the UN sources. Given the systematic nature of the exchange rate deviation index (ratio of *PPP* to *ER*), use of exchange rate converted per capita GDP is likely to magnify differences in per capita incomes.

Table 1 Parameter estimates under alternative specifications -Y05 model

	Regression		State Space Model Y05							
	Without spatial errors (Panel 1)		Spatial PC weights (Panel 2)		Spatial TW weights (Panel 3)		Spatial EW weights (Panel 4)		No spatial (Panel 5)	
Variable	Estimate	S.E.	Estimate	S.E.	Estimate	S.E.	Estimate	S.E.	Estimate	S.E.
Intercept			2.078	1.338	3.780	1.325	2.462	1.329	3.200	1.172
dum75_79	−0.199	0.063	0.868	0.341	1.191	0.350	1.675	0.354	0.030	0.153
dum80_84	−0.128	0.238	0.668	0.331	0.977	0.337	1.011	0.338	−0.130	0.155
dum85_89	−0.621	0.239	−0.040	0.324	0.105	0.327	0.257	0.327	−0.865	0.154
dum90_92	−0.282	0.244	−0.504	0.376	−0.742	0.378	−0.671	0.379	−0.937	0.168
dum93_95	−0.451	0.244	−1.059	0.375	−1.212	0.377	−1.012	0.378	−1.275	0.170
dum96_98	−0.445	0.241	−1.571	0.376	−1.808	0.378	−1.716	0.381	−1.538	0.174
dum99_01	−0.692	0.246	−1.484	0.382	−1.602	0.387	−1.280	0.391	−1.624	0.180
dum02_04	−0.786	0.238	−1.650	0.389	−1.828	0.399	−1.480	0.405	−1.770	0.185
dum05	−0.540	0.237	−3.306	0.582	−4.752	0.591	−4.761	0.594	−2.693	0.233
D_anz	−0.770	0.221	−0.443	0.394	−0.519	0.362	−0.328	0.380	−0.395	0.400
D_asean	0.016	0.080	0.075	0.281	0.137	0.267	0.010	0.269	0.010	0.250
D_cac	−0.029	0.155	0.221	0.278	0.203	0.277	0.311	0.292	0.372	0.255
D_cafrica	0.101	0.116	0.033	0.321	0.077	0.314	−0.025	0.311	0.329	0.306
D_eafrica	0.118	0.094	0.090	0.283	0.105	0.280	−0.007	0.277	0.321	0.264
D_euro	0.092	0.045	0.104	0.170	0.169	0.174	0.065	0.170	0.164	0.170
D_mena	0.045	0.073	−0.041	0.194	−0.103	0.190	−0.193	0.187	−0.032	0.184
D_mercsr	−0.081	0.082	0.720	0.274	0.846	0.267	0.817	0.272	1.182	0.244
D_nafta	−0.243	0.086	−0.023	0.305	0.034	0.311	−0.025	0.285	0.032	0.284
D_safrica	0.066	0.122	−0.052	0.302	−0.085	0.290	−0.190	0.289	0.171	0.284
D_scucar	0.228	0.148	0.293	0.261	0.159	0.266	0.281	0.264	0.432	0.254
D_spr	0.632	0.206	0.925	0.302	0.810	0.290	0.696	0.287	0.833	0.269
D_usd	0.073	0.069	0.569	0.138	0.576	0.138	0.589	0.133	0.571	0.137
D_wafrica	0.256	0.089	−0.551	0.269	−0.526	0.263	−0.666	0.259	−0.326	0.253
Agedep	0.365	0.174	−0.258	0.571	−0.344	0.560	−0.226	0.560	−0.458	0.556

Table 1 (continued)

Variable	Regression: Without spatial errors (PANEL 1) Estimate	S.E.	STATE SPACE MODEL Y05: Spatial PC weights (PANEL 2) Estimate	S.E.	Spatial TW weights (PANEL 3) Estimate	S.E.	Spatial EW weights (PANEL 4) Estimate	S.E.	No spatial (PANEL 5) Estimate	S.E.
AGVAGUN	–0.009	0.002	–0.019	0.007	–0.021	0.007	–0.020	0.007	–0.019	0.007
TRACTORPW	0.094	0.061	0.159	0.245	0.216	0.245	0.253	0.239	0.308	0.242
LABPOP	–0.003	0.003	–0.013	0.011	–0.015	0.011	–0.012	0.011	–0.025	0.011
LIFE	–0.006	0.004	–0.007	0.012	–0.012	0.012	–0.008	0.012	–0.013	0.011
LITERATE	2.1E-04	1.4E-04	–4.0E-04	4.2E-04	–4.0E-04	4.1E-04	–4.8E-04	4.2E-04	– 2.4E-04	4.1E-04
NTRVAG2	–0.004	0.003	–0.012	0.008	–0.013	0.008	–0.016	0.008	–0.010	0.008
EXPG	–0.002	0.003	–0.006	0.006	–0.008	0.006	–0.006	0.006	–0.005	0.006
PHONES	0.001	1.8E-04	0.003	0.001	0.003	0.001	0.003	0.001	0.003	0.001
RADPCCN	5.0E-06	7.0E-06	–5.5E-05	2.3E-05	–4.8E-05	2.3E-05	–6.7E-05	2.3E-05	–6.6E-05	2.3E-05
RURPOP	–0.004	0.001	–0.004	0.005	–0.004	0.005	–0.005	0.005	–0.004	0.005
SECENR	3.3E-05	5.4E-05	–9.5E-05	1.7E-04	–1.1E-04	1.6E-04	–9.5E-05	1.6E-04	–1.9E-04	1.7E-04
TRADEGUN	–2.1E-04	0.002	2.4E-05	3.2E-03	1.4E-04	0.003	–8.0E-05	3.2E-03	–0.002	0.003
MANUFEXP	–2.4E-04	0.001	0.002	0.002	0.002	0.002	0.001	0.002	0.003	0.002
MANUFIMP	0.003	0.001	0.004	0.004	0.003	0.004	0.006	0.004	0.004	0.004
R^2	0.737									
lnL			–1.30e+07		–1.37e+07		–1.33e+07		–1.30e+07	

Table 1 (continued)

Variable	Regression		State Space Model Y05							
	Without spatial errors (Panel 1)		Spatial PC weights (Panel 2)		Spatial TW weights (Panel 3)		Spatial EW weights (Panel 4)		No spatial (Panel 5)	
	Estimate	S.E.	Estimate	S.E.	Estimate	S.E.	Estimate	S.E.	Estimate	S.E.
σ_η^2			7.00		7.01		7.00		7.00	
σ_u^2			4.50		4.50		4.50		4.50	
σ_ξ^2			0.79		0.80		0.80		1.00	
ϕ			0.70		0.70		0.69		0.00	

Table 2 Parameter estimates under alternative specifications - No05 model

Variable	Regression Without spatial errors (Panel1) Estimate	S.E.	State Space Model No05 Spatial PC weights (Panel 2) Estimate	S.E.	Spatial TW weights (Panel 3) Estimate	S.E.	Spatial EW weights (Panel 4) Estimate	S.E.	No spatial (Panel 5) Estimate	S.E.
Intercept			2.864	1.474	4.294	1.460	3.339	1.460	3.553	1.334
dum75_79	–0.197	0.064	0.137	0.311	0.188	0.311	0.429	0.311	–0.435	0.188
dum80_84	–2.296	0.540	–2.005	0.306	–1.965	0.305	–1.966	0.305	–2.538	0.191
dum85_89	–2.810	0.544	–2.839	0.300	–2.914	0.299	–2.809	0.299	–3.330	0.189
dum90_92	–2.507	0.548	–3.192	0.337	–3.480	0.337	–3.392	0.339	–3.419	0.206
dum93_95	–2.697	0.551	–3.697	0.338	–3.951	0.338	–3.796	0.340	–3.782	0.209
dum96_98	–2.686	0.549	–4.157	0.340	–4.501	0.341	–4.347	0.345	–4.069	0.214
dum99_01	–2.957	0.554	–4.166	0.347	–4.415	0.350	–4.184	0.354	–4.159	0.222
dum02_04	–3.078	0.556	–4.485	0.335	–4.742	0.342	–4.559	0.347	–4.418	0.221
D_anz	–0.792	0.302	–0.491	0.457	–0.538	0.426	–0.358	0.445	–0.451	0.468
D_asean	0.034	0.097	0.092	0.307	0.153	0.295	0.062	0.298	0.063	0.285
D_cac	–0.054	0.158	0.274	0.307	0.295	0.308	0.351	0.321	0.361	0.293
D_cafrica	0.324	0.151	0.348	0.353	0.392	0.347	0.319	0.345	0.543	0.347
D_eafrica	0.274	0.113	0.322	0.311	0.329	0.308	0.254	0.305	0.473	0.300
D_euro	0.119	0.052	0.173	0.201	0.212	0.206	0.135	0.203	0.193	0.201
D_mena	0.123	0.096	0.063	0.219	0.025	0.215	–0.024	0.213	0.048	0.212
D_mercsr	0.095	0.103	0.881	0.304	0.991	0.300	0.955	0.304	1.262	0.282
D_nafta	–0.243	0.094	–0.011	0.352	0.020	0.363	0.024	0.337	0.025	0.336
D_safrica	–0.017	0.155	–0.002	0.338	–0.017	0.327	–0.095	0.326	0.130	0.326
D_scucar	0.184	0.158	0.279	0.295	0.194	0.301	0.260	0.299	0.387	0.293
D_spr	0.584	0.288	0.896	0.326	0.848	0.317	0.761	0.314	0.844	0.305
D_usd	0.112	0.078	0.550	0.159	0.559	0.159	0.556	0.154	0.560	0.160
D_wafrica	0.542	0.118	–0.139	0.297	–0.118	0.291	–0.211	0.288	–0.010	0.289
AGEDEP	0.962	0.251	0.272	0.641	0.104	0.631	0.217	0.632	0.136	0.632
AGVAGUN	–0.006	0.003	–0.014	0.007	–0.016	0.007	–0.016	0.007	–0.014	0.007

Table 2 (continued)

Variable	Regression Without spatial errors (Panel1) Estimate	S.E.	State Space Model No05 Spatial PC weights (Panel 2) Estimate	S.E.	Spatial TW weights (Panel 3) Estimate	S.E.	Spatial EW weights (Panel 4) Estimate	S.E.	No spatial (Panel 5) Estimate	S.E.
TRACTORPW	0.026	0.072	0.071	0.281	0.108	0.280	0.141	0.276	0.158	0.278
LABPOP	0.001	0.004	–0.013	0.013	–0.016	0.013	–0.016	0.013	–0.023	0.012
LIFE	0.013	0.006	0.009	0.013	0.004	0.013	0.007	0.013	0.005	0.013
LITERATE	1.4E-04	1.9E-04	–3.9E-04	4.6E-04	–3.6E-04	4.6E-04	–3.9E-04	4.6E-04	–2.3E-04	4.6E-04
NTRVAG2	–0.003	0.003	–0.012	0.009	–0.013	0.009	–0.015	0.009	–0.010	0.009
EXPG	0.008	0.005	0.003	0.007	0.002	0.007	0.003	0.007	0.004	0.007
PHONES	0.001	2.2E-04	0.002	0.001	0.003	0.001	0.003	0.001	0.003	0.001
RADPCCN	5.0E-06	8.0E-06	–4.4E-05	2.6E-05	–4.3E-05	2.6E-05	–5.4E-05	2.6E-05	–5.4E-05	2.6E-05
RURPOP	–0.003	0.002	–0.005	0.005	–0.005	0.005	–0.004	0.005	–0.005	0.005
SECENR	1.3E-04	7.4E-05	6.0E-06	1.9E-04	1.2E-05	1.9E-04	–1.2E-05	1.9E-04	–6.5E-05	1.9E-04
TRADEGUN	–0.005	0.003	–0.004	0.004	–0.004	0.004	–0.005	0.004	–0.005	0.004
MANUFEXP	0.001	0.001	0.003	0.003	0.003	0.003	0.003	0.003	0.004	0.003
MANUFIMP	0.005	0.001	0.005	0.004	0.004	0.004	0.006	0.004	0.005	0.004
R^2	0.753									
lnL			–1.39e+04		–1.41e+04		–1.42e+04		–1.39e+04	
σ_η^2			7.00		6.98		7.01		8.00	
σ_u^2			6.50		6.50		6.50		6.50	
σ_ξ^2			0.80		0.80		0.81		0.80	
ϕ			0.55		0.56		0.55		0.00	

Table 3 LM tests for spatial correlation

W_t	PCW		TW		EW	
	LM	P-VALUE	LM	P-VALUE	LM	P-VALUE
LS RESID 1985	**0.122**	0.726	**0.069**	0.793	**2.791**	0.095
LS RESID 2005	**4.48**	0.034	**4.208**	0.040	**14.54**	0.0001

- The log-likelihood value is highest for the model without spatial correlation (Panel 5, when $\phi = 0$) and that with PC weights (Panel 2). This is the case for both models (Y05 and No05).
- The estimate of the spatial parameter is around 0.7 for the Y05 model and 0.55 for the No05 model.
- The computed LM statistic for the null hypothesis of no spatial correlation (the reader is referred to Florax and Graaff (2004) for a description of this test) provides evidence of spatial autocorrelation. The residuals of the multiple regression in Panel 1 of Table 1 are used for the testing. Table 3 presents the computed values. The LM statistics is computed for the least squares residuals corresponding to the year 1985 (56 observations) and those corresponding to the year 2005 (110 observations) and for each of the alternative specification of the $\boldsymbol{W}_t$ matrix. The null is rejected in all cases when the 2005 residuals are used, but only rejected in one case (EW) when the 1985 residuals are used.

In summary, the variables included in the price level regression seem to explain over 70% of the variance of the deviation of *PPPs* from *ERs;* there is evidence that the errors of the price level regression are spatially correlated although the likelihood value of the models with and without spatial errors do not differ substantially. Since the main objective of the RRD method is to produce a panel of *PPPs,* the next section evaluates the prediction accuracy of the constructed panels.

5.2.2 PPP Predictions and Prediction Performance

As an illustration of the *PPPs* series that are produced by the method, Tables 4, 5 and 6 present the constructed series for three countries (Spain, China and India) for the combinations, Y05_PCW, Y05_TW, Y05_EW, No05_PCW, No05_TW and No05_EW. Table 7 presents the computed *mean percent absolute deviation (MPAD)* for each combination, where *MPAD* is defined as follows:

$$MPAD = \frac{1}{110}\left(\sum_{i=1}^{110}\left|\frac{P\tilde{P}P_i - P\hat{P}P_i}{P\tilde{P}P_i}\right| \times 100\right) \tag{14}$$

where,

$P\tilde{P}P_i$ is the ICP benchmark observation for country i in the 2005 round.
$P\hat{P}P_i$ is the RRD predicted PPP value for country i in 2005.

Table 4 *PPP* series for spain (implied price movements preserved)

YEAR	ER	ICP	Y05				No05				
			PC weights	Trade weights	Equal weights	No spatial	PC weights	Trade weights	Equal weights	No spatial	PWT6.2[(a)]
1971	0.4175		0.1763	0.1763	0.1763	0.1760	0.1763	0.1768	0.1765	0.1724	32.9789
1972	0.3863		0.1835	0.1836	0.1836	0.1833	0.1836	0.1840	0.1838	0.1795	34.1987
1973	0.3502		0.1945	0.1945	0.1945	0.1942	0.1945	0.1950	0.1947	0.1902	36.2469
1974	0.3467		0.2068	0.2068	0.2068	0.2065	0.2068	0.2073	0.2070	0.2023	39.0841
1975	0.3450	**0.2540**	0.2206	0.2207	0.2207	0.2203	0.2207	0.2212	0.2209	0.2158	41.4173
1976	0.4021		0.2430	0.2431	0.2431	0.2427	0.2431	0.2437	0.2433	0.2377	45.7205
1977	0.4565		0.2820	0.2820	0.2820	0.2816	0.2820	0.2827	0.2823	0.2758	52.9058
1978	0.4608		0.3178	0.3179	0.3179	0.3174	0.3179	0.3187	0.3182	0.3109	58.9438
1979	0.4034		0.3431	0.3432	0.3432	0.3427	0.3432	0.3440	0.3435	0.3357	63.1692
1980	0.4309	**0.3825**	0.3566	0.3566	0.3566	0.3561	0.3567	0.3575	0.3570	0.3488	66.3191
1981	0.5549		0.3662	0.3663	0.3663	0.3657	0.3663	0.3672	0.3667	0.3582	69.6305
1982	0.6603		0.3921	0.3921	0.3921	0.3915	0.3922	0.3931	0.3925	0.3835	74.7093
1983	0.8620		0.4220	0.4220	0.4220	0.4214	0.4221	0.4231	0.4225	0.4128	81.3117
1984	0.9662		0.4509	0.4510	0.4510	0.4503	0.4510	0.4521	0.4514	0.4411	86.6931
1985	1.0220	**0.5728**	0.4751	0.4751	0.4751	0.4744	0.4752	0.4763	0.4756	0.4647	90.8481
1986	0.8417		0.5153	0.5153	0.5153	0.5146	0.5154	0.5166	0.5159	0.5040	96.2679
1987	0.7421		0.5312	0.5313	0.5313	0.5305	0.5314	0.5326	0.5319	0.5197	98.5311
1988	0.7001		0.5441	0.5442	0.5442	0.5433	0.5442	0.5455	0.5448	0.5322	100.2722
1989	0.7115		0.5604	0.5604	0.5604	0.5596	0.5605	0.5618	0.5610	0.5481	102.9329
1990	0.6126	**0.6581**	0.5790	0.5791	0.5791	0.5782	0.5792	0.5805	0.5797	0.5664	105.1833
1991	0.6245		0.5983	0.5983	0.5983	0.5974	0.5984	0.5998	0.5990	0.5852	108.4699
1992	0.6153		0.6241	0.6241	0.6241	0.6232	0.6242	0.6257	0.6248	0.6104	112.8342
1993	0.7649	**0.7032**	0.6377	0.6378	0.6378	0.6368	0.6379	0.6394	0.6385	0.6238	116.3522
1994	0.8051		0.6488	0.6488	0.6488	0.6479	0.6489	0.6505	0.6495	0.6346	119.0284
1995	0.7494		0.6672	0.6672	0.6672	0.6662	0.6673	0.6689	0.6680	0.6526	121.9997
1996	0.7613	**0.7433**	0.6774	0.6774	0.6774	0.6764	0.6775	0.6791	0.6781	0.6626	124.1070
1997	0.8800		0.6821	0.6822	0.6822	0.6811	0.6823	0.6839	0.6829	0.6672	125.8340

Table 4 (continued)

YEAR	ER	ICP	Y05				No05				
			PC weights	Trade weights	Equal weights	No spatial	PC weights	Trade weights	Equal weights	No spatial	PWT6.2[(a)]
1998	0.8979		0.6913	0.6914	0.6914	0.6904	0.6915	0.6932	0.6922	0.6763	127.7315
1999	0.9386	**0.7493**	0.6994	0.6995	0.6995	0.6984	0.6996	0.7012	0.7002	0.6841	0.7773
2000	1.0854		0.7081	0.7082	0.7082	0.7071	0.7083	0.7100	0.7089	0.6927	0.7923
2001	1.1175		0.7203	0.7204	0.7204	0.7193	0.7205	0.7222	0.7212	0.7046	0.8002
2002	1.0626	**0.7428**	0.7393	0.7394	0.7394	0.7383	0.7395	0.7413	0.7402	0.7232	0.8121
2003	0.8860		0.7532	0.7533	0.7533	0.7522	0.7534	0.7552	0.7541	0.7368	0.8220
2004	0.8054		0.7637	0.7638	0.7638	0.7626	0.7639	0.7657	0.7646	0.7470	0.8361
2005	0.8041	**0.7700**	0.7700	0.7700	0.7700	0.7689	0.7701	0.7720	0.7709	0.7532	
SE			**0.0063**	**0.0063**	**0.0063**	**0.0070**	**0.0339**	**0.0340**	**0.0339**	**0.0352**	

[(a)] 1971–1998 in Pesetas, 1999 onwards in Euros

Table 5 *PPP* series for china (Implied Price Movements Preserved)

YEAR	ER	ICP	Y05				No05				
			PC weights	Trade weights	Equal weights	No spatial	PC weights	Trade weights	Equal weights	No spatial	PWT6.2
1971	2.4600		3.0921	3.0908	3.0910	3.0687	3.2156	2.8455	2.8976	0.7692	1.8079
1972	2.2500		2.9659	2.9647	2.9650	2.9435	3.0844	2.7294	2.7794	0.7379	1.7330
1973	1.9900		2.8133	2.8121	2.8123	2.7920	2.9257	2.5889	2.6363	0.6999	1.6537
1974	1.9600		2.5862	2.5852	2.5854	2.5667	2.6896	2.3800	2.4235	0.6434	1.4735
1975	1.8600		2.3354	2.3345	2.3347	2.3178	2.4288	2.1492	2.1885	0.5810	1.3468
1976	1.9400		2.2042	2.2033	2.2035	2.1875	2.2923	2.0284	2.0655	0.5484	1.2158
1977	1.8600		2.0953	2.0944	2.0946	2.0794	2.1790	1.9282	1.9635	0.5212	1.1718
1978	1.6800		1.9837	1.9829	1.9831	1.9687	2.0630	1.8255	1.8590	0.4935	1.1341
1979	1.5500		1.8967	1.8960	1.8961	1.8824	1.9725	1.7455	1.7774	0.4719	1.0551
1980	1.5000		1.8046	1.8038	1.8040	1.7909	1.8767	1.6606	1.6910	0.4489	0.9971
1981	1.7000		1.6877	1.6870	1.6871	1.6749	1.7551	1.5531	1.5815	0.4199	0.9372
1982	1.8900		1.5877	1.5871	1.5872	1.5757	1.6512	1.4611	1.4879	0.3950	0.9002
1983	1.9800		1.5436	1.5430	1.5431	1.5320	1.6053	1.4205	1.4465	0.3840	0.8881
1984	2.3200		1.5606	1.5599	1.5600	1.5488	1.6229	1.4361	1.4624	0.3882	0.8779
1985	2.9400		1.6676	1.6669	1.6671	1.6550	1.7342	1.5346	1.5627	0.4149	0.9262
1986	3.4500		1.7063	1.7056	1.7058	1.6934	1.7745	1.5703	1.5990	0.4245	0.9518
1987	3.7200		1.7446	1.7439	1.7440	1.7314	1.8143	1.6055	1.6349	0.4340	0.9810
1988	3.7200		1.8912	1.8905	1.8906	1.8769	1.9668	1.7404	1.7723	0.4705	1.1012
1989	3.7700		1.9826	1.9818	1.9819	1.9676	2.0618	1.8245	1.8579	0.4932	1.1724
1990	4.7800		2.0171	2.0163	2.0164	2.0018	2.0977	1.8562	1.8902	0.5018	1.1418
1991	5.3200		2.0802	2.0793	2.0795	2.0644	2.1633	1.9143	1.9493	0.5175	1.1610
1992	5.5100		2.1940	2.1931	2.1933	2.1774	2.2817	2.0190	2.0560	0.5458	1.2352
1993	5.7600		2.4955	2.4945	2.4947	2.4766	2.5952	2.2965	2.3385	0.6208	1.4751
1994	8.6200		2.9476	2.9464	2.9466	2.9253	3.0654	2.7125	2.7622	0.7333	1.7552
1995	8.3500		3.2855	3.2841	3.2844	3.2606	3.4167	3.0235	3.0788	0.8173	1.9755
1996	8.3100		3.4316	3.4302	3.4305	3.4056	3.5687	3.1579	3.2157	0.8537	2.0745
1997	8.2900		3.4262	3.4249	3.4251	3.4003	3.5631	3.1530	3.2107	0.8524	2.0825

Table 5 (continued)

YEAR	ER	ICP	Y05				No05				
			PC weights	Trade weights	Equal weights	No spatial	PC weights	Trade weights	Equal weights	No spatial	PWT6.2
1998	8.2800		3.3595	3.3581	3.3584	3.3341	3.4937	3.0916	3.1482	0.8358	2.0595
1999	8.2800		3.2701	3.2688	3.2690	3.2454	3.4008	3.0093	3.0644	0.8135	1.9973
2000	8.2800		3.2664	3.2650	3.2653	3.2417	3.3969	3.0059	3.0609	0.8126	1.9630
2001	8.2800		3.2550	3.2537	3.2539	3.2304	3.3850	2.9954	3.0502	0.8098	1.9597
2002	8.2800		3.2177	3.2164	3.2166	3.1934	3.3463	2.9611	3.0153	0.8005	1.9411
2003	8.2800		3.2359	3.2345	3.2348	3.2114	3.3652	2.9778	3.0323	0.8050	1.9769
2004	8.2800		3.3711	3.3697	3.3700	3.3456	3.5058	3.1022	3.1590	0.8386	2.1447
2005	8.1943	**3.45**	3.4546	3.4532	3.4535	3.4285	3.5926	3.1791	3.2373	0.8594	
SE			**0.0925**	**0.0924**	**0.0924**	**0.1026**	**2.3203**	**2.0502**	**2.0880**	**0.6014**	

Table 6 *PPP* series for India (implied price movements preserved)

YEAR	ER	ICP	Y05				No05				
			PC weights	Trade weights	Equal weights	No spatial	PC weights	Trade weights	Equal weights	No spatial	PWT6.2
1971	7.4900		4.3430	4.3441	4.3399	4.3146	3.6719	3.8755	3.7932	2.2284	2.8198
1972	7.5900		4.6204	4.6216	4.6172	4.5903	3.9065	4.1231	4.0355	2.3708	2.9330
1973	7.7400		5.1595	5.1608	5.1559	5.1259	4.3623	4.6041	4.5063	2.6474	3.2572
1974	8.1000		5.5225	5.5239	5.5186	5.4865	4.6692	4.9280	4.8234	2.8337	3.6413
1975	8.3800	**2.5940**	4.9675	4.9687	4.9640	4.9351	4.1999	4.4328	4.3386	2.5489	3.2869
1976	8.9600		4.9774	4.9786	4.9739	4.9449	4.2083	4.4416	4.3473	2.5540	3.1441
1977	8.7400		4.9415	4.9427	4.9380	4.9093	4.1779	4.4096	4.3159	2.5356	3.1191
1978	8.1900		4.7335	4.7347	4.7302	4.7027	4.0021	4.2240	4.1343	2.4289	3.0267
1979	8.1300		5.0607	5.0620	5.0572	5.0277	4.2787	4.5160	4.4201	2.5967	3.0674
1980	7.8600	**3.1045**	5.1732	5.1746	5.1696	5.1395	4.3739	4.6164	4.5183	2.6545	3.0990
1981	8.6600		5.2145	5.2158	5.2108	5.1805	4.4087	4.6532	4.5543	2.6756	3.1755
1982	9.4600		5.2947	5.2960	5.2910	5.2602	4.4765	4.7247	4.6244	2.7168	3.1664
1983	10.1000		5.5457	5.5472	5.5419	5.5096	4.6888	4.9488	4.8437	2.8456	3.3516
1984	11.4000		5.7418	5.7433	5.7378	5.7044	4.8546	5.1238	5.0150	2.9462	3.4445
1985	12.4000	**4.6670**	5.9716	5.9731	5.9674	5.9326	5.0488	5.3288	5.2156	3.0641	3.5533
1986	12.6000		6.2370	6.2386	6.2326	6.1963	5.2733	5.5656	5.4474	3.2003	3.6601
1987	13.0000		6.6284	6.6301	6.6237	6.5852	5.6042	5.9149	5.7893	3.4011	3.7893
1988	13.9000		6.9401	6.9419	6.9352	6.8949	5.8677	6.1931	6.0615	3.5611	4.0469
1989	16.2000		7.2445	7.2463	7.2394	7.1972	6.1250	6.4647	6.3274	3.7173	4.2150
1990	17.5000		7.7103	7.7123	7.7049	7.6600	6.5189	6.8803	6.7342	3.9563	4.4348
1991	22.7000		8.4792	8.4814	8.4733	8.4239	7.1690	7.5665	7.4058	4.3508	4.8928
1992	25.9000		9.0215	9.0238	9.0152	8.9627	7.6275	8.0504	7.8795	4.6291	5.2393
1993	30.5000		9.6550	9.6575	9.6482	9.5921	8.1631	8.6157	8.4327	4.9542	5.4891
1994	31.4000		10.3713	10.3740	10.3640	10.3037	8.7687	9.2549	9.0584	5.3217	6.0479
1995	32.4000		11.0759	11.0787	11.0681	11.0037	9.3645	9.8837	9.6738	5.6832	6.5390
1996	35.4000		11.6557	11.6587	11.6475	11.5797	9.8546	10.4010	10.1802	5.9807	6.8681
1997	36.3000		12.2106	12.2137	12.2020	12.1310	10.3238	10.8962	10.6648	6.2655	7.0947

Table 6 (continued)

YEAR	ER	ICP	Y05				No05				
			PC weights	Trade weights	Equal weights	No spatial	PC weights	Trade weights	Equal weights	No spatial	PWT6.2
1998	41.3000		13.0289	13.0322	13.0198	12.9440	11.0157	11.6264	11.3795	6.6854	7.5103
1999	43.1000		13.4420	13.4454	13.4326	13.3544	11.3649	11.9951	11.7403	6.8973	7.7342
2000	44.9000		13.6184	13.6219	13.6089	13.5297	11.5141	12.1525	11.8944	6.9879	7.8449
2001	47.2000		13.7141	13.7176	13.7045	13.6247	11.5950	12.2379	11.9780	7.0370	8.0032
2002	48.6000		14.0024	14.0060	13.9926	13.9111	11.8388	12.4952	12.2298	7.1849	8.1162
2003	46.6000		14.2440	14.2476	14.2340	14.1511	12.0430	12.7107	12.4408	7.3088	8.1461
2004	45.3000		14.4862	14.4899	14.4760	14.3917	12.2478	12.9268	12.6523	7.4331	
2005	44.2725	**14.6700**	14.6851	14.6889	14.6748	14.5894	12.4160	13.1044	12.8261	7.5352	
SE			**0.5038**	**0.5039**	**0.5034**	**0.5593**	**7.0325**	**7.4189**	**7.2084**	**4.5895**	

Table 7 Performance in the prediction of the 2005 ICP benchmark by alternative spatial specifications

Mean percent absolute deviation	PPP PCW	PPP TW	PPP EW	PPP NOSPT
Model Estimated Without 2005 ICP Data	34.04	33.58	34.72	38.40
Model Estimated With 2005 ICP Data	0.45	0.44	0.47	0.76

Thus, for each case the *MPAD* value in the body of the table can be read as a per cent. For example, the average deviation in the *PPP* predictions for the 110 countries for the year 2005 obtained from the No05_PCW model is 34.04%, which decreases to 0.45% when the 2005 ICP data are incorporated (Y05_PCW). The predictions from the "Y05_" models are *in-sample* predictions given that the 2005 data are used to estimate the model and therefore expected to be much closer to the ICP observed values although not necessarily identical given the model accounts for some measurement error. Using the *MPAD* we can also evaluate the performance of the models with and without a spatial error. By treating 2005 as a non-benchmark year, it is clear that the predictions made by the model without spatial errors are worse than those made using any of the alternative spatial specifications. The spatial specifications PCW and TW have an *MPAD* of 34%, the EW of 35% and the NoSpt of 38%. What is interesting is that this pattern still exists when the *MPAD* of the Y05 model are considered. The *MPAD* of the PCW is 0.45%, TW is 0.44%, EW is 0.47% and NoSpt is 0.76%.

The use of averages allows an overall comparison, however, it can conceal the extent of the deviations for some countries. For instance, China's predictions from the No05 models (Table 5) show that the *PPP* predictions from the model with spatial errors (either of them) were significantly closer to the ICP measurement than that of the model without spatial errors. The ICP value was Yuan 3.45, the predictions of the spatial models were Yuan 3.59 from the No05_PWC, Yuan 3.18 from the No05_TW, Yuan 3.24 from the No05_EW and Yuan 0.86 from the No05_NoSpt. China had never participated in an ICP exercise before 2005, and thus, the predictions of the spatial models came remarkably close to the ICP benchmark of 2005. Further, the PWT6.2 includes predictions up to the year 2004 and so we can compare the predicted value from the No05 models for the year 2004. The spatial versions of RRD predict: Yuan 3.51, Yuan 3.10, and Yuan 3.16 (PWC, TW, EW, respectively), while PWT6.2 is Yuan 2.15. A set of similar arguments can be made for India (see Table 6).

From the discussion so far we conclude that the use of a spatial error structure produces a substantial improvement in the prediction of *PPPs* both on average and for the less developed countries in the sample even when the 2005 data are incorporated in the estimation. However, this might not be the case for OECD countries that have been involved in frequent comparisons since the early 1990's. Table 4 presents the predicted series for Spain. Spain has been involved in the EUROSTAT/OECD comparisons of 1990, 1993, 1996, 1999, 2002 and 2005. It is however interesting to note that while the spatial versions of the No05 model predicted Euro 0.77 for

2005 (ICP value for 2005 is 0.77), the non-spatial version predicted 0.75, a small deviation. For the Y05 model, all predictions are Euro 0.77. This pattern repeats across all the EUROSTAT/OECD countries (available from the authors). Thus, the use of a spatial error does not change the predictions in any significant way. This is reassuring, as the comparisons of EUROSTAT/OECD are of high quality and thus the model should not adjust observed international comparison's values. Finally, we note that the PWT6.2 for 2004 comes higher than any of the predictions made from any of the versions of the RRD method (Euro 0.84).

6 Conclusions

The paper builds on the earlier work of Rao et al. (2008, 2009) where a comprehensive econometric approach to the construction of consistent time-space extrapolations of PPPs was proposed. The Rao et al. approach makes use of a particular specification for the spatially correlated errors and in this paper we focus on alternative specifications of the spatial weights matrix. An approach based on principal components analysis is used to estimate a common factor for each country as a measure of the pairwise distance to all other countries in the sample. This measure is converted into spatial weights and compared to alternative weights matrices based only on bilateral trade information. The empirical analysis reported in the paper makes use of the data set compiled by Rao et al. spanning the period 1970–2005 and 141 countries. The results clearly indicate the need to adequately model spatial autocorrelation in predicting PPPs for non-benchmark countries and years. The trade weights matrix and the matrix based on principal components appear to perform equally well. Though the empirical results presented here focus on selected countries, detailed results for all the countries included in the analysis are available from the authors.

Appendix: The Construction of the Economic Distance Measure used in the PCW

A principal components approach is used to construct the measure. This technique allows the representation of the variance-covariance (or correlation) structure of a set of variables through a small number of linear combinations called components. Variables are grouped into components by their cross-correlations (Johnson & Wichern, 2002).

Variables Included in the Construction

The distance between countries is measured using a range of variables. The variables are chosen considering trade closeness, geological proximity, cultural closeness. To provide the reader with an example of the data used, summary statistics for a small group of countries are presented in Appendix DA.4.

a) Trade closeness:
   We use the trade share between countries which is expressed as the volume of trade (export plus import) with each trading partner in the sample as a percentage of total trade of a particular country. This is constructed for every five year intervals ; 1970, 1975, 1980, 1985, 1990, 1995, 2000 and 2005
b) Geographical proximity:
   Two dummies are used to consider geographical proximity, regional membership and close neighbours.

1. Regional membership: Asia pacific region, Europe, South America, North America, Central America and the Caribbean, Saharan Africa (except North Africa), North Africa and Middle East.
2. Close neighbours: We consider both land and sea proximities: e.g. Canada and the US are bordering countries. Sri Lanka and India as well as Singapore and Malaysia are also close neighbours. This dummy captures some aspects of proximity not caputred by the regional dummies. For instance, Venezuela and Trinidad and Tobago are close neighbours, but they are classified in different regions. Further, Japan and Australia are not close in terms of distance, but they are in the same region.

c) Cultural and colonial closeness:

1. Common language: A dummy is used for each language. In addition, the closeness of the dialect is also considered. For instance, Danish is very close to Swedish and Norwegian.
2. Common colonies: This dummy indicates the colonial relationship of countries. We do not totally fix to the standard definition for a colony but include protectorates and other types of foreign rules and consider countries under different types of foreign rules and their colonial occupants (such as UK, Spain etc.)to construct the dummy. Some countries had several colonial relationships over their history. For instance, Sri Lanka was partly or fully under the rules of England, Portugal and Holland. To avoid this complexity, we construct the dummy considering the last colonial power of the country or the colonial rule that made a significant impact on the particular country.

Estimation

The steps involved in the PC estimation procedure are summarized in the main text and some more details are presented here.

Step 1
A separate principal components model is estimated for each country using the variables for each time period. Therefore, 141 models are estimated. For each country there could be up to nine variables defined: Trade period, region, border, language1, language2,. . ., language5, colonial. We will refer to the number of variables by m.
Step 2
The m principal components are orthogonal linear combinations of the m variables. The estimate of the weight of a given variable on a principal component is known as the "loading." We select the first component (corresponding to the largest eigenvalue) as the common factor. During the procedure we drop variables and re-estimate the model if, a) it has a negative loading on the first component, b) its loading is below 0.4. Although this choice of "significance" of the variables is rather ad hoc, it is a common rule of thumb in the principal components literature. For each country we use the same model (that is, the same subset of variables) to estimate the principal components model for all eight time periods.

We present the example of France to illustrate. The factor loadings for common colony and common language are not significant in the first component as the magnitude of the loading is less than 0.4 (see Table). These variables, common colony and Lang 1, are significantly loaded on the second component.

France (all available variables)		
	Components	
	1	2
Trade75	.865	.036
Border	.906	–.006
Comcol	.005	.934
Lang 1	.176	.924
Region	.743	–.258

Thus, in this case the principal component model is re-estimated without Comcol and Lang1 and shown on the next table.
Thus, the common factor for France for the period 1975–1979 is given by

$$cfFr7579 = 0.86 \times Trade + 0.902 \times Border + 0.763 \times Region$$

Interested readers can consult the authors for the complete set of results.

France (reduced set of variables)	
	1
Trade75	.860
Border	.902
Region	.763

Table A.1 Descriptive statistics for selected countries

Country		Bilateral trade share									Border	Colony	Common language				Region
		70	75	80	80	85	90	95	00	05			Lang1	Lang2	Lang3	Lang4	
US	No.	110	127	127	127	127	138	139	140	140	5	57	67	18			21
	$\overline{X}$	0.01	0.01	0.01	0.01	0.01	0.01	0.01	0.01	0.01	0.04	0.4	0.48	0.13			0.15
	SD	0.02	0.02	0.02	0.02	0.02	0.02	0.02	0.02	0.02	0.19	0.49	0.5	0.33			0.36
	min	0	0	0	0	0	0	0	0	0	0	0	0	0			0
	max	0.24	0.22	0.17	0.17	0.21	0.2	0.21	0.21	0.2	1	1	1	1			1
UK	No.	129	132	140	140	140	138	137	138	138	7	57	67				24
	$\overline{X}$	0.01	0.01	0.01	0.01	0.01	0.01	0.01	0.01	0.01	0.05	0.4	0.48				0.17
	SD	0.02	0.02	0.02	0.02	0.02	0.02	0.02	0.02	0.02	0.22	0.49	0.5				0.38
	min	0	0	0	0	0	0	0	0	0	0	0	0				0
	max	0.13	0.09	0.11	0.11	0.13	0.14	0.14	0.15	0.13	1	1	1				1
Belgium	No.	118	125	132	132	126	133	135	134	135	5	3	34	2	9		24
	$\overline{X}$	0.01	0.01	0.01	0.01	0.01	0.01	0.01	0.01	0.01	0.04	0.02	0.24	0.01	0.06		0.17
	SD	0.03	0.03	0.03	0.03	0.03	0.03	0.03	0.03	0.03	0.19	0.14	0.43	0.12	0.25		0.38
	min	0	0	0	0	0	0	0	0	0	0	0	0	0	0		0
	max	0.24	0.23	0.21	0.21	0.21	0.23	0.22	0.17	0.19	1	1	1	1	1		1
Denmark	No.	128	132	138	138	134	133	133	134	134	6	3	67	2	9		24
	$\overline{X}$	0.01	0.01	0.01	0.01	0.01	0.01	0.01	0.01	0.01	0.04	0.02	0.48	0.01	0.06		0.17
	SD	0.03	0.02	0.02	0.02	0.02	0.02	0.02	0.02	0.02	0.2	0.14	0.5	0.12	0.25		0.38
	min	0	0	0	0	0	0	0	0	0	0	0	0	0	0		0
	max	0.17	0.17	0.19	0.19	0.19	0.22	0.23	0.21	0.2	1	1	1	1	1		1
France	No.	118	129	130	130	131	136	137	137	135	8	23	34				24
	$\overline{X}$	0.01	0.01	0.01	0.01	0.01	0.01	0.01	0.01	0.01	0.06	0.16	0.24				0.17
	SD	0.02	0.02	0.02	0.02	0.02	0.02	0.02	0.02	0.02	0.23	0.37	0.43				0.38
	min	0	0	0	0	0	0	0	0	0	0	0	0				0
	max	0.2	0.17	0.16	0.16	0.16	0.18	0.17	0.16	0.18	1	1	1				1
		70	75	80	80	85	90	95	00	05		Colony	Lang1	Lang2	Lang3	Lang4	
Domincan Rep	No.	27	32	72	72	68	68	29	106	93	13	17	18				21
	$\overline{X}$	0.01	0.01	0.01	0.01	0.01	0.01	0.01	0.01	0.01	0.09	0.12	0.13				0.15

Table A.1 (continued)

Country		Bilateral trade share									Border	Colony	Common language				Region
	SD	0.07	0.06	0.04	0.04	0.05	0.04	0.05	0.06	0.05	0.29	0.33	0.33				0.36
	min	0	0	0	0	0	0	0	0	0	0	0	0				0
	max	0.8	0.69	0.47	0.47	0.51	0.49	0.51	0.71	0.61	1	1	1				1
Ecuador	No.	58	71	46	46	47	47	54	96	104	2	2	18				11
	$\overline{X}$	0.01	0.01	0.01	0.01	0.01	0.01	0.01	0.01	0.01	0.01	0.01	0.13				0.08
	SD	0.04	0.04	0.04	0.04	0.05	0.04	0.03	0.03	0.03	0.12	0.12	0.33				0.27
	min	0	0	0	0	0	0	0	0	0	0	0	0				0
	max	0.43	0.45	0.41	0.41	0.57	0.48	0.38	0.36	0.39	1	1	1				1
El Salvador	No.	44	46	45	45	44	64	73	78	81	4	17	18				21
	$\overline{X}$	0.01	0.01	0.01	0.01	0.01	0.01	0.01	0.01	0.01	0.03	0.12	0.13				0.15
	SD	0.03	0.03	0.03	0.03	0.04	0.04	0.03	0.05	0.04	0.17	0.33	0.33				0.36
	min	0	0	0	0	0	0	0	0	0	0	0	0				0
	max	0.25	0.29	0.32	0.32	0.4	0.4	0.38	0.58	0.51	1	1	1				1
Guatemala	No.	57	59	59	59	80	62	68	95	99	6	17	18				21
	$\overline{X}$	0.01	0.01	0.01	0.01	0.01	0.01	0.01	0.01	0.01	0.04	0.12	0.13				0.15
	SD	0.03	0.03	0.03	0.03	0.03	0.04	0.04	0.04	0.04	0.2	0.33	0.33				0.36
	min	0	0	0	0	0	0	0	0	0	0	0	0				0
	max	0.31	0.3	0.33	0.33	0.39	0.41	0.41	0.4	0.44	1	1	1				1
China	No.	91	109	53	53	121	130	134	135	137	9	1	4				25
	$\overline{X}$	0.01	0.01	0.01	0.01	0.01	0.01	0.01	0.01	0.01	0.06	0.01	0.03				0.18
	SD	0.02	0.03	0.03	0.03	0.03	0.04	0.03	0.03	0.02	0.25	0.08	0.17				0.38
	min	0	0	0	0	0	0	0	0	0	0	0	0				0
	max	0.23	0.31	0.27	0.27	0.33	0.39	0.23	0.2	0.18	1	1	1				1
India	No.	115	115	128	128	100	132	133	132	134	6	57	67	2	3	2	25
	$\overline{X}$	0.01	0.01	0.01	0.01	0.01	0.01	0.01	0.01	0.01	0.04	0.4	0.48	0.01	0.02	0.01	0.18
	SD	0.03	0.02	0.02	0.02	0.02	0.02	0.02	0.02	0.02	0.2	0.49	0.5	0.12	0.14	0.12	0.38
	min	0	0	0	0	0	0	0	0	0	0	0	0	0	0	0	0
	max	0.25	0.21	0.15	0.15	0.16	0.14	0.14	0.16	0.13	1	1	1	1	1	1	1

Step 3
A factor score is computed for each pair of countries using the common factors, for example *cfFr7579* is used to compute the scores for France for the years 1975–1979. These factor scores are rescaled to prepare the proximity matrix using the formula presented in (13).

Descriptive Statistics for the Variables used to Construct the PCW. Selected Countries Shown.

The table below shows descriptive statistics for eleven countries in the sample for each of the variables used to estimate a common factor for each country. The label "Lang #" stands for *language number*. The maximum number of languages considered for any country was five.

References

ADB. (2005). Purchasing power parities and real expenditures. URL www.adb.org/Documents/Reports/ICP-Purchasing-Power-Expenditures/default.asp.

Ahmad, S. (1996). *Regression estimates of per capita GDP based on purchasing power parities* (pp. 237–264). Volume International Comparisons of Prices, Output and Productivity: Contributions to Economic Analysis Series, Chapter II. Elsevier.

Barro, R. J., & SalaiMartin, J. (2004). *Economic growth* (2nd ed.). Cambridge, MA: MIT Press.

Bergstrand, J. H. (1991). Structural determinants of real exchange rates and national price levels: Some empirical evidence. *American Economic Review,* 81(1), 325–334.

Bergstrand, J. H. (1996). *Productivity, factor endowments, military expenditures, and national price levels* (pp. 297–317). Volume International Comparisons of Prices, Output and Productivity: Contributions to Economic Analysis Series, Chapter II, Elsevier Science Publishers.

Castles, I., & Henderson, D. (February, 2003). Ipcc issues: A swag of documents. URL http://www.lavoisier.com.au/articles/climate-policy/economics/castles-henderson2003-2.php.

Clague, C. (1988). *Explanations of national price levels* (pp. 237–262). Volume World Comparisons of Incomes, Prices and Product: Contributions to Economic Analysis Series, Chapter III. North Holland: Elsevier.

Durbin, J., & Koopman, S. J. (2001). *Time series analysis by state space methods.* Oxford: Oxford University Press.

Durlauf, S. N., Johnson, P., & Temple, J. (2005). Growth Econometrics. In *Handbook of economic growth* (Vol. 1, pp. 555–677). Elsevier.

Florax, R. J. G. M., & de Graa, T. (2004). *The performance of diagnostic tests for spatial dependence in linear regression models: A meta-analysis of simulation studies* (pp. 29–65). Chapter 2. Germany: Springer-Verlag.

Harvey, A. C. (1989). *Forecasting, structural time series models and the Kalman filter.* Cambridge: Cambridge University.

Johnson, R. A., & Wichern, D. W. (2002). *Applied multivariate statistical analysis* (5th ed.). New Jersey: Prentice Hall.

Krämer, W. (2005). Finite sample power of clifford type tests for spatial disturbance correlation in linear regression. *Journal of Statistical Planning and Inference, 128,* 489–496.

Kravis, I. B., & Lipsey, R. E. (1983). *Toward an explanation of national price levels.* Princeton, NJ: Princeton Studies in International Finance No. 52.

Kravis, I. B., & Lipsey, R. E. (1986, April). *The assessment of national price levels*. Philadelphia: Eastern Economic Association Meetings.

Maddison, A. (1995). *Monitoring the world economy, 1820–1992*. Paris and Washington, DC: Development Centre of the Organisation for Economic Co-operation and Development.

Maddison, A. (2007). *Contours of the world economy 1–2030 AD: Essays in macro-economic history*. Oxford: Oxford University Press.

McKibbin, W. J., & Stegman, A. (2005). Convergence and per capita carbon emissions. (Working paper in international economics 4.05). Sydney: Lowey Institute for International Policy.

Milanovic, B. (2002). True world income distribution, 1988 and 1993: First calculation based on household surveys alone. *The Economic Journal, 112*, 51–92.

Ord, K. (1965). Estimation methods for models of spatial interaction. *Journal of the American Statistical Association, 70*, 120–126.

Rao, D. S. P., Rambaldi, A. N., & Doran, H. E. (2008). A method to construct world tables of purchasing power parities and real incomes based on multiple benchmarks and auxiliary information: Analytical and empirical results. (CEPA Working Paper Series, ISSN No. 19324398 WP05/08). School of Economics, University of Queensland.

Rao, D. S. P., Rambaldi, A. N., & Doran, H. E. (2009). Extrapolation of purchasing power parities using multiple benchmarks and auxiliary information: A new approach, *The Review of Income and Wealth,* page forthcoming.

Rose, A. (2004, January). Bilateral trade data set. URL http://faculty.haas.berkeley.edu/arose/RecRes.htm.

SalaiMartin, X. (2002). *15 years of new growth economics: What have we learnt?* (Discussion Papers 0102-47). Department of Economics, Columbia University.

Summers, R., & Heston, A. (1991, May). The penn world talbe (mark 5): An expanded set of international comparisons, 1950–1988. *The Quarterly Journal of Economics, 106*(2), 327–368.

World Bank. (2006). World development indicators 2006, the international bank for reconstruction and development. Washington, DC.

Price Indexes across Space and Time and the Stochastic Properties of Prices

Matteo M. Pelagatti*

1 Introduction

Many institutions responsible for the production of official price statistics are moving towards automatic procedures for surveying the relevant data. The first Italian city to start this process was the City of Milan, which since the end of the Nineties receives price data directly from the databases of large-scale retailers.

This work is part of a research project in which the Statistical Office of the City of Milan aims at expanding the information base gathered from large-scale retailers in order to have actual (cash) sale prices and quantities. At the moment, few supermarkets are providing us with monthly sales data which include total monthly sold values and quantities for every single product on their shelves.

The availability of this kind of data is extremely valuable since it allows the comparison of the cost of living across space and time. Supermarket products are, indeed, rather homogeneous over both dimensions (space and time) and represent an important share of personal consumptions (in Italy 77% of food and 88% of beverage and grocery expenditures according to AC Nielsen).

A natural problem that arises when a system of space-time price indexes is to be computed is the choice of the formulas. Indeed, while there is a general consensus on the choice of bilateral indexes, the debate on multilateral comparisons is still open and will probably remain such in the future. In fact, it is well known that in a system of multilateral price-quantity indexes the property of transitivity, essential for building a coherent system, is incompatible with the property of characteristicity

M.M. Pelagatti (✉)
Dipartimento di Statistica, Università degli Studi di Milano-Bicocca Via Bicocca degli Arcimboldi 8, I-20126, Milano

*In this work I use data that the Statistical Office of the City of Milan has obtained from supermarkets expressly for this research. I thank Flavio Necchi (Comune di Milano), supermarket managers I cannot mention for preserving data anonymity and my colleague Marco Fattore for our frequent discussions on the theme and for sharing with me and Flavio the effort of fostering our project and convincing the other partners.

L. Biggeri, G. Ferrari (eds.), *Price Indexes in Time and Space*, Contributions to Statistics, DOI 10.1007/978-3-7908-2140-6_5, © Springer-Verlag Berlin Heidelberg 2010

and consequently with the proportionality of either the price indexes or their respective cofactor. This amounts to the impossibility of factorizing all the value indexes in a coherent system of proportional price indexes times proportional quantity indexes. This is maybe not surprising as index numbers are an extremely synthetic way of summarizing huge masses of information. For thorough discussions of this theme the reader should refer to Balk (1996); van Veelen (2002); Balk (2008, Chap. 7).

The consequence of the above consideration is that whoever wants to build a transitive system of price and quantity indexes needs to express a preference on the properties the system and the single pairs of price and quantities indexes should satisfy. Now, for our type of data (supermarket scanner data) it is natural to ask price indexes rather than quantity indexes to pass all possible tests. Indeed, price indexes would contain important information for comparing the cost of living in different places, while quantity indexes would just indicate that a supermarket sells more or less than another supermarket. On the contrary, if the interest is comparing the wealth of a country with respect to other countries through their national accounts, then quantity indexes are probably more informative. The fact that products may change through time and space brings about further considerations that have to be accounted for in the construction of the system of indexes.

In this paper, I address the problem of building a coherent system of space-time price indexes that should be the best possible compromise, when all the issues that our data pose have been considered. The novelty of the approach I pursue in this work is the consideration also of the stochastic properties that prices and sold quantities of each product seem to conform to in real-world data. In this light, the new properties of *cointegration-preservation* and the slightly more general *stability-preservation* are defined and discussed. I do not seek with this paper the production of a complete theory of cointegration- or stability-preserving indexes, since, as discussed in Sect. 3, "less linear" definitions of integrated and cointegrated processes are needed (Ermini & Granger 1993,Granger 1995) given the non-linearity of index number functions.

2 Existing Literature

While there is some literature on space-time harmonization of official CPI and PPP statistics (Rao, 2001; Ferrari, Laureti, & Mostacci, 2005; Eurostat, 1995, Sect. 7), the issue of building a coherent system of indexes over space and time starting from raw data (to the best of my knowledge) has been considered only by Hill (2004). Martini and Zavanella (1995) considered the problem as well, but supposing that quantities are not available and found that the only solution compliant with the axiomatic theory is a system of direct Jevons indexes.

Thus, the starting point of this work is the article by Hill (2004), but as will stand clear from the next sections, my recommendations concerning the appropiate formulas for the index system will differ significantly from Hill's.

Hill (2004) considers the following classes of multilateral formulas

- Geary-Khamis (Geary 1958, Khamis 1972),
- GEKS (Gini 1924, Eltetö and Köves, 1964; Szulc, 1964),
- Minimum Spanning-Tree methods (Martini 1992; Zavanella 1996),
- the Weighted Country Product Dummy (WCPD) method (Rao 2005),

under different graph configurations and possible alternatives to join sub-graphs into a single graph. The different formulas and graph configurations are evaluated according to the following criteria:

Temporal Fixity (TF): old indexes are not affected when information concerning a new time period is added;

Spatial Fixity (SF): old indexes are not affected when information concerning a new country is added;

Temporal Consistency (TC): the temporal indexes for each country do not depend on the other countries in the comparison;

Spatial Consistency (SC): spatial results for each time point do not depend on the other time periods in comparison;

Temporal Displacement (TD): maximum temporal displacement of all the bilateral spatial comparisons, where the temporal displacement of a bilateral comparison is the time span of the data used to compute it; if there are T time points, the temporal displacement ranges between 0 (only contemporaneous data used) and $T-1$ (data from time T and form time 1 used).

As Hill notices, in a transitive system of indexes temporal and spatial consistency cannot coexist if quantities are to be taken into account.

In his considerations on the choice of a method for space-time comparisons Hill suggests "that one should prefer methods that maintain temporal fixity and temporal consistency". In order to reduce temporal displacement Hill suggests that temporal consistency could be periodically broken in order to reconcile spatial and temporal price indexes and proposes different method to do this. It is to be stressed that Hill draws his conclusions with in mind the comparison of European Union countries based on the Harmonized Indexes of Consumer Prices and OECD spatial price indexes. So his ideal setup is somewhat different from mine.

As the next sections will show, according to the nature of our data the property of time consistency turns out to lead to indexes with very undesirable properties. Furthermore, in addition to the families of multilateral indexes considered by Hill, also the performances of the method used by the Economic Commission for Latin America and the Caribbean (ECLAC, 1978) and Gerardi's 1982 Unit-Country-Weight (UCW) system will be analyzed.

3 Price Behaviour and Price Indexes

Before being able to express a rational preference for a particular transitive system of multilateral indexes, it is important to consider how prices and quantities tend to behave both in space and time. The best way to describe the behaviour of prices and quantities in time is probably through stochastic models.

A reasonable model for the dynamics of (log) prices in discrete time is the following integrated process

$$\log p_{n,k,t} = \log p_{n,k,t-1} + \delta_{n,k} + \eta_{n,k,t}, \tag{1}$$

where $p_{n,k,t}$ denotes the price of commodity $n \in \{1,\dots,N\}$ in country $k \in \{1,\dots,K\}$ at time $t \in \{1,\dots,T\}$, $\delta_{n,k}$ is a time-invariant drift parameter and $\{\eta_{n,k,t}\}_{t=1,\dots,T}$ is a mean-zero stationary sequence. Notice that, since $\log(1+x) \approx x$ for small x, then, for moderate price increases,

$$\frac{p_{n,k,t} - p_{n,k,t-1}}{p_{n,k,t-1}} \approx \log \frac{p_{n,k,t}}{p_{n,k,t-1}} = \delta_{n,k} + \eta_{n,k,t},$$

and $\delta_{n,k}$ turns out to be approximately the mean increment rate of the price of product n in country k with respect to one time unit. If the hypothesis of a constant mean increment rate is too strong, the drift $\delta_{n,k}$ may be made time-dependent with the possible specification

$$\delta_{n,k,t} = \delta_{n,k,t-1} + \zeta_{n,k,t}, \tag{2}$$

where $\zeta_{n,k,t}$ is, again, a mean-zero stationary sequence. The process (1) is commonly referred to as integrated process of order 1, or in short I(1), while the pair (1)–(2) form an integrated process of order 2, or I(2). More generally, a process I(d) is a nonstationary process that needs (at least) d differences to become stationary.

Just as an illustrative example, let's consider the monthly prices (in US$ per MT[2]) of the Thai rice (B garde, FOB), which is easy to have access to, being frequently traded in international commodity markets (Fig. 1). Applying unit root tests[3] to the log-prices we cannot reject the null of a unit root at any commonly used size, while the same tests applied to the increments (differences) of log-prices reject the hypothesis of integration. Indeed an ARIMA(1,1,0) process seems to fit the rise log-prices quite well.

---

[2]MT stands for *measurement ton* a unit of volume used for measuring the cargo of a ship, truck, train, or other freight carrier, equal to exactly 40 cubic feet, or approximately 1.1326 cubic meters.

[3]The ADF, and Elliot, Rothenberg and Stock (1996) DF-GLS and point-optimal tests were applied. The stationarity test KPSS leads to the same conclusions.

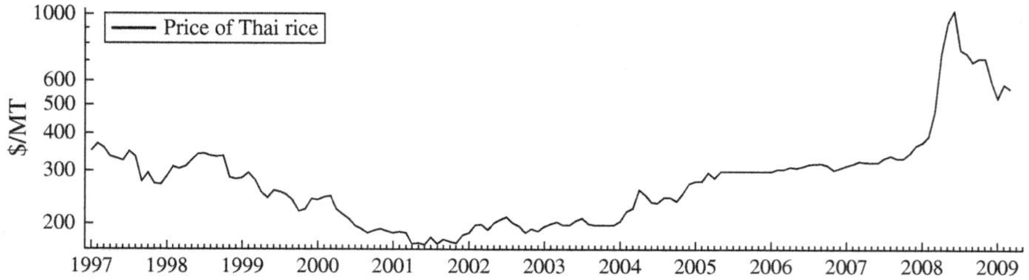

Fig. 1 Price of Thai rice in log scale from January 1997 to March 2009

If we consider the nature of supermarket data within a country, then it is very likely that the same product across different stores will have in the long-run approximately proportional prices[4]. In fact it is hard to believe that the prices of the same good across different stores may drift apart without bound, while it is not uncommon to observe the same goods differently priced across a country or a city. Turning these concepts in formulas, we have

$$p_{n,k,t} \approx b_{n,k,j} p_{n,j,t} \quad \Leftrightarrow \quad \log p_{n,k,t} \approx \beta_{n,k,j} + \log p_{n,j,t} \tag{3}$$

with $k \neq j$, $\beta_{n,k,j} = \log b_{n,k,j}$. This condition may be stated in a more statistically meaningful fashion as

$$\log p_{n,k,t} = \beta_{n,k,j} + \log p_{n,j,t} + \varepsilon_{n,k,t}, \tag{4}$$

where $\varepsilon_{n,k,t}$ is a mean-zero stationary process. If the price time series are well described by the processes (1)–(2) above, then this condition (4) amounts to say that the prices of good n in the two countries are *cointegrated*[5]. It is trivial to prove that cointegration between pairs of price series is a transitive property: if A is cointegrated with B and B is cointegrated with C, then A is cointegrated with C.

Now, if all the commodities in two countries of a panel are cointegrated, then the corresponding space index should be stationary with respect to the time parameter t. In fact, if this does not happen, the index reveals price divergence even though all price-pairs are not diverging.

The aforementioned property for space-time index numbers could be named *cointegration-preservation* if we are ready to accept the integration/cointegration approach, but could also be named *stability-preservation* if we want to leave the concepts of divergence and co-divergence of prices open, where by divergence it

[4]The *Law of One Price* implies that prices of tradable goods should be equal in the long run, but in order to allow some kind of *intra-country Balassa-Samuelson effect* we relax the identity to proportionality.

[5]The reader not acquainted with the concept of cointegration should refer to any recent text on time series econometrics. A thorough and mathematically rigourous treatment of cointegration in the framework of vector autoregressive processes may be found in Johansen (1995).

is meant that the price of a commodity does not have an *attractor*[6] and by co-divergence we mean that the ratio of the prices of the same good in two countries does have an attractor.

Even if we concentrate on the somewhat smaller world of cointegration-preserving indexes, the development of a complete theory is rather complex, as the concepts of integration and cointegration are intrinsically linear, while price index formulas are not necessarily linear in the logarithm of price ratios. The analysis of nonlinear functions of integrated and cointegrated processes needs the development of alternative concepts and tools and goes beyond the scope of this paper. Ermini and Granger (1993) and Granger (1995) are the only papers I am acquainted with that deal with this issue, and could be a starting point for building a complete theory of cointegration-preserving indexes.

From Eq. (4) it is clear that in order to recover the discrepancy between $p_{n,j,t}$ and $p_{n,k,t}$ the index should be built as a function of the elementary price ratios

$$\frac{p_{n,k,t}}{p_{n,j,t}} = b_{n,k,j} \exp\left(\varepsilon_{n,k,t}\right),$$

while other type of functions of the the N prices will in general drift apart even when all the prices of the same commodity are cointegrated across country pairs. Surprisingly this condition seems to be only sufficient for the construction of cointegration-preserving indexes, in fact the ratio of values evaluated with fixed quantities

$$\frac{\Sigma_{n=1}^{N} p_{n,k,t} q_n}{\Sigma_{n=1}^{N} p_{n,j,t} q_n} \tag{5}$$

seems to enjoy the property, even though a formal proof of this fact can be sought only with a more specific definition of *attraction* to a fixed value or *mean-reversion*.

Figure 2 depicts simulated time series and their ratios for every time point. Three pairs of integrated and (country-wise) cointegrated (log-price) time series are generated for $n = 1, 2, 3$ and $t = 1, \ldots, 100$ from

$$\log p_{n,1,t} = \sum_{s=1}^{t} \varepsilon_{n,1,s}, \quad \varepsilon_{n,1,s} \sim \text{NID}(0,1) \tag{6}$$

$$\log p_{n,2,t} = \log p_{n,1,t} + \eta_{n,t} \quad \eta_{n,t} \sim \text{NID}(0,1) \tag{7}$$

where $\text{NID}(\mu,\sigma^2)$ stands for normally independently distributed with mean μ and variance σ^2. Panel a) of Fig. 2 shows the three ratios $p_{n,2,t}/p_{n,1,t}$ on a log scale. Since the three ratios tend to move around an attractor, so should the price index

[6] In order to leave the concept open enough we can take the definition of Granger (1995) "a process may be said to have an attractor if there is some mechanism that produces an inclination to return to some value – usually its mean".

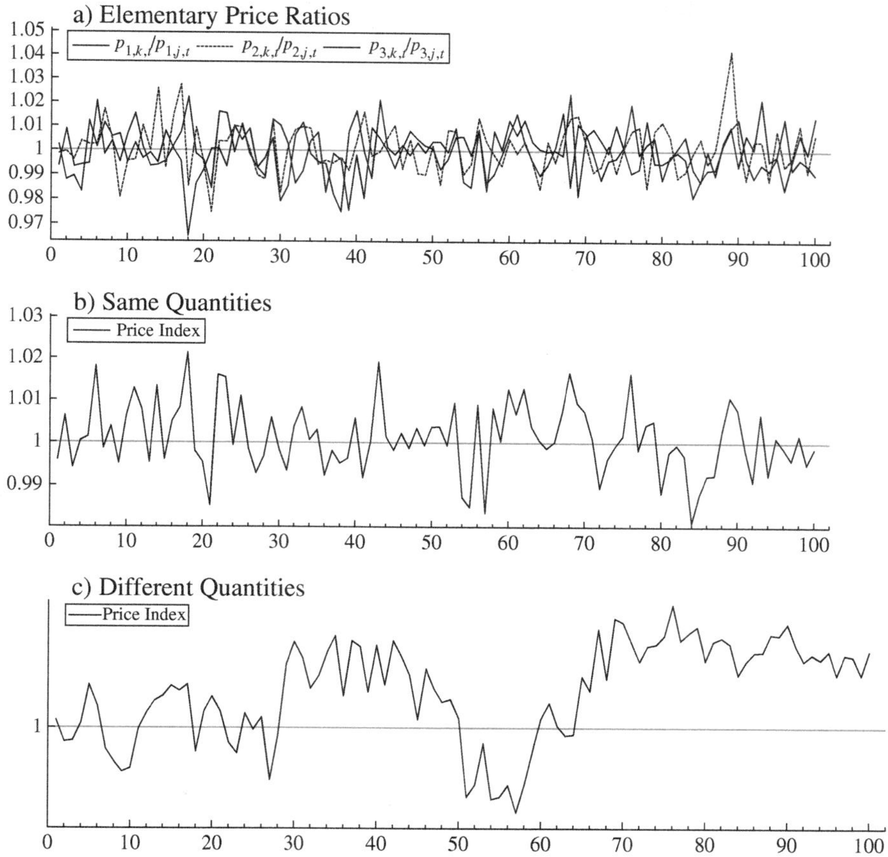

Fig. 2 **a**) price ratios, **b**) ratio of values based on same quantities, **c**) ratio of values based on different quantities. All values are on a log-scale

between the two countries. Panel b) represents the index defined in Eq. (5) with quantities vector $q = (1,10,100)$. Panel c) reports the values of an index built as ratio of values where the quantities for the numerator are $q_2 = (100,1,10)$ and those for the denominator are $q_1 = (1,10,100)$. While the index in panel b) tends to move around the attractor 1, the index in panel c) does not seem to enjoy this property.

The index-number-theory competent reader would certainly object that an index of the type depicted in panel c) is not a proper index (at least axiomatically), but many economists test the *purchasing power parity theory* using this type of indexes. In fact, in many empirical papers unit root tests are applied to (the logarithm of) real exchange rates (RER) built using consumer price indexes (CPI) and exchange rates of country pairs: $\text{RER}_t = E_t \times \text{CPI}_{2,t}/\text{CPI}_{1,t}$, with E_t nominal exchange rate. Now, without loss of generality suppose that the exchange rate is constantly equal to one, then it is straightforward to see that RARs are built exactly as those indexes of panel c)[7]. Applying a unit root test[8] to our simulated data, we get a rejection (mean-reversion) for panel b) and a non-rejection (integration) for panel c) data.

Thus, the first (empirical) result of the above reasoning is that the so called *purchasing power parity puzzle*, that is, the fact that real exchange rates turn out to be integrated or at least too persistent if compared to what the theory prescribes, may be ascribed to the (wrong) choice of the index number formula (ratio of CPIs) for building real exchange rates. Even though this result may be considered a byproduct of our research, its relevance could be far-reaching if the reader considers that the exact sequence of words "purchasing power parity puzzle" produces 7650 entries in Google and 1810 in Google Scholar.

Recall that in a system of multilateral price indexes the properties of transitivity and characteristicity are incompatible. Since we want to preserve transitivity, the indexes of the system we are going to build won't be characteristic. From index number theory (Martini 2001, Sect. 6.2), we know that if we have the price vectors of two situations under comparison and *non-characteristic values* $\{v_n\}$, then the only index that satisfies proportionality (PR), commensurability (C), homogeneity (H), monotonicity (M) and base reversal (B) has form

$$\prod_{n=1}^{N}\left(\frac{p_{n,k}}{p_{n,j}}\right)^{wn}, \quad \text{with} \quad w_n = \frac{v_n}{\Sigma_{n=1}^{N} v_n}. \tag{8}$$

If, instead, only non characteristic quantities $\{q_n\}$ are available, then the only index with properties PR,C, H, M, B has form

$$\frac{\Sigma_{n=1}^{N} p_{n,k}q_n}{\Sigma_{n=1}^{N} p_{n,j}q_n}. \tag{9}$$

It is interesting to notice that, from our above reasoning, these two type of indexes are those compliant with the property of cointegration preservation. The fact that the values v_n and the quantities q_n may evolve with time further complicates the theoretical analysis of cointegration-preserving or stability-preserving indexes.

Thus, living the formal analysis of systems of cointegration-preserving indexes to future research, we try to draw some conclusions through a simulation experiment

---

[7] Recall that the CPI index for country k has generally the form

$$\mathrm{CPI}_k = \sum_n \frac{p_{n,k,t}}{p_{n,k,0}} w_n$$

for some weights $\{w_n\}$. Thus, if log-prices are modelled as integrated processes then $\log p_{n,k,t} = \Sigma_{s=1}^{t} \varepsilon_{n,k,s} + \log p_{n,k,0}$, and $\log(p_{n,k,t}/p_{n,k,0})$ is an integrated process that starts form zero (at time $t = 0$), exactly as in our simulation.

[8] A Dickey-Fuller test was applied imposing non deterministic regressors, since under the alternative of mean reversion the mean of the log of the price indexes should be 0, being the price pairs almost identical.

considering different widely used multilateral systems of price indexes and various graph configurations with respect to time and space.

4 A Simulation Experiment

The experiment is based on simulated prices for $N = 3$ commodities, $K = 3$ sites and $T = 100$ time points. The prices have been generated using the exponential of Gaussian I(1) processes and are cointegrated across only two countries, while prices in the third country follow idiosyncratic I(1) processes. The simulated prices are depicted in the first column of Fig. 3 and the elementary price ratios for each non-redundant pair of sites are shown in the second column of the same figure. From this graph it is clear that a well-behaved system of indexes should reveal that the prices of places 2 and 3 have stable ratios, while the prices of place 1 tend to diverge with respect to the other two sites' prices.

The quantities were obtained from the exponential of nine independent random walks with Student's t increments. The choice of Student's distribution is based on the consideration that in a supermarket, due to promotions, advertisement and openings of neighbouring competitors, quantities should move more erratically than

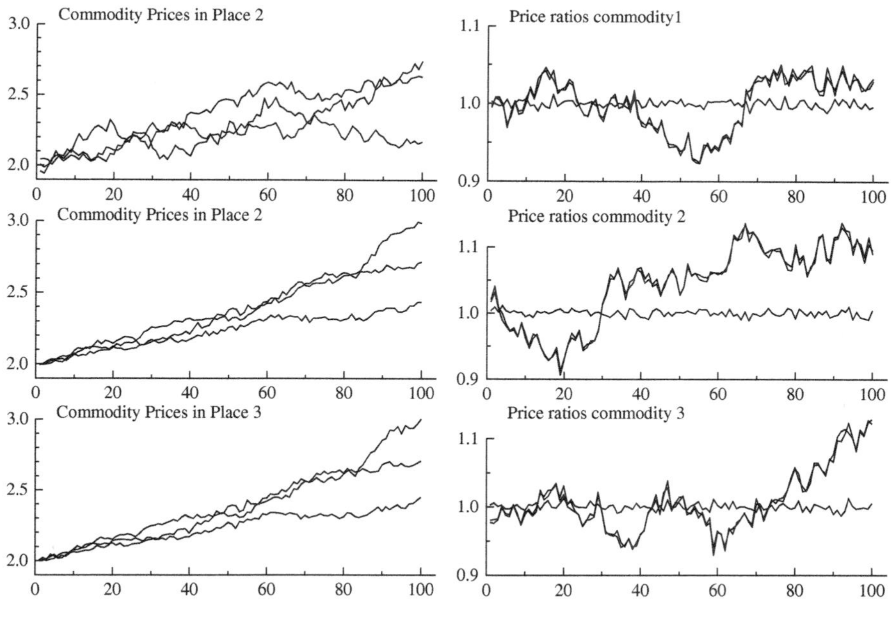

Fig. 3 *First column*: time series of simulated prices for 3 commodities in 3 sites. The first site prices are not cointegrated with the prices of the same commodity in the other sites, while prices in the second and third site have cointegrated prices for the same commodity. *Second column*: time series of price ratios across space for each commodity

prices. However, the choice of the distribution of the increments does not affect the conclusions of the experiment.[9]

The index formulas applied to the simulated data are the following. Let S be the number of the situations to be compared and denote with $P_{a,b}$ the generic index for comparing the situation b with situation a (base).

GEKS (Gini, 1924; Eltetö and Köves, 1964; Szulc, 1964)]
$P_{a,b} = \prod_{s=1}^{S} (F_{a,s} \cdot F_{s,b})^{1/s}$, where $F_{a,b}$ is a Fisher index with base a.

UCW (Gerardi, 1982)
$P_{a,b} = (\Sigma_n p_{n,b} \cdot q_n)/(\Sigma_n p_{n,a} \cdot q_n)$ with $q_n = \prod_s q_{n,s}^{1/s}$.

ECLAC (ECLAC, 1978)
$P_{a,b} = (\Sigma_n p_{n,b} \cdot q_n)/(\Sigma_n p_{n,a} \cdot q_n)$ with $q_n = \Sigma_s q_{n,s}$.

GK (Geary, 1958; Khamis, 1972)
$P_{a,b} = P_b/P_a$ with

$$\begin{cases} P_s = \left(\Sigma_n p_{n,s} \cdot q_{n,s}\right) / \left(\Sigma_n \pi_n \cdot q_{n,s}\right) \\ \pi_n = \left(\Sigma_{s=1}^{S} p_{n,s} \cdot q_{n,s}/P_s\right) / \left(\Sigma_{s=1}^{S} q_{n,s}\right). \end{cases}$$

WCPD (Summers, 1973; Rao, 2005)
$P_{a,b} = \exp(\pi_b - \pi_a)$ where $\pi_1 = 0$ and for $s = 2, \dots, S$ the π_s are those which solve the weighted least squares problem

$$\min_{\pi,\beta} \sum_{r=2}^{S} \sum_{m=1}^{N} w_{m,r} \left[\log p_{m,r} - \sum_{s=2}^{S} \pi_s D_{m,r}^{(s)} - \sum_{n=1} \beta_n E_{m,r}^{(n)} \right]^2,$$

with $w_{m,r} = (p_{m,r} \cdot q_{m,r})/(\Sigma_{m=1}^{N} p_{m,r} \cdot q_{m,r})$ expenditure share of product m in country r, and

$$D_{m,r}^{(s)} = \begin{cases} 1 & \text{for } r = s, \\ 0 & \text{otherwise} \end{cases} \qquad E_{m,r}^{(n)} = \begin{cases} 1 & \text{for } m = n, \\ 0 & \text{otherwise.} \end{cases}$$

As proved by Rao (2005), this system is equivalent to the one proposed by Rao (1990), defined by $P_{a,b} = P_b/P_a$, where the P_s's solve the system

$$\begin{cases} P_s = \prod_{n=1}^{N} \left(\frac{p_{n,s}}{\pi_n}\right)^{w_{n,s}} \\ \pi_s = \prod_{s=1}^{S} \left(\frac{p_{n,s}}{P_s}\right)^{w_{ij}^*}, \end{cases}$$

with $w_{n,s}^* = w_{n,s}/\Sigma_{s=1}^{S} w_{n,s}$.

[9]For brevity's sake, the quantities and the exact processes' formulas are not reported in the paper, but the software developed for carrying out all the simulations is available from the author on request. The code is written in Ox, so the reader interested in running the software should download the Ox interpreter from www.doornik.com.

MST (Martini, 1992; Zavanella, 1996; Hill, 1999)]
Let $D = \{d_{i,j}\}_{i,j \in \{1,\dots,S\}}$ be a matrix of (symmetric) measures of discrepancy between situation i and situation j, based on either prices or quantities, or on both; and let G_D be the minimum spanning tree graph derived from matrix D. Then, denoting with $F_{a,b}$ the Fisher price index with base a,

$$P_{a,b} = \begin{cases} F_{a,b} & \text{when } a \text{ and } b \text{ are directly linked} \\ P_{a,i_1} \cdot P_{i_1,i_2} \cdot \dots \cdot P_{i_M,b} & \text{otherwise,} \end{cases}$$

where $\{(a,i_1),(i_1,i_2),\dots,(i_M,b)\}$ is any set of vertices with edges linking a with b. There are several ways of measuring the closeness (usually to proportionality) of any two situations. Since the main concern of this work is on price indexes, then the following discrepancy measure based on price vectors proposed by Diewert (2002) has been implemented:

$$d_{i,j} = \sum_{n=1}^{N} \left[\left(\frac{w_{n,i} + w_{n,j}}{2} \right) \left(F_{i,j}^{-1} \frac{p_{n,j}}{p_{n,i}} + F_{i,j} \frac{p_{n,i}}{p_{n,j}} - 2 \right) \right]$$

where $w_{n,s}$ is the expenditure share of commodity n in situation s defined above.

This set of transitive multilateral systems has been applied in three configurations:

Ensemble the systems have been applied directly to all space and time situations;
Time-based the multilateral systems have been applied only at time $t = 1$, then chained time indexes have been computed for each space situation and space indexes have been derived indirectly;
Space-based for every time point t the systems have been applied with respect to space and the time indexes for each space situation have been derived using a "world" price index chained across time.

4.1 Ensemble Indexes

In this configuration all combinations of space and time points are simply treated as situations without differentiating the treatment of the two dimensions. All the properties proposed by Hill (2004) are generally violated and the temporal displacement is maximum $(T - 1)$. MST methods could be forced to satisfy temporal fixity and space fixity by joining new space or time situations to the closest situations already in the graph. This would amount to the loss of optimality (minimality), but in general the approximation will be acceptable, especially when new time points are added. In fact, it is reasonable to believe the prices in the present are closer to prices in the near past.

As far as cointegration-preservation is concerned, the property should be satisfied only by the UCLAC and the UCW systems, which are in the form specified in Eq. (9). Indeed, none of the other index formulas is either in the form (8) or (9). If an opportune discrepancy function is chosen, the MST could be forced to chose the paths that minimize the divergence of the index numbers. At the moment, no such function has been studied, but I expect the one used here (Diewert, 2002) to do a reasonable job, being price proportionality (identity) related to a particular form of cointegration-preservation.

Now, from Fig. 3 it is clear how the prices of one of the three sites tend to diverge form those in the other two places starting from around time $t = 65$. This is probably a consequence of the choice of generating random walks with correlated increments also for the non-cointegrated prices. The choice was based on the consideration that it is very unlikely that the same commodity's price increments are uncorrelated across space. The result is that the divergence starts to become evident after a while.

So, what a good system of price indexes should indicate is that two places share almost the same prices (indexes close to one) and one place has prices that around observation $t = 65$ start diverging.

By observing Fig. 4 it appears clear how according to the GEKS, GK and WCDP systems all price indexes tend to diverge (particularly after $t = 65$), erroneously indicating that in all the considered places prices are drifting apart. On the contrary, we know that in two countries the prices of the same commodity are almost identical. There seems to be some kind of *dragging effect* of the indexes computed for site-pairs where prices are actually diverging.

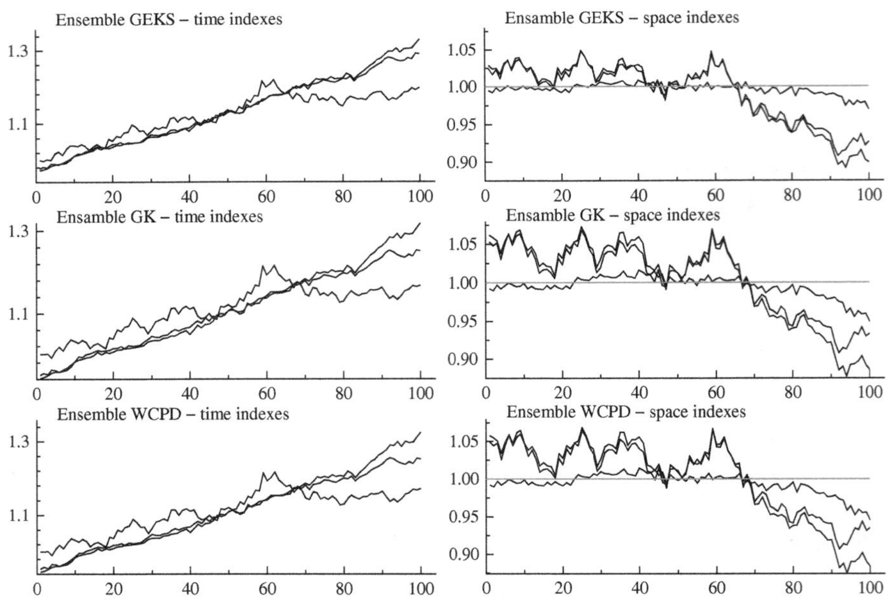

Fig. 4 Ensemble indexes not preserving cointegration

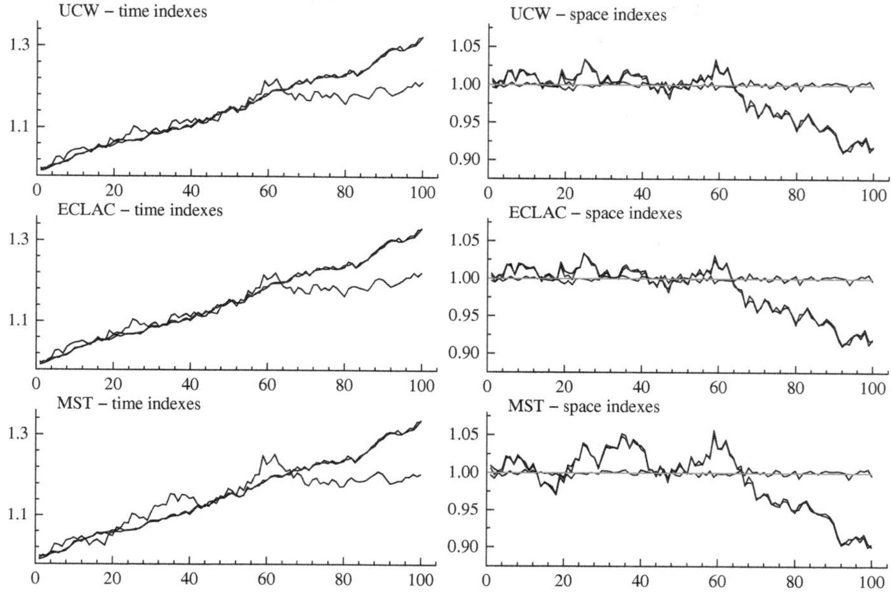

Fig. 5 Ensemble indexes preserving cointegration. The MST system is not cointegration-preserving, but it seems to be a well approximation to such a system

As expected, the indexes series depicted in Fig. 5 demonstrate that the UCW and UCLAC systems are preserving the cointegration for the two places where prices are cointegrated, while comparisons with the place with non-cointegrated prices stress their divergence.

The very good behaviour of the MST system is striking, but there are two facts to be taken into account. On the one hand, for two of the three places we simulated prices that are almost proportional (actually almost identical), where proportionality is the target of the discrepancy measure proposed by (Diewert, 2002), on the other hand a better discrepancy measure with cointegration as objective could be designed for those situations where prices are pairwise cointegrated but not almost proportional and this would probably yield a performance similar to the one we are observing here.

4.2 Time-Based Indexes

Time-based indexes are built as chained Fisher price indexes across time for every single site in sample, initialised at time $t = 1$ with one of the six multilateral systems. Time consistency and time fixity hold by construction, while space fixity depends on the choice of the multilateral system used for initializing the indexes: among the six multilateral systems considered in this work none preserve space

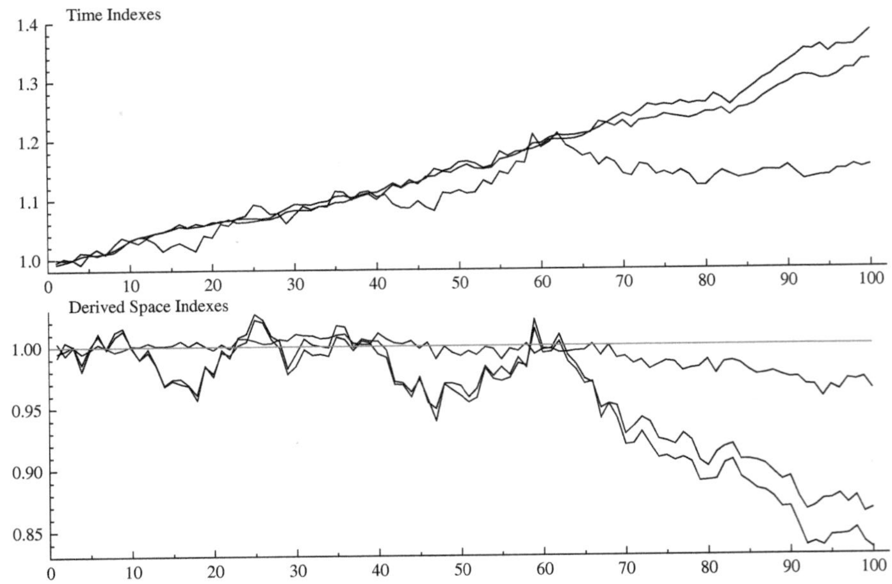

Fig. 6 Time-based indexes with space index at time $t = 1$ being GEKS

fixity, even though the MST graph may be forced to host a new situation losing its optimality only marginally.

As noted before, these indexes are of the type (erroneously) used in many papers that test the purchasing power parity theory and they do not preserve cointegration. The evolution of the time-indexes depends on the choice of the starting system of space-indexes only by a proportionality factor, so in Fig. 6 only the GEKS based indexes are shown. The figure confirms what expected: all three bilateral comparisons diverge and this is particularly evident starting from observation $t = 65$.

4.3 Space-Based Indexes

In this configuration, for every t an idiosyncratic system of space-indexes has been computed and the time dimension has been obtained by multiplying the space-indexes by a "world" time-index built applying chained Fisher indexes to all the prices and quantities in the sample.

Without a formal definition of persistence, it is hard to say if the results of this computation preserves cointegration or stability. In fact, since the weights used for the price indexes change with time, the use of the concept of stationarity would certainly lead to a negative answer, while working with less strong concept of some form of weak dependence (e.g. mixing or Granger's (1995) short memory in mean) may suggest conditions under which stability preservation may hold. Figure 7 presents strong evidence in support of the cointegration preservation property. We

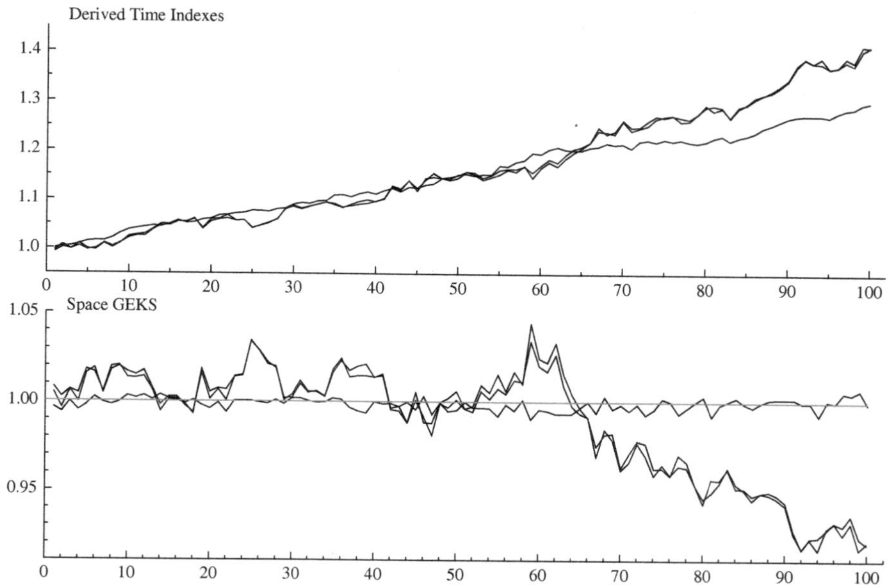

Fig. 7 Space-based EKS indexes updated across time using chained "world" Fisher indexes

report only the space-based GEKS system, being the other systems very similar as far as the cointegration preserving behaviour is concerned.

5 Application to Real Data

We applied the same scheme of the above simulation experiment to real data concerning prices and values of goods sold in two supermarkets and two hypermarkets placed in different areas of Milan. The time series represent total monthly sold values and quantities dating from January 2006 to December 2007. Prices have been recovered by dividing values by quantities, thus, if prices have changed within a month the derived prices are quantity-weighted means. In order to avoid adding a further problem (i.e. the substitution bias) to those considered in this paper, only goods present in the $K = 4$ stores at each of the $T = 24$ months have been used. The total number of goods entering all the indexes is $N = 913$.

Figure 8 shows a selection of the multilateral systems computed on these data. Recall that now there are four time series of price indexes across time and $K(K-1)/2 = 6$ non redundant comparisons across space. The first thing that can be noticed is that there are two pairs of stores with similar prices. Indeed, the two supermarkets and the two hypermarkets tend to have close prices, while supermarkets tend to be more expensive than hypermarkets (up to some 10%).

In the considered 24 months the price excursions were modest, so it is hard to notice much from the plots. Nonetheless, the time-based GEKS reveals a divergence

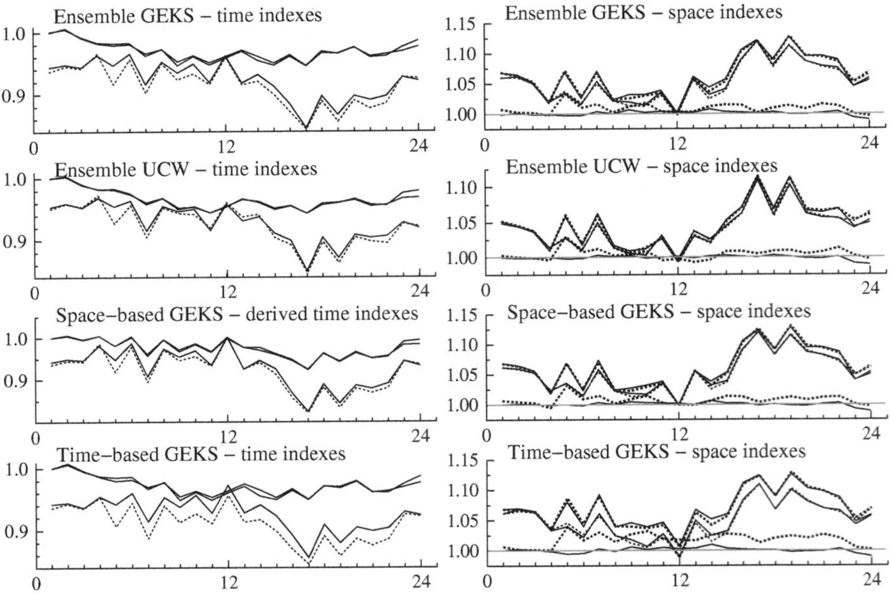

Fig. 8 A selection of space-time indexes for four department stores in Milan

of price levels for the two supermarkets that does not show up in the two systems that are expected to be cointegration-preserving (ensemble UCW and space-based GEKS). A much smaller dragging effect can be noted also in the two panels with ensemble GEKS indexes.

These results confirm that time-based indexes represent a bad choice even though the dragging effect for these data was not too strong. The best choices seem to be the ensemble UCW (as well as ensemble UCLAC not shown here) and space-based indexes (the ones not reported had a very similar behaviour). It will be interesting to repeat this application when four or more years of time series become available.

6 Conclusion

The issue of the construction of coherent price indexes across space and time has been somewhat neglected in the scientific literature. A likely motivation for this was the scarce availability, at least in the past, of sufficiently comparable data across both dimensions. These limitations have been partially overcome by scanner data from large-scale retailers.

With this work I propose a stability-preservation property that space-time indexes should satisfy in order to avoid misleading conclusions. In its seminal work on this theme Hill (2004) considers many relevant tests, but the omission of the property proposed here leads him to endorse systems of indexes where the importance of the time dimension prevails on the space dimension. On the contrary, our evidence suggests that space-based indexes should form the basis for well-behaving system of space-time indexes. Nonetheless, Hill proposes a periodic correction of time-based indexes by means of space indexes as well. This surely limits the possible divergence of indexes that should not diverge, but at the cost of introducing a spurious periodicity into the index time series.

Our simulation results confirmed by the empirical application should foster further research towards the construction of a complete theory of cointegration-preserving indexes, that should overcome the intrinsic linearity of the definitions of integration and cointegration, possibly extending the work of Granger (1993) and Granger (1995) or borrowing from the theory of mixing processes.

References

Balk, B. M. (1996). A comparison of ten methods for multilateral international price and volume comparison. *Journal of Official Statistics, 12*(2), 199–222.

Balk, B. M. (2008). *Price and quantity index numbers. Models for measuring aggregate change and difference.* New York: Cambridge University Press.

Diewert, W. E. (2002). *Similarity and dissimilarity indexes: An axiomatic approach* (Discussion Paper No. 02-10). Department of Economics, University of British Columbia.

ECLAC. (1978). Series Históricas del Crecimiento de Améerica Latina. Santiago, Chile: Economic Commission for Latin America and the Caribbean.

Elliot, G., Rothenberg, T. J., & Stock, J. H. (1996). Efficient tests for an autoregressive unit root. *Econometrica, 64*(4), 813–836.

Eltetö, O., & Koves, P. (1964). On an index number computation problem in international comparison (in Hungarian), *Statisztikai Szemle, 42*, 507–518.

Ermini, L., & Granger, C. W. J. (1993). Some generalizations on the algebra of I(1) processes. *Journal of Econometrics, 58*, 369–384.

Eurostat. (1995, December 18–20). *Improving the quality of price indices.* Proceedings of the International Seminar held in Florence, Commission of the European Community.

Ferrari, G., Laureti, T., & Mostacci, F. (2005). Time-space harmonization of consumer price indexes in Euro-zone countries. *International Advances in Economic Research, 11*, 359–378. DOI: 10.1007/s11294-005-2275-7

Geary, R. C. (1958). A note on the comparison of exchange rates and PPPs between countries. *Journal of the Royal Statistical Society Series A, 121*(1), 97–99.

Gerardi, D. (1982). Selected problems of inter-country comparisons on the basis of the experience of the EEC. *Review of Income and Welth, 28*, 253–259.

Gini, C. (1924). Quelques Considéretions au Sujet de la Construction des Nombres.

Indices des Prix et des Questions Analogues. *Metron, 4*, 3–162.

Granger, C. W. J. (1995). Modelling nonlinear relationships between extended-memory variables. *Econometrica, 63*(2), 265–279.

Hill, R. J. (1999). Comparing price levels across countries using minimum spanning trees. *Review of Economics and Statistics, 81*(1), 135–142.

Hill, R. J. (2004). Constructing price indexes across space and time: the case of the European union. *American Economic Review, 94*(5), 1379–1410.

Johansen, S. (1995). *Likelihood-based inference in cointegrated vector autoregressive models.* Oxford: Oxford University Press.

Khamis, S. H. (1972). A new system of index numbers for national and international purposes. *Journal of the Royal Statistical Society, Series A, 135*(1), 96–121.
Martini, M. (1992). *I numeri indice in un apporccio assiomatico*. Milan: Giuffré Editore.
Martini, M. (2001). *Numeri indice per il confronto nel tempo e nello spazio*. Milan: CUSL.
Martini, M., & Zavanella, B. (1995). *Problems of space and time consistency between spatial and temporal indeces*. Proceedings of the Internartional Seminar Improving the Quality of Price Indeces: CPI & PPP, Commission of the European Communities, 371–390.
Rao, D. S. P. (1990). A system of log-change index numbers for multilateral comparisons. In J. Salazar-Carrillo & D. S. P. Rao (Eds.), *Comparisons of prices and real products in Latin America*, Contribution to Economic Analysis Series. Amsterdam: North Holland.
Rao, D. S. P. (2001). *Integration of CPI and PPP: Methodological issues, feasibility and recommendations*, Joint World Bank-OECD Seminar on PPP (pp. 1–26). Washington, DC.
Rao, D. S. P. (2005). On the equivalence of weighted country-product-dummy (CPD) method and the Rao-system for multilateral comparison. *Review of Income and Wealth, 51*(4), 571–580.
Summers, R. (1973). International comparisons based upon incomplete data. *Review of Income and Wealth, 19*(1), 1–16.
Szulc, B. (1964). Index numbers of multilateral regional comparison (in Polish). *Przeglad Statysticzny, 3*, 239–254.
van Veelen, M. (2002). An impossibility theorem concerning multilateral international comparison of volumes. *Econometrica, 70*(1), 369–375.
Zavanella, B. (1996). International comparisons and construction of optimal graphs. In S. Camiz & S. Stefani (Eds.), *Proceedings of the conferences on matrices and graphs: Theory and applications to economics*, University of Brescia, Italy, 8th June 1993, 22 June 1995 (pp. 176–199). River Edge NJ: World Scientific.

Intra-National Price Level Differentials: The Italian Experience

Rita De Carli

1 Data Source and Survey Framework

The Italian research project was started in order to answer to the growing demand for official estimates of purchasing power parities (PPPs) at an intra-national level. With specific reference to household consumption, it led to the making of evaluations of difference in real consumption expenditures, salaries, revenues, and so on. Actually, these specific consumer PLIs are very useful in economic analysis and will provide valuable information to policy makers and economists interested in the dynamic adjustments taking place within countries.[1]

For the time being very few National Statistical Institutes (NSIs) have calculated these indices.[2] Taking this international experience into account, the methodology adopted was largely oriented towards respecting the comparability principle, requiring that prices to be compared be collected on identical or equivalent products, with reference to their physical and economic characteristics. From an economic point of view, this assumption, also referred to as the "like with like" principle, imposes that purchasers are indifferent towards the baskets collected in the different areas, as they meet the same needs. This principle also ensures that the differences in the prices collected in the different areas reflect the actual price gaps, so that comparability requires pricing to constant quality.

The comparability principle is essentially assured by product specifications that fully define the selected basket in terms of item and transaction characteristics that are supposed to influence prices. Item specifications are essentially the brand and the

R. De Carli (✉)
ISTAT, Rome, Italy
e-mail: decarli@istat.it

[1] In this article I present the first results of a project to calculate the differential level of consumer prices between twenty selected major Italian cities, jointly developed by Istat, Unioncamere and "Istituto Guglielmo Tagliacarne", with the cooperation of Municipal Statistical Offices which are in charge of most of the CPI survey. See also Istat (2008).

[2] Cambridge Econometrics Department conducted a specific survey for EUROSTAT to calculate regional PLIs, using some price data collected for the CPI and integrating them with other specific survey data (Cambridge Econometrics, 2002).

L. Biggeri, G. Ferrari (eds.), *Price Indexes in Time and Space*, Contributions to Statistics, DOI 10.1007/978-3-7908-2140-6_6,

model, or even tight defined clusters of these. Clusters of generic products, like non-processed food or brandless clothing and furniture products, in which only technical parameters are considered to make price differences, have been then used to make separate comparisons.

The comparability requirement takes advantage of facilitating the interpretation of results. At an operational level, however, difficulties emerge regarding product identification and item availability, which affect data requirements and the survey framework.

1.1 The Usefulness of Existing CPI Data

In respecting a general rule of parsimony, the potentiality of the existing CPI dataset was first investigated, in order to use the available data on prices as much as possible. As the product definitions were mostly exhaustive in terms of ensuring comparability only for food and non-alcoholic beverages, and not for clothing, and furniture, two supplementary surveys were carried out for these two latter expenditure chapters.

Even if the product specification within CPI data for food and non-alcoholic beverages make collected items fully identifiable in terms of their characteristics, their complete usefulness is not automatically guaranteed, as the choice of product items inside each BH is decided by the Municipal Statistical Offices which are in charge of most of CPI surveys. In particular, in the common CPI survey framework, the price collectors have to take the prices of the most sold items in each sampled outlet. Therefore, the collected items list could be not identical between municipalities, even though it refers to a common product list.

If we look at the at CPI survey framework itself, it is constructed to calculate indices with reference to over 500 BHs (there were 562 in 2006), some of which consist of several products (for example, the BH "Vegetables" includes 21 different types of vegetables, and fruit refers to 16 different products); the resulting price indices are published with reference to 205 groups of BHs.[3]

From an operational point of view, it is organised in two different sections: (i) a centralised survey, directly performed by ISTAT on some products that exhibit common characteristics within the country or whose prices are set in a common way, and (ii) local surveys, conducted by each municipality independently, but with strict regulations and procedures established by ISTAT.

The centralised survey concerns some 20% of the total expenditure basket. Anyway, most of the price quotations collected centrally do not vary between regions, because their value is usually set by central government, as happens for health and communications services. In order to estimate spatial price differentials, it is not necessary make direct use of any price quotations for these latter products, as they do not exhibit any geographical difference in price levels. On the other hand,

[3]More in detail, with reference to the COICOP classification system, the Consumer Price indices are published for 12 expenditure chapters, 38 product categories, 106 product groups, and 205 group of BHs.

the remaining price data is useful for PLI purposes, and their peculiarities must be detected centrally, in order to check their correspondence with PLI requirements.

The local surveys refer to a basket that covers about 80% of total household expenditure. They are currently performed by a total of 84 municipalities, including the 20 regional capitals, which we wish to investigate in this experiment. They run monthly in nearly 40,000 outlets, and in around 10,000 households for rents. The shops sample plan is set apart by each municipality, in order to reflect as much as possible the structure of local distribution. Even the number of shops interviewed is related to the magnitude of the town. In each shop, price quotations are taken on the most sold items. As a result, nearly 400,000 price quotations are collected this way each month.[4]

For the purpose of our intra-national PLIs calculations, selection of products was fixed in order to ensure their comparability. Taking the CPI basket in terms of BHs as a constraint, each item has been then submitted to many checks, in order to detect its degree of comparability. In fact, the real usefulness of this CPI data depends on the possibility of carrying out a preliminary job of refiling and recoding these items, which are locally stored. Indeed, drawing up a unique national item coding system to be used for this purpose was one of the more practical goals of this project. Moreover, the processed food items by CPI dataset were identified and grouped into more homogeneous subsets in terms of brands, models and packages, while the unprocessed food quotes were cleaned of their seasonality. As a result, out of 790,754 price quotations collected on Food and Beverages throughout the year 2006 by the local CPI surveys in the 20 analysed municipalities, a basket of 1,337 products was identified in terms of homogeneous characteristics that could guarantee comparability, such as brand, model, package etc, for a total of 544,414 selected price quotations.[5]

1.2 The Direct Integrating Surveys

Supplementary direct surveys were only implemented by the Municipal Statistical Office for the purposes of intra-national PLIs, to make up for the lack of

---

[4] For some products that show a bigger price volatility, the temporal cadence could vary. More precisely, the price collecting frequency is defined as follows: • twice a month, every 1st and 12th of each month for fruit and vegetables and fish, every 1st and 15th for car fuel; • monthly for other foods like fresh meat, processed food, for non durable goods such as house cleaning products, for semi-durable goods such as clothing, for some durable goods such as appliances, for certain services such as cinema, and for utility bills (gas, water, etc.); • quarterly in the months of February, May, August and November for some durable goods such as furniture and for some services, especially medical services; in the months of January, April, July and October for rents and for housing.
In 2008 there are 450 BHs the prices of which are recorded monthly, 61 for which they are collected quarterly, and 22 for which collection is once a fortnight.

[5] The remaining 246,340 available price quotations were not equipped with sufficient information as to their basic characteristics so as to permit the exact identification of the collected products, and so they were not used in the PPP calculations.

identification criteria ensuring comparability in some specific products such as clothing and footwear, and furniture products.[6]

These two direct surveys were conducted thanks to the experience gained in the current international PPP surveys. Using this experience as a benchmark, the item list was designed in conformity with the international basket, although it was organized to better fit the Italian economy.

The survey framework was indeed crucial here, as the sample design was set to better portray the Italian market, while their full comparability was ensured from the start at the moment of their inclusion in the basket. The item characteristics in terms of brand, model, and packaging were adapted to the Italian context only after having consulted market experts.

In order to ensure comparability, item definitions were accompanied by full technical and aesthetic features, and for some products photos were also included in the two lists. The sample design of outlet in terms of typology, was defined locally by analogy with the CPI survey. As a result, over a total of 400 products, 249 of Clothing and Footwear and 151 of Furniture, nearly 20 thousand (19,923) price quotations were collected in this way.

Caution was also used in the setting of the period of price collection and the frequency of collection. Although the PLI index is a structural indicator, the price collection period was defined infra-annually, with a collection frequency changing from product to product. The choice was made with particular caution, especially with reference to the seasonality that usually affects some food and clothing items, which might not be available or which might have different price levels in specific periods of the year. For these reasons, a frequency of 10 days was established for fruit and vegetables, with a monthly one for processed food, a semi-annual one for clothing, and an annual one for furnishing.

The exact number of quotes analysed depends on to the item definition and on the city analysed. Moreover, the larger the definition of the product – in terms of brand, model and technical characteristics – the greater the number of quotes. As a result, out of a total of 564,337 price quotations analysed, most of them (544,414) were collected monthly by the CPI surveys in 2006. The remaining 19,923 price quotations were directly collected semi-annually for clothing (13,226 quotes) and annually for furniture (6,697 quotes), in the same cities. Considering that only 19 municipalities[7] out of the 20 selected performed the survey for clothing while 18 of them did it for furniture,[8] we calculated that we collected on average over 2.5 price quotations per product in each municipality (2.8 for Clothing and Footwear and 2.5 for Furniture), while only 1.7 annual average price quotation per Food and

[6] Similar choices have already been measured in other countries. For the United Kingdom see Wingfield et al. (2005).

[7] The municipality of Bari did not do the direct survey on Clothing and Footwear. The municipalities of Cagliari and Campobasso only did one of the two biannual surveys.

[8] The municipality of Bari and Cagliari did not the direct survey on Furniture.

Non-Alcoholic products were selected on average in each of the 20 municipalities analysed.

2 Basic Methodology

In accordance, to some extent, with internationally shared methodology in PPP calculation, the Italian intra-national consumer PLIs were calculated to measure the cost of purchasing a common selected basket of goods in each of the twenty selected major cities. It expresses the cost relative to buying the same basket nationally, i.e. how much more (or less) does it cost to buy the basket in one particular city compared with the average cost of it in Italy.

The indices were calculated using the EKS formula introduced by Elteto Koves Szulc in the early 1960s. It was used to impose transitivity to the bilateral indexes calculated between each pair of cities, where the bilateral related to two specific cities, say k and b, taken from a larger contest of $j = 1.2\ldots$ J cities, involving a set of $i = 1.2\ldots$, l products, is defined as follows:

$$\mathrm{J}_{kb} = \left(\frac{\prod_{i=1}^{I} {}_{i}p_{k}}{\prod_{i=1}^{I} {}_{i}p_{b}} \right)^{{}_{i}q_{b}} \tag{1}$$

J_{kb} being the Jevons index, while ${}_{i}p_{k}$ and ${}_{i}q_{k}$ are respectively quotes and the relative quantity of the item i in the city k, so that $\Sigma_i\, {}_{i}q_{k} = 1$.

In practice, when calculating PLIs at BHs, a pseudo-weighted geometric mean of differences in comparable products belonging to each BH is first performed, as product weights are generally unknown and estimated as proxies. Due to the lack of any official information on the market shares of items, the collection frequency was set as weight. Results were then weighted together by groups of BHs, using the same expenditure structure as the CPIs.

Overall, the PLIs for the three chapters of expenditure are then calculated following these steps: first the geometric mean of binary Jevons indices are calculated for each pair of regions using pseudo-weights at item level, and transitivity is imposed using EKS formula, then the resulting indices are weighted, summed, and averaged using expenditure weights.

In formulas, considering the $i = 1,\ldots, I$ comparable products inside the $\mathrm{BH_h}$, we first calculate the binary relatives for regions k and b using the respecting $w_i..w_I$ pseudo-weights, estimated by collection frequencies

$${}_{h}\mathrm{J}_{kb} = \prod_{i}^{I} \left(\frac{{}_{i}p_{k}}{{}_{i}p_{b}} \right)^{\frac{w_i}{\sum w_i}} \tag{2}$$

Then transitivity is imposed using the EKS formula, to obtain PLIs at each $h = 1..J$ BHs. The resulted indices aim to synthesize the geographical differences between the prices of all comparable items of a single product: in the case of durum wheat pasta for instance, the calculation is carried out on two well known brands of locally marked items, using their collection frequency as weights on all kinds of locally collected pasta products.

At a higher level of aggregation, the results achieved at BHs are used instead of price quotations in calculating the Jevons indices, while the local expenditure structure adopted in the CPIs is used as a weight.[9]

3 Main Results

The results achieved seem to outline an articulated framework. In particular, the results for Food and Non Alcoholic Beverages tend to show higher price levels in the northern cities than in the southern ones, while products for clothing and furniture show significantly different profiles (see Table 1).

Among the northern cities, Milan, relatively one of the most expensive cities for food and furniture, only exhibits price levels closer to the national average with regard to clothing. Similarly, Reggio Calabria only stands as the most expensive city if as regards clothing, while its price levels for the remaining two chapters are below the national average of around 1 and 6% respectively for furniture and food.

Grouping up the cities in three sets on the basis of their deviation from the national average, we get the following breakdown:

Food and beverages: cities with price levels more than 2% higher than the average are Turin, Aosta, Genoa, Milan, Bolzano, Venice, Trieste, and Bologna; cities with price levels between +2% and –2% compared to the average are Ancona and Perugia; cities with price levels more than 2% lower than the average are Florence, Rome, Naples, L'Aquila, Campobasso, Bari, Potenza, Reggio Calabria, Palermo, and Cagliari

Clothing and footwear: cities with price levels more than 2% higher than the average are Bolzano, Venice, Trieste, Perugia, Reggio Calabria, and Cagliari; cities with price levels between +2% and –2% compared to the average are Turin, Genoa, Milan, Bologna, Ancona, Florence, Rome, and Potenza; cities with price levels more than 2% lower than the average are Aosta, Naples, L'Aquila Campobasso, and Palermo.

Furniture: cities with price levels more than 2% higher than the average are Turin, Aosta, Genoa, Milan, Bolzano, Bologna, Florence, and Rome; cities with price levels between +2% and –2 % compared to the average are

[9]Unlike the traditional procedure, the aggregation procedure implemented here makes use of the geometric mean of the lower-level indices weighted by the local structure of consumption, instead of the arithmetic mean.

Table 1 Intra-national price level indices (PLIs), for food and non alcoholic beverages, clothing and footwear, furniture. 20 Cities = 100. Year 2006*

Cities	Food and non alcoholic beverages			Clothing and footwear			Furniture		
	Processed	Non processed	Total	Well known brand	Brandless	Total	Well known brand	Brandless	Total
Turin	99.8	108.3	103.3	99.9	96.7	98.3	109.7	97.6	103.9
Aosta	104.	111.1	106.9	94.8	85.7	90.2	103.5	117.7	110.2
Genoa	102.6	113.	107.1	100.1	102.3	101.2	100.8	118.7	108.6
Milan	100.8	126.0	111.2	103.4	95.2	99.3	114.1	140.3	125.8
Bolzano	106.6	123.4	113.3	102.9	107.4	105.1	105.3	103.5	104.5
Venice	100.6	117.5	107.6	108.6	102.2	105.4	94.1	94.3	94.3
Trieste	104.8	115.6	109.4	103.	107.8	105.4	101.	100.	100.6
Bologna	98.2	119.7	107.	102.6	98.8	100.7	99.2	108.5	103.5
Ancona	99.6	103.8	101.4	96.8	104.8	100.7	94.9	87.4	91.3
Florence	97.9	94.9	96.6	104.7	96.7	100.7	108.6	105.1	107.
Perugia	99.6	103.	101.	100.	107.1	103.4	100.2	90.3	95.4
Rome	98.8	94.2	96.7	100.4	96.9	98.6	106.3	120.4	112.8
Naples	95.7	78.8	88.	97.2	92.9	95.1	86.9	90.3	88.6
L'Aquila	–	97.4	97.8	95.8	99.9	97.8	91.4	98.5	94.7
Campobasso	98.4	83.6	91.8	92.6	100.1	96.2	93.8	62.8	77.2
Bari	98.5	80.5	91.	–	–	–	–	–	–
Potenza	97.4	86.4	92.5	97.8	99.9	98.9	100.8	100.3	100.6
Reggio Calabria	98.8	87.	93.5	100.3	113.3	106.5	95.1	103.2	98.9
Palermo	97.9	87.5	93.2	101.5	89.4	95.4	98.4	85.2	91.9
Cagliari	100.8	89.7	95.8	99.1	106.8	102.8	–	–	–
min	*95.7*	*78.8*	*88.*	*92.6*	*85.7*	*90.2*	*86.9*	*62.8*	*77.2*
max	*106.6*	*126.*	*113.3*	*108.6*	*113.3*	*106.5*	*114.1*	*140.3*	*125.8*
Italy	**100.**	**100.**	**100.**	**100.**	**100.**	**100.**	**100.**	**100.**	**100.**

*The city of L'Aquila was not included in processed food calculations because of the low number of comparable items. Bari did not perform direct surveys; the cities of Cagliari and Campobasso only did one of the two half-yearly Clothing and Footwear surveys.
Source: ISTAT-Unioncamere-Istituto Guglielmo Tagliacarne.

Trieste, Potenza and Reggio Calabria; cities with price levels more than 2% lower than the average are Venice, Ancona, Perugia, Naples, L'Aquila, Campobasso, and Palermo.

The results achieved in the breakdown, made on a comparability basis, show a growing differential both for less comparable products, such as unprocessed food (mainly bread, meat, fish, vegetables and fruit), and for brandless products, such as clothing and furniture.

Considering the PLI index estimations on these two subgroups of food, price differentials are relatively smaller for processed products, with variations of less than 5% in absolute value, and wider for unprocessed products, with Milan exhibiting a

Table 2 Products surveyed in calculating price level indices (PLIs), for food and non alcoholic beverages, clothing and footwear, furniture*. Year 2006

	Food and non alcoholic beverages			Clothing and footwear			Furniture		
Cities	Processed	Non processed	Total	Well known brand	Brandless	Total	Well known brand	Brandless	Total
Turin	438	169	607	144	101	245	92	58	150
Aosta	249	97	346	125	101	226	72	56	128
Genoa	466	136	601	122	101	223	77	52	129
Milan	698	156	854	136	101	237	78	57	135
Bolzano	250	99	349	121	101	222	77	55	132
Venice	394	122	516	133	98	231	84	54	138
Trieste	277	119	396	124	97	221	66	43	109
Bologna	343	129	472	131	101	232	78	53	131
Ancona	240	81	321	116	89	205	81	51	132
Florence	379	120	499	121	96	217	73	53	126
Perugia	220	93	313	75	79	154	47	29	76
Rome	454	165	619	135	100	235	88	59	147
Naples	337	121	458	140	101	241	76	58	134
L'Aquila	–	99	99	121	97	218	81	56	137
Campobasso	124	105	229	65	54	119	58	42	100
Bari	255	117	372	–	–	–	–	–	–
Potenza	218	92	310	98	100	198	67	53	120
Reggio Calabria	369	111	480	96	97	193	75	58	133
Palermo	395	117	512	130	99	229	78	54	132
Cagliari	231	106	337	43	36	79	–	–	–
min	*124*	*81*	*229*	*43*	*36*	*79*	*47*	*29*	*76*
max	*698*	*169*	*854*	*144*	*101*	*245*	*92*	*59*	*150*
Italy	**1,106**	**231**	**1,337**	**148**	**101**	**249**	**92**	**59**	**151**

*The city of L'Aquila was not included processed food calculations because of the low number of comparable items. Bari did not perform direct surveys; cities of Cagliari and Campobasso only did one of the two half-yearly Clothing and Footwear surveys.
Source: ISTAT-Unioncamere-Istituto Guglielmo Tagliacarne.

variation of +26% and Naples one of –21% compared to the national average. We presume that the greater variability of price levels observed on the latter product category to some extent reflects the traditional market typology of unprocessed food (the open market), as well as certain aspects related to the localization and the nature of the products sold.

As for the other expenditure chapters, similar differences emerge between the well-known brand products and generic products, the prices of the latter being slightly more variable than the former.

Tables 4, 5 and 6 show the bilateral PLI matrix calculated on the three expenditure chapters. Within the matrix, the bilateral PLIs index is reported with reference to each pair of cities. The cities listed at the side (first column of the table) form the

Table 3 Price quotations used in calculating price level indices (PLIs), for food and non alcoholic beverages, clothing and footwear, furniture*. Year 2006

Cities	Food and non alcoholic beverages			Clothing and footwear			Furniture		
	Processed	Non processed	Total	Well known brand	Brandless	Total	Well known brand	Brandless	Total
Turin	16,765	41,019	57,784	457	569	1,026	329	278	607
Aosta	5,315	5,064	10,379	228	560	788	92	162	254
Genoa	15,795	15,072	30,867	229	478	707	133	184	317
Milan	39,176	41,931	81,107	339	594	933	184	197	381
Bolzano	5,142	6,008	11,150	309	693	1,002	202	281	483
Venice	10,177	12,500	22,677	226	407	633	210	167	377
Trieste	7,192	10,576	17,768	232	367	599	137	118	255
Bologna	16,625	23,792	40,417	299	617	916	200	194	394
Ancona	7,325	4,799	12,124	208	242	450	236	120	356
Florence	13,129	20,153	33,282	300	412	712	200	180	380
Perugia	5,935	4,737	10,672	153	255	408	90	50	140
Rome	22,688	36,949	59,637	399	504	903	264	221	485
Naples	12,407	25,898	38,305	385	645	1,030	204	195	399
L'Aquila		6,853	6,853	208	379	587	171	214	385
Campobasso	2,616	6,838	9,454	131	192	323	118	121	239
Bari	7,361	14,558	21,919						
Potenza	7,086	5,478	12,564	155	392	547	127	154	281
Reggio Calabria	11,821	12,427	24,248	166	368	534	260	271	531
Palermo	12,627	15,894	28,521	298	672	970	170	263	433
Cagliari	5,937	8,749	14,686	72	86	158			
min	*2,616*	*4,737*	*6,853*	*72*	*86*	*158*	*90*	*50*	*140*
max	*39,176*	*41,931*	*81,107*	*457*	*693*	*1,030*	*329*	*281*	*607*
Italy	**225,119**	**319,295**	**544,414**	**4,794**	**8,432**	**13,226**	**3,327**	**3,370**	**6,697**

*The city of L'Aquila was not included in processed food calculations because of the low number of comparable items. Bari did not do direct surveys; the cities of Cagliari and Campobasso only did one of the two half-yearly Clothing and Footwear surveys.
Source: ISTAT-Unioncamere-Istituto Guglielmo Tagliacarne.

basis of comparison: each cell then contains the deviation of the price levels of the city given in the heading (first row of the table) compared to those of the city at the side, which are set to 1. For instance, if you look at the twelfth row, first column of Table 4, you will find that price levels in Turin are 6.9% higher than in Rome (CPP = 106.9); on the other hand, you will see that price levels in Rome are 6.4% lower than in Turin (first row, twelfth column: CPP = 93.6). Differences in the two bilateral indices are due to different weights in terms of expenditure consumption in the two cities.

The same information could be indirectly deduced from Table 2: there it is shown that in Turin we have price levels which are 3.3% higher than the average (PLI = 103.3), while in Rome we have price levels which are 3.3% under the

Table 4 Bilateral price level indices (PLIs) matrix, for food and non alcoholic beverages[1]. 20 Cities*= 100. year 2006

	Turin	Aosta	Genoa	Milan	Bozen	Venice	Trieste	Bologna	Ancona	Florence
Turin	**100.0**	103.5	103.6	107.6	109.7	104.1	105.9	103.6	98.2	93.4
Aosta	96.6	**100.0**	100.2	104.0	106.0	100.6	102.3	100.1	94.9	90.3
Genoa	96.5	89.8	**100.0**	103.8	105.8	100.4	102.1	99.9	94.7	90.1
Milan	92.9	96.2	96.3	**100.0**	101.9	96.7	98.4	96.2	91.2	86.8
Bozen	91.2	94.4	94.5	98.1	**100.0**	94.9	96.5	94.4	89.5	85.2
Venice	96.0	99.4	99.6	103.4	105.3	**100.0**	101.7	99.5	94.3	89.7
Trieste	94.5	97.7	97.9	101.7	103.6	98.3	**100.0**	97.8	92.7	88.3
Bologna	96.6	99.9	100.1	103.9	105.9	100.5	102.2	**100.0**	94.8	90.2
Ancona	101.9	105.4	105.6	109.6	111.7	106.1	107.8	105.5	**100.0**	95.2
Florence	107.0	110.7	110.9	115.2	117.4	111.4	113.3	110.8	105.1	**100.0**
Perugia	102.3	105.8	106.0	110.1	112.2	106.5	108.3	105.9	100.4	95.6
Rome	106.9	110.6	110.8	115.0	117.2	111.3	113.1	110.7	104.9	99.8
Naples	117.4	121.5	121.7	126.3	128.7	122.2	124.3	121.6	115.2	109.7
L'Aquila	105.7	109.3	109.5	113.7	115.9	110.0	111.8	109.4	103.7	98.7
Campobasso	112.6	116.5	116.7	121.2	123.5	117.3	119.2	116.6	110.6	105.2
Bari	113.5	117.5	117.7	122.2	124.5	118.2	120.2	117.6	111.4	106.1
Potenza	111.7	115.6	115.8	120.2	122.5	116.3	118.3	115.7	109.7	104.4
Reggio Calabria	110.5	114.3	114.5	118.9	121.1	115.0	116.9	114.4	108.4	103.2
Palermo	110.8	114.7	114.9	119.3	121.5	115.4	117.3	114.8	108.8	103.5
Cagliari	107.9	111.6	111.8	116.1	118.3	112.3	114.2	111.7	105.9	100.8
Italy	**103.3**	**106.9**	**107.1**	**111.2**	**113.3**	**107.6**	**109.4**	**107.0**	**101.4**	**96.6**

Table 4 (continued)

	Perugia	Rome	Naples	L'Aquila	Campobassc	Bari	Potenza	Reggio Calal	Palermo	Cagliari
Turin	97.8	93.6	85.2	94.6	68.8	88.1	89.5	90.5	90.2	92.7
Aosta	94.5	90.4	82.3	91.5	85.8	85.1	86.5	87.5	87.2	89.6
Genoa	94.3	90.3	82.2	91.3	85.7	85.0	86.4	87.3	87.1	89.4
Milan	90.8	87.0	79.2	87.9	82.5	81.9	83.2	84.1	83.8	86.1
Bozen	89.1	85.3	77.7	86.3	81.0	80.3	81.6	82.5	82.3	84.5
Venice	93.9	89.9	81.8	90.9	85.3	84.6	86.0	86.9	86.7	89.0
Trieste	92.3	88.4	80.5	89.4	83.9	63.2	84.6	85.5	85.2	87.6
Bologna	94.4	90.4	82.2	91.4	85.7	85.1	86.4	87.4	87.1	89.5
Ancona	99.6	95.3	86.8	96.4	90.5	89.7	91.2	92.2	91.9	94.4
Florence	104.6	100.2	91.2	101.3	95.0	94.3	95.8	96.9	96.6	99.2
Perugia	100.0	**95.7**	87.1	96.8	90.8	90.1	91.6	92.6	92.3	94.8
Rome	104.5	100.0	**91.0**	101.1	94.9	94.1	95.7	96.7	96.4	99.1
Naples	114.8	109.9	100.0	**111.1**	104.2	103.4	105.1	106.3	105.9	108.8
L'Aquila	103.3	96.9	90.0	100.0	**93.8**	93.1	94.6	95.6	95.3	98.0
Campobasso	110.1	105.4	95.9	106.6	100.0	**99.2**	100.8	102.0	101.6	104.4
Bari	111.0	106.2	96.7	107.4	100.8	100.0	**101.6**	102.8	102.4	105.2
Potenza	109.2	104.5	95.2	105.7	99.2	98.4	100.0	**101.1**	100.8	103.6
Reggio Calabria	108.0	103.4	94.1	104.6	98.1	97.3	989	100.0	**99.7**	102.4
Palermo	106.3	103.7	94.4	104.9	96.4	97.6	99.2	1Q0.3	100.0	**102.7**
Cagliari	105.4	100.9	91.9	102.1	95.8	95.0	96.6	97.6	97.3	**100.0**
Italy	**101.0**	**96.7**	**88.0**	**97.8**	**91.8**	**91.0**	**92.5**	**93.5**	**93.2**	**95.6**

*The city of L'Aquial was not included in processed food calculations of the low number of comparable items. Bari did not do direct surveys; the cities of Cagliari and Campobasso only did one of the two half-yearly Clothing and Footwear surveys.
Source: Istat-Unioncamere-lstituto Gluglielmo Tagliacarm.

Table 5 Bilateral price level indices (PLIs) matrix, for clothing and footwear. 20 Cities*= 100. Year 2006.

	Turin	Aosta	Genoa	Milan	Bozen	Venice	Trieste	Bologna	Ancona	Florence
Turin	**100.0**	91.8	103.0	101.0	106.9	107.2	107.2	102.5	102.4	102.4
Aosta	109.0	**100.0**	112.2	110.0	116.5	116.8	116.8	111.6	111.6	111.6
Genoa	97.1	89.1	**100.0**	98.1	103.9	104.1	104.1	99.S	99.5	99.5
Milan	99.0	90.9	101.9	**100.0**	105.9	106.2	106.1	101.5	101.4	101.4
Bozen	93.5	85.8	96.3	94.4	**100.0**	100.3	100.2	95.8	95.8	95.8
Venice	93.3	85.6	96.0	94.2	99.7	**100.0**	100.0	95.6	95.5	95.5
Trieste	93.3	65.6	96.1	94.2	99.8	100.0	**100.0**	95.6	95.6	95.6
Bologna	97.6	89.6	100.5	98.6	104.4	104.6	104.6	**100.0**	100.0	100.0
Ancona	97.6	89.6	100.5	98.6	104.4	104.7	104.7	100.0	**100.0**	100.0
Florence	97.6	89.6	100.5	98.6	104.4	104.7	104.7	100.0	100.0	**100.0**
Perugia	95.1	87.2	97.9	96.0	101.6	101.9	101.9	97.4	97.4	97.4
Rome	99.7	91.5	102.6	100.6	106.6	106.8	106.8	102.1	102.1	102.1
Naples	103.4	94.9	106.4	104.4	110.5	110.8	110.8	105.9	105.9	105.9
L'Aquila	100.5	92.3	103.5	101.5	107.5	107.8	107.8	103.0	103.0	103.0
Campobasso	102.2	93.8	105.2	103.2	109.3	109.6	109.6	104.7	104.7	104.7
Bari	–	–	–	–	–	–	–	–	–	–
Potenza	99.4	91.3	102.4	100.4	106.3	106.6	106.6	101.9	101.9	101.8
Reggio Calabria	92.3	84.7	95.0	93.2	98.7	99.0	98.9	94.6	94.5	94.5
Palermo	103.1	94.6	106.1	104.1	110.2	110.5	110.5	105.6	105.6	105.6
Caglrari	95.6	87.8	98.5	96.6	102.3	102.5	102.5	98.0	97.9	97.9
Italy	**98.3**	**90.2**	**101.2**	**99.3**	**105.1**	**105.4**	**105.4**	**100.7**	**100.7**	**100.7**

Table 5 (continued)

	Perugia	Rome	Naples	L'Aquila	Campobassc	Bari	Potenza	Reggio Calal	Palermo	Cagliari
Turin	105.2	100.3	96.7	99.5	97.8	–	100.6	108.3	97.0	104.6
Aosta	114.6	109.3	105.4	108.4	106.6	–	109.6	118.1	105.7	113.9
Genoa	102.2	97.5	94.0	96.6	95.0	–	97.7	105.2	94.2	101.6
Milan	104.2	99.4	95.8	98.5	96.9	–	99.6	107.3	96.1	103.5
Bozen	98.4	93.8	90.5	93.0	91.5	–	94.0	101.3	90.7	97.8
Venice	98.1	93.6	90.2	92.8	91.2	–	93.8	101.1	90.5	97.5
Trieste	98.2	93.6	90.3	92.8	91.3	–	93.8	101.1	90.5	97.6
Bologna	102.7	97.9	94.4	97.1	95.5	–	98.1	105.7	94.7	102.1
Ancona	102.7	98.0	94.5	97.1	95.5	–	98.2	105.8	94.7	102.1
Florence	102.7	98.0	94.5	97.1	95.5	–	98.2	105.8	94.7	102.1
Perugia	100.0	**95.4**	92.0	94.6	93.0	–	95.6	103.0	92.2	99.4
Rome	104.8	100.0	**96.4**	99.1	97.5	–	100.2	108.0	96.7	104.2
Naples	108.7	103.7	100.0	**102.8**	101.1	–	103.9	112.0	100.3	108.1
L'Aquila	105.8	100.9	97.3	100.0	**98.4**	–	101.1	108.9	97.5	105.1
Gampobasso	107.5	102.6	98.9	101.7	100.0	–	102.8	110.8	99.2	106.9
Bari	–	–	–	–	–	–	–	–	–	–
Polenza	104.6	99.8	96.2	98.9	97.3	–	100.0	**107.7**	96.5	104.0
Reggio Calabria	97.1	92.6	89.3	91.8	90.3		92.8	100.0	**89.5**	96.5
Palermo	108.4	103.4	99.7	102.5	100.8	–	103.6	111.7	100.0	**107.8**
CagHari	100.6	96.0	92.5	95.1	93.6	–	96.2	103.6	92.8	100.0
Italy	**103.4**	**98.6**	**95.1**	**97.8**	**96.2**		**98.9**	**106.5**	**95.4**	**102.8**

*The city of L'Aquial was not included in processed food calculations of the low number of comparable items. Bari did not do direct surveys: the cities of Cagliari and Campobasso only did one of the two half-yearly Clothing and Footwear surveys.

Source: Istat-Unioncamere-lstituto Gluglielmo Tagliacarm.

Table 6 Bilateral price level indices (PLIs) matrix, for funiture. 20 Cities*= 100. Year 2006.

	Turin	Aosta	Genoa	Milan	Bozen	Venice	Trieste	Bologna	Ancona	Florence
Turin	**100.0**	106.1	104.6	121.1	100.6	90.8	96.8	99.6	87.9	103.0
Aosta	94.3	**100.0**	98.6	114.2	94.8	85.5	91.3	93.9	82.9	97.1
Genoa	95.6	101.5	**100.0**	115.8	96.2	86.8	92.6	95.3	84.1	98.5
Milan	82.6	87.6	86.3	**100.0**	83.0	74.9	79.9	82.2	72.6	95.0
Bozen	99.4	105.5	103.9	120.4	**100.0**	90.2	96.2	99.0	67.4	102.4
Venlce	110.2	116.9	115.2	133.5	110.8	**100.0**	106.7	109.8	96.9	113.5
Trieste	103.3	109.6	108.0	125.1	103.9	93.7	**100.0**	102.9	90.8	106.4
Bologna	100.4	106.5	105.0	121.6	101.0	91.1	97.2	**100.0**	88.3	103.4
Ancona	113.7	120.6	118.9	137.8	114.4	103.2	110.1	113.3	**100.0**	117.2
Florence	97.1	103.0	101.5	117.6	97.6	88.1	94.0	96.7	85.4	**100.0**
Perugia	108.9	115.5	113.9	131.9	109.5	98.8	105.4	108.5	95.7	112.2
Rome	92.1	97.7	96.3	111.5	92.6	83.6	89.1	91.7	81.0	94.9
Naples	117.3	124.4	122.6	142.1	118.0	106.5	113.6	116.9	103.1	120.8
L'Aquila	109.6	116.3	114.6	132.8	110.3	99.5	106.1	109.2	96.4	112.9
Campobasso	134.5	142.7	140.6	162.9	135.3	122.1	130.2	134.0	118.3	138.6
Bari	–	–	–	–	–	–	–	–	–	–
Polenza	103.2	109.5	108.0	125.1	103.9	93.7	100.0	102.9	90.8	106.4
Reggio Calabria	105.0	111.4	109.8	127.2	105.6	95.3	101.7	104.6	92.3	108.2
Palemio	113.0	119.9	118.2	136.9	113.7	102.6	109.4	112.6	99.4	116.4
Cagliari	–	–	–	–	–	–	–	–	–	–
Italy	**103.9**	**110.2**	**108.6**	**125.8**	**104.5**	**94.3**	**100.6**	**103.5**	**91.3**	**107.0**

Table 6 (continued)

	Perugia	Rome	Naples	L'Aquila	Campobassc	Bari	Potenza	Reggio Calal	Palermo	Cagliari
Turin	91.8	106.6	85.3	91.2	74.3	–	96.9	95.2	88.5	–
Aosta	66.6	102.4	80.4	86.0	70.1	–	91.3	89.8	83.4	–
Genoa	87.8	103.9	81.5	87.2	71.1	–	92.6	91.1	84.6	–
Milan	75.8	89.7	70.4	75.3	61.4	–	80.0	78.6	73.1	–
Bozen	91.3	108.0	84.8	90.7	73.9	–	96.3	94.7	68.0	
Venice	101.2	119.7	93.9	100.5	81.9	–	106.7	104.9	97.5	–
Trieste	94.9	112.2	88.1	94.2	76.8	–	**100.0**	98.4	91.4	–
Bologna	92.2	109.0	85.6	91.6	74.8	–	97.2	95.6	88.8	–
Ancona	104.4	123.5	97.0	103.7	84.5	–	110.1	108.3	100.6	–
Florence	89.2	105.4	82.8	66.6	72.2	–	94.0	92.4	85.9	
Perugla	**100.0**	118.3	92.8	99.3	80.9	–	105.5	103.7	96.4	–
Rome	84.6	**100.0**	78.5	84.0	88.4	–	89.2	07.7	81.5	
Naples	107.7	127.4	**100.0**	107.0	87.2	–	113.6	111.7	103.8	–
L'Aquila	100.7	119.1	93.5	**100.0**	61.5	–	106.2	104.4	97.0	–
Campobasso	123.5	148.1	114.7	122.7	**100.0**	–	130.3	128.1	119.0	–
Bari	–	–	–	–	–	–	–	–	–	–
Potenza	94.8	112.1	88.0	94.2	76.8	–	**100.0**	98.3	91.4	–
Reggio Calabria	96.4	114.0	89.5	95.8	78.1	–	101.7	**100.0**	92.9	–
Palermo	103.8	122.7	96.3	103.1	84.0	–	109.4	107.6	**100.0**	–
Cagliari	–	–	–	–	–	–	–	–	–	–
Italy	**95.4**	**112.8**	**88.6**	**94.7**	**77.2**		**100.6**	**96.9**	**91.9**	

*The city of L'Aquial was not included in processed food calculations of the low number of comparable items. Bari did not do direct surveys: the cities of Cagliari and Campobasso only did one of the two half-yearly Clothing and Footwear surveys.

Source: Istat-Unioncamere-Istituto Gluglielmo Tagliacarm.

average (PLI = 96.7). As a consequence, the difference between the two indices equals the above mentioned bilateral parities (100 × [103.3/96.7] = 1.068; 100 × [96.7/103.3] = 93.6). Note that, the geometric average index has been given in the last row of the bilateral price parity matrix, and it is equal to the data reported in Table 2.

Therefore, compared with the information contained in Table 2, the bilateral price parity matrix has the sole advantage of making the comparison between each pair of cities immediate.

References

Biggeri, L., De Carli, R., & Laureti, T. (2008). The interpretation of the PPPs and the integration with the CPIs work at regional level: a method for measuring the factors that affect the differences. Paper presented at the joint UNECE/ILO Meeting on Consumer Price Indexes. Geneva.

Cambridge Econometrics (2002, 23 December). *Provision of Statistical Services, CPI and Ppp Application* (A final report for EUROSTAT). Unit of Price Comparison and Correction Coefficient.

EUROSTAT/OECD (2006). *Methodological manual on purchasing power parities*, ISSN 1725-0048. Retrieved from http://epp.eurostat.ec.europa.eu/pls/portal/url/page/PGP_MISCELLANEOUS/PGE_DOC_DETAIL?p_product_code=KS-BE-06-002

Istat, U. T. (2008, 22 aprile). Le differenze nel livello dei prezzi tra i capoluoghi delle regioni italiane per alcune tipologie di beni. Anno 2006. *Note per la stampa.*

Wingfield D., Feniwick D., & Smith K. (2005, February). Relative regional consumer price levels in 2004, Office for National Statistics, Economic Trends 615. Retrieved from http://www.statistics.gov.uk/articles/economic_trends/ET615Wingfield.pd

Part III
Subindexes

Consumer Price Indexes: An Analysis of Heterogeneity Across Sub-Populations

Raffaele Santioni, Isabella Carbonaro, and Margherita Carlucci

1 Introduction

The soundness of Consumer Price Index (CPI) as a measure of inflation has long been debated, focusing on the different sources of bias inherent to the use of a constant basket of goods and services.[1] At the end of the twentieth century the Boskin Commission's report (Boskin, Dulberger, Gordon, Griliches, & Jorgenson, 1996) showed an overstatement of 1.1% per year from CPI data, mainly due to substitution bias and inadequate treatment of quality changes. On the contrary, following the euro cash changeover in 2002, evidence of wide gaps between consumers' perceptions and official measures of inflation was found in many countries of the euro area,[2] showing understatements of more than 3.5% per year from CPI data in 2003.

The gap between perceived and official inflation could lead to serious consequences for economic policy: erosion of the euro's public acceptance and its institutional framework (Del Giovane & Sabbatini, 2006), distorted inflation expectations influencing prices and wages; and even questioning the credibility of monetary policy (Brachinger, 2006).

According to literature, there are many possible explanations for this gap.

Basically, official measures of inflation rate follow a methodology established at international level, based on a subset of actual final consumption of resident households, derived from national accounts statistics. So the index is not representative

R. Santioni (✉)
Bank of Italy, Economic Research Unit, Rome, Italy
e-mail: raffaele.santioni@bancaditalia.it

[1]This research was funded by PRIN 2005. The Authors thank Martino Lo Cascio and the participants in the International Workshop on Price Index Numbers in Time and Space, held in Florence, September 29th 2008, for useful comments. The views expressed in this paper are those of the authors and do not involve the responsibility of the Bank of Italy.

[2]See, for instance, ECB (2007), Jungermann, Brachinger, Belting, Grinberg, and Zacharias (2007) for Germany; Kurri (2006) for Finland; Fluch and Stix (2005) for Austria; Álvarez González et al. (2004) for Spain; Del Giovane and Sabbatini (2006) and Del Giovane, Fabiani, & Sabbatini (2008) for Italy. Assessment of perceived inflation derives from qualitative opinion surveys, such as European Commission's Consumer Survey.

L. Biggeri, G. Ferrari (eds.), *Price Indexes in Time and Space*, Contributions to Statistics, DOI 10.1007/978-3-7908-2140-6_7, © Springer-Verlag Berlin Heidelberg 2010

of a particular social group but of the community. Otherwise, inflation perceived by consumers appears conditioned by the individual patterns of consumption, and also by subjective factors, such as asymmetries in the perception of price movements upwards and downwards.

CPI can only be an average of differentiated price increases of individual products and in different territories. Therefore the index, summarizing the spending behavior of all households, might not match the inflation rate experienced by each individual family as a result of its specific consumption basket.

Thus experimental consumer price indexes have been proposed for targeted sub-populations specifically relevant for social and economic policy, that are likely to have consumption patterns different from the average, i.e. poor (Garner, Johnson, & Kokoski, 1996) and elderly (Stewart, 2008) households. Experimental indexes are a reweighting of the CPI elementary indexes using expenditure weights from households belonging to the sub-population of interest (BLS, 1997).

Moreover, the goods and services whose prices enter in CPI do not include items relevant in family expenditures that are not considered as consumption but as family's investment or inter-institutions transfers. The first is the case of house prices, not covered in the domain of consumer prices, but whose increase substantially affects the perceived loss of purchasing power among lower-income households (Ranyard, Missier, Bonini, Duxbury, & Summers, 2008). The second is the case of insurance premiums, valued in CPI net of claims and surrenders: thus increases in their prices enter the index with weights much smaller than their actual burden on family's expenditures.

On the other hand, due to asymmetry in perceptions, consumers seem to form their idea of general inflation on the basis of a smaller commodity basket than that of the CPI. Price increases in items with high purchase frequency and out-of-pocket payment, like food, petrol, coffee services and hairdressing, count much more than price reductions in infrequently purchased items, paid for with credit card or via automatic bank transfer, like home electronics.

Speculative behavior of price setters in "grey zones" (Wunder, Schwarze, Krug, & Herzog, 2008): non-competitive markets of the service sector, such as restaurants (Adriani, Marini, & Scaramozzino, 2009), cinema tickets and dry cleaning, more influential costs of the adjustment to the new currency for items paid for in cash, varying accuracy of consumers' memory of prices, may also play a role.

These findings suggest that an "out-of-pocket" subsample of CPI basket could provide an index better suited to catch perceived inflation rate of growth (Lyziak, 2009; ECB, 2003; contrasting findings are reported in Antonides, 2008).

Aim of this work is to investigate the possibilities offered by alternative methods of construction of consumer price indices in order to capture the variability of inflation rates facing different households' types.

2 Definition of Experimental Price Indexes

CPI measures the changes in the price of a fixed basket of goods and services purchased by an average – or representative – consumer. The increase in prices which individual households actually faces is, however, a phenomenon depending on the specific family's consumption profile.

In fact, factors contributing to erosion of real purchasing power of a given amount of spending can be attributed to two main sources of heterogeneity (Schultze & Mackie, 2002):

1. Households differently allocate their budget spending in various categories of goods and services, with consumer spending patterns mainly related to households' profile. These typical differences in the allocation of budget shares, or *across-strata heterogeneity*, are the rationale behind the construction of BLS experimental indexes based on specific weighting systems;
2. Households' consumption patterns differ not only in expenditure budget shares, but also in retail outlets and consumption items actually purchased. This one is called *within-stratum heterogeneity*. If the hypothesis that all families face the same set of retail place is to be dismissed, and if there are reasons to believe that the elderly or the poor face higher prices, these groups could suffer more rapid inflation rates than the entire population (Rao, 2000).

Official statistics' data only allow to control for differences in spending patterns across households, since Household Budget Surveys (HBS) currently provide detailed data on family expenditures' shares, while price quotes by outlet and items are not published. Thus, official data do not allow estimates of price dispersion[3] faced by different households.

Following Pollak (1998), another way to look at heterogeneity involves answering three questions: "How many indices?"; "Beer or champagne?"; and "What type of group indices?". The first one refers to heterogeneity in households' consumption patterns, calling for different indexes for population subgroups (across-strata heterogeneity). The second one raises the problem of appropriate selection of items and outlets included in the inflation index (within-stratum). The third one introduces an issue not yet analyzed.

CPI is a weighted average of products' price indices, with weights given by their respective shares in aggregate consumption expenditure. Assuming that all households face the same lower-level price indexes, this means that each household receives a weight in accordance to its total expenditure. Since expenditure increases with income, this approach (the so-called *plutocratic* approach) gives greater influence to the consumption patterns of rich households than to poor ones. Thus, since the families who spend more contribute more to national expenditure shares compared to less affluent ones, the wealthiest families have a greater say in determining CPI value (the "one dollar one vote" criterion). Empirical evidence shows that the ideal "average" consumer of CPI is rather a rich one. For the United States in 1990,

[3]It is worthwhile noting that this kind of price heterogeneity also affects CPI via formulas used for calculating elementary indexes, since the use of geometric (as in Italy and several other countries) or arithmetic unweighted means of prices' ratios lead to different values depending on the variances of the logarithms of sampled prices (Silver & Heravi, 2007).

Deaton (1998) located this consumer in the 75 percentile of the distribution of consumption expenditure; for Spain during the 1990s Izquierdo, Ley, and Ruiz-Castillo (2003) in the 61 percentile, while household data from four Latin American countries (Brazil, Colombia, Mexico, and Peru) in a period spanning from 1984 to 2003 showed that standard CPI weights typically reflect those of a consumer located in the 80–90 percentile (Goni, Lopez, & Servén, 2006).

On the contrary, if we decide to accord equal weight to each household, the aggregate price index follows the *democratic* formula (the "one household one vote" criterion).

One reason for official statistics to prefer plutocratic indexes (Kokoski, 2003) lies in that only information on prices and aggregated expenditure shares are needed, while for a democratic index one must first construct the price index for each individual household, then average them to produce an aggregate index.

To show the differences between plutocratic and democratic formulas, we define the Laspeyres price index ${}^{h}L^{t}_{t-1}$ for each individual household, h, as follows:

$$ {}^{h}L^{t}_{t-1} = \sum_{i=1}^{n} \left(\frac{p_{i,t}}{p_{i,t-1}} \right) {}^{h}s_{i,t-1} = \sum_{i=1}^{n} P^{t}_{i,t-1} \, {}^{h}s_{i,t-1} \tag{1} $$

where ${}^{h}s_{i,t-1}$ is the product i's share of household h's total expenditure in the base period t–1 and $P^{t}_{i,t-1}$ is product i's elementary price index. The aggregate price index for all H households could be expressed by:

$$ {}^{H}L^{t}_{t-1} = \sum_{h \in H} {}^{h}w_{t-1} \, {}^{h}L^{t}_{t-1} = \sum_{h \in H} {}^{h}w_{t-1} \sum_{i=1}^{n} P^{t}_{i,t-1} \, {}^{h}s_{i,t-1} \tag{2} $$

where ${}^{h}w_{t-1}$ is the weight given to household h in the aggregate index.

In the democratic approach each household h counts with the same weight ${}^{h}w_{t-1} = 1/H$ and the corresponding index is the unweighted arithmetic mean of the H individuals households' price indexes.

$$ {}^{H}_{d}L^{t}_{t-1} = \frac{1}{H} \sum_{h \in H} {}^{h}L^{t}_{t-1} = \frac{1}{H} \sum_{h \in H} \sum_{i=1}^{n} P^{t}_{i,t-1} \, {}^{h}s_{i,t-1} \tag{3} $$

In the plutocratic approach weights are proportional to each household's total expenditure ${}^{h}E_{t-1}$ at the base period,

$$ {}^{h}w_{t-1} = \frac{{}^{h}E_{t-1}}{\sum_{h \in H} {}^{h}E_{t-1}} \tag{4} $$

The corresponding index equals to:

$$ {}^{H}_{p}L^{t}_{t-1} = \sum_{h \in H} {}^{h}L^{t}_{t-1} \frac{{}^{h}E_{t-1}}{\sum_{h \in H} {}^{h}E_{t-1}} = \sum_{h \in H} \sum_{i=1}^{n} P^{t}_{i,t-1} \frac{{}^{h}p_{i,t-1}\,{}^{h}q_{i,t-1}}{{}^{H}E_{t-1}} \tag{5} $$

It is straightforward to show that ${}^{H}_{p}L^{t}_{t-1}$ is the current CPI index.

Due to the combined effect of differences in consumption patterns across individual households (i.e. rich households buying luxuries and poor ones buying necessities) and in inflation rates across goods (i.e. with luxury goods or necessities experiencing higher than average inflation), the two approaches lead to different estimates of overall inflation rates. The gap[4] between inflation measured according to CPI and according to a democratic index is called *plutocratic bias*. Depending on whether prices behave in an anti-rich or an anti-poor manner the bias will be positive or negative. In any case, the more different family consumption patterns are, the less CPI could represent inflation rate suffered by poorer households.

Therefore, the use of a single price index to adjust the income of a collectivity has a redistributive effect. Individuals whose price index variations are lower than those experienced by the official index used to deflate nominal income receive an unexpected benefit, as opposed to those who face inflation rates higher than "average" one, which in fact suffer a net loss. As a result, it is desirable to use specific price indices for subgroups of the population when analyzing time changes of households' purchasing power and inequality.[5] Some attempts have already been performed in Italy (ISTAT, 2007), even if there are many conceptual and operational drawbacks (Biggeri & Leoni, 2003).

In this study we used ISTAT Households Budget Survey (HBS) data for the years 1999–2005 and ISTAT monthly price indexes (CPI) by representative positions for 2000–2006.

We first developed sub-indexes breaking down Italian families' by equivalised expenditure decile. Then, we focused our attention on selected households that are likely both to have specific consumption behavior and to be particularly sensitive to the loss of purchasing power:

- single person with dependent children;
- head of family workless through unemployment or economic inactivity;
- elderly.

[4]It is worthwhile noting that the assumption of identical lower-level price indexes may lead to substantial underestimation of this gap (Izquierdo et al., 2003; Kokoski, 2003).

[5]The impact on inequality depends on how much households' expenditure profiles differ across income-groups (see Son & Kakwani, 2006).

To control[6] for across-strata heterogeneity we constructed sub-indexes defined applying to lower-level price indexes specific family types' weights according to their expenditures patterns resulting from HBS.

We also checked the effect of changes in the sample of goods and services, developing a Laspeyres price index for a basket of frequently purchased goods and services, defined on the basis of empirical evidence provided by previous studies (Carlucci & Zelli, 1998) as common "Necessities" of any type of households, regardless the phase of family life cycle they are within, related to daily basic personal needs and housing maintenance. This reference basket includes food and non-alcoholic beverages, electricity and household fuels, household cleaning products, communication charges.

3 Data

3.1 Harmonisation of HBS and CPI Data

To build up weights for sub-populations we used HBS micro data on family characteristics and expenditures, broken down in nearly 300 categories (corresponding to five-digit level Classification Of Individual COnsumption by Purpose, COICOP–HBS) for the years 1999–2005.

These data are not fully correspondent to CPI requirements either in coverage or in classification criteria. CPI measures the changes over time in the prices of a basket of goods and services representative of all those purchased for final consumption of household. It means that CPI does not include investment items, compensative and non-monetary transactions, recorded as families' expenditures in HBS. Thus we firstly had to remove HBS headings such as mortgages, life insurances, repayments of loans and outlays, services engaged for major maintenance and repairs or for extensions and conversions of dwellings (capital formation), purchases of second-hand vehicles (compensative) and imputed rentals of owner-occupied houses (non-monetary). We also removed expenditures on game of chance and expenditures abroad, not covered in the CPI sample.

Merging the remaining HBS expenditures and CPI price indexes is not a straightforward operation, since the headings of the two sources do not always coincide. In building the transition matrix to match CPI indexes to HBS categories, we had to copy with the fact that HBS commodity headings are not only more aggregated than CPI representative positions but also follow different grouping criteria than the ones used for CPI. We operated at the most disaggregated level we could, namely at the voice of product's level. Only when it was not possible we matched headings

[6] With the caveat that weights used for sub-groups derive from sub-samples of HBS whose size may affect estimates' precision (BLS, 1997).

at group of product's level.[7] In few cases the HBS heading corresponded to different CPI levels, so we matched CPI of representative positions with voices of product' ones to cover the HBS definition.[8] In the end, we defined a one-to-one match between about 280 HBS expenditure items and 155 consumer price indexes, exhausting the entire reference domain of CPI listing.

3.2 *Weights' Identification*

Expenditures of each household in the HBS samples have been broken down in the 155 groups for which we have matched CPI elementary price indexes. For household h we computed the vector of weights at time t, $t = 1999, \ldots, 2005$, namely ${}^h s_t = \left({}^h s_{1,t}, {}^h s_{2,t}, \ldots, {}^h s_{155,t}\right)$, where ${}^h s_{i,t} = \frac{{}^h E_{i,t}}{\sum_{i=1}^{155} {}^h E_{i,t}}$, ${}^h E_{i,t}$ is the expenditure for the group i, $i = 1, 2, \ldots, 155$, and $\sum_{i=1}^{155} {}^h E_{i,t}$ total expenditure.[9]

It is worthwhile noting that even if National Accounts (NA) estimates of consumption – that provide weights for official CPI – heavily rely on HBS data, the two sources lead to different values, specially for selected items, such as items that are purchased at a frequency lower than one month. However, we may assume that, in terms of quality and reliability, the weighting coefficients obtained from the HBS represent a good approximation of NA ones and therefore can be used for the transition from a structure of common weights for all family types to a set of weights for subgroups.

Hence, we evaluated monthly (calculation base) consumer price indexes for each h-th family as:

$$ {}_{cb}^{h}L^{m,t} = \sum_{i=1}^{155} {}^{12,t-1}P_i^{m,t}\, {}^h s_{i,t-1} \quad h,m = 1,\ldots,12 \text{ and } t = 2000,\ldots,2006 \qquad (6) $$

where ${}^{12,t-1}P_i^{m,t}$ is the national price index of the i-th group between month m of year t and December of the year t–1 (calculation base), as published by ISTAT.

---

[7] This means that in these cases our lower-level price ratios are themselves price indexes (see Hill, 2004) but at this level of detail we are fairly confident that this should not have important effects on our final results.

[8] For instance, to match the HBS heading "Bread and other bakery products" we had to aggregate the index of group of products "Bread" with the indexes of the two representative positions "Crackers" and "Bread sticks".

[9] Total expenditure here refers only to the HBS expenditures covered by CPI domain of observation (see Sect. 3.1).

Chained indexes (reference base December 1999) have finally been calculated by

$$ {}^{h}_{rb}L^{m,t} = \prod_{k=2000}^{t-1} \frac{{}^{h}_{cb}L^{12,k}}{100}\, {}^{h}_{cb}L^{m,t} \quad h,m = 1,\ldots,12 \text{ and } t = 2000,\ldots,2006 \tag{7} $$

3.3 Identification of Sub-Groups

For ethical and thence policy aims, the most important issue to investigate is to what extent poorer families are liable to suffer inflation rates higher than average. In this context, families' wellbeing can be measured in terms of the amount of goods and services for households' consumption. Thus, economic condition for each family has been defined in terms of its HBS expenditures, excluding only those which are not included in consumption, as contributing to capital formation (see above, Sect. 3.1). Since we wanted a proxy for economic welfare, we considered also non monetary flows that accrue the amount of available goods and services, such as services of owner-occupied dwellings, measured as imputed rentals.

We applied the concept of equivalent total consumption expenditure of the family in order to compare families of different size. It has been determined by dividing total expenditure of the family for appropriate deflators, the so-called equivalence scales, that take into account the economies of scale in family consumption. In literature different equivalence scales have been proposed, without a general consensus about which one to use in the different situations. Since the choice could affect the results,[10] we decided to use the official scale used in Italy by ISTAT to assess poverty, the Carbonaro's scale (1985).

We have therefore separated households into deciles of the distribution of equivalent total consumption expenditure, with decile 1 comprising the poorest[11] 10% of the households, decile 10 the richest 10% and so on. The thresholds used for the different years are given below, Table 1.

In addition to select population's subgroups on the basis of equivalent expenditure deciles, other types of families were taken into consideration based on the characteristics of household's members (relationship with the reference person, age, occupation). As noted before (see Sect. 2), these family types are as follows:

- single person with dependent children (less than 18 years old);
- head of the family workless through unemployment or economic inactivity[12];

[10]For instance, Engel's equivalence scales measured with food share could overestimate the needs of large families, while Rothbarth's ones may underestimate the needs of families with children (see Carlucci & Zelli, 1998).

[11]"Poorest" and "richest" refer to our definition of wellbeing in terms of goods and services available for consumption.

[12]Excluding unemployed single person with dependent children.

Table 1 Equivalent expenditure deciles: lower boundary of groups (*current euro*)

Year	Second decile	Third decile	Fourth decile	Fifth decile	Sixth decile	Seventh decile	Eighth decile	Ninth decile	Highest 10%
1999	431.33	558.22	661.48	763.15	880.03	1106.50	1171.28	1407.38	1858.50
2000	451.06	574.61	685.46	796.43	915.15	1052.44	1233.39	1487.99	1990.37
2001	460.45	584.20	693.66	805.13	917.93	1067.95	1254.07	1516.27	1998.48
2002	477.07	603.92	713.60	825.85	950.78	1093.81	1267.46	1512.90	1979.53
2003	510.68	643.20	757.59	875.17	1003.92	1154.36	1343.60	1611.66	2113.61
2004	521.50	666.03	791.98	918.58	1058.45	1222.22	1424.44	1709.52	2234.62
2005	541.82	686.49	816.18	950.26	1087.28	1248.31	1446.60	1744.88	2234.01

- head of the family retired;
- elderly, regardless their occupational status.

These groups of households are characterized by substantial differences, as can easily be seen in Table 2, derived from HB Survey for the year 2005.

Data for the year 2005 support our choice of these families as the most sensitive to the risk of poverty.

Table 2 Characteristics of households by sub-group (HBS, 2005)

Household	Average monthly household expenditure (current euro)	Equivalent household expenditure (current euro)	Average household members	Number of households in the HBS sample	Grossed number of households
Single person with children (age < 18)	2,209	1,183	2.4	389	422,951
Head of the family retired	1,983	1,252	2.0	9,373	8,841,900
Head of the family unemployed	1,853	985	2.8	523	533,547
Other households	2,594	1,351	2.8	13,822	13,469,311
Elderly households	1,531	1,225	1.4	5,366	5,534,829
Other households	2,589	1,326	2.8	18,741	17,732,880
Lowest 10%	909	429	2.9	2,526	2,328,541
Second decile	1,267	618	2.8	2,472	2,325,176
Third decile	1,537	751	2.8	2,444	2,326,603
Fourth decile	1,757	884	2.7	2,491	2,326,978
Fifth decile	1,998	1,019	2.6	2,406	2,326,638
Sixth decile	2,216	1,167	2.5	2,408	2,327,465
Seventh decile	2,447	1,343	2.4	2,337	2,326,364
Eighth decile	2,763	1,582	2.2	2,342	2,328,638
Ninth decile	3,314	1,961	2.1	2,259	2,324,709
Highest 10%	5,169	3,266	2.0	2,422	2,326,597

All the selected groups spent less than the other households, with households whose head is unemployed and single parents showing major[13] shortcomings. Thus, even if their limited sample size prevent from drawing quantitative conclusions, we thought that qualitative indication about possible inflation differentials for these groups could be of some interest for policy aims.

4 Results

4.1 Weights' Effects

Based on the considerations developed in Sects. 3.2 and 3.3, consumer price indexes were determined for each subpopulation considered using both the democratic and the plutocratic approaches. Chain price indexes[14] for households sorted by equivalent expenditure decile are presented in Table 3.

Poorer households seem to have experienced higher inflation than richer ones, as shown both by plutocratic and democratic indexes. Chained price indexes (reference

Table 3 Plutocratic and democratic price indexes by families' decile (chained indexes reference base, values at December of each year)

Year	Lowest 10%		Second decile		Third decile		Fourth decile		Fifth decile	
	P	D	P	D	P	D	P	D	P	D
2000	103.58	103.58	103.48	103.48	103.42	103.43	103.44	103.47	103.34	103.37
2001	105.91	105.98	105.70	105.73	105.53	105.60	105.59	105.64	105.41	105.46
2002	108.84	108.82	108.63	108.57	108.44	108.43	108.53	108.52	108.32	108.32
2003	111.81	111.83	111.53	111.48	111.29	111.32	111.33	111.35	111.08	111.12
2004	113.72	113.50	113.57	113.35	113.40	113.29	113.50	113.40	113.32	113.28
2005	116.46	116.17	116.32	116.06	116.16	116.06	116.23	116.13	116.06	116.05
2006	119.50	119.29	119.05	118.83	118.74	118.70	118.76	118.71	118.47	118.51
Year	Sixth decile		Seventh decile		Eighth decile		Ninth decile		Highest 10%	
	P	D	P	D	P	D	P	D	P	D
2000	103.28	103.31	103.23	103.26	103.18	103.22	103.13	103.16	102.70	102.79
2001	105.34	105.38	105.27	105.28	105.18	105.19	105.14	105.16	104.67	104.78
2002	108.26	108.26	108.17	108.14	108.08	108.07	107.97	108.01	107.24	107.36
2003	110.95	111.00	110.85	110.87	110.72	110.75	110.56	110.65	109.53	109.76
2004	113.23	113.19	113.17	113.12	113.07	113.07	112.97	113.04	112.16	112.38
2005	115.88	115.86	115.84	115.83	115.64	115.68	115.44	115.57	114.65	114.89
2006	118.27	118.29	118.15	118.19	117.97	118.05	117.77	117.91	116.74	117.05

Note: (P) Plutocratic Index; (D) Democratic Index.

[13] Here, the use of an equivalence scale of Engel's type could have brought to an underestimation of older families' disadvantage versus the other targeted groups.

[14] Calculation base and reference base monthly indexes are available at request.

base) display a regularly decreasing pattern by decile during the entire period 2000–2006. Moreover, indexes' differential between the poorest and the richest households slightly increases over time.

Analysis of trend inflation gives similar results, with families in the lowest tenth facing on the whole higher inflation rates, except for the periods August 2004–September 2005 and April–August 2002 (Fig. 1).

It is interesting to note that periods of anti-poor price behaviour coincide with periods of higher gaps between official and perceived inflation in Italy, as reported by Del Giovane and Sabbatini (2006).

Inflation rates by subgroups defined by conditions of the head of the family (single parent or out-of-work) do not significantly differ from the other households (Table 4), supporting the hypothesis that price dispersion within demographic group is so great to make across strata heterogeneity negligible (Garner et al., 1996).

This is also the case of older families that on average do not seem to have suffered faster inflation growth than the other households (Table 4). A slow-down of elderly' inflation rate of growth had been reported for the same period by Stewart (2008), caused primarily by changes in the relative inflation rates of medical care compared with overall inflation.

These data seem to suggest that income disposable for consumption is the only variable that could affect so much expenditures' patterns as to determine a grade of across-strata heterogeneity detectable even in presence of marked price heterogeneity within groups.

These differences in consumptions' profiles are synthesized in Fig. 2, reporting expenditures' shares for the 12 COICOP divisions, for the lowest and the highest deciles. The expenditures of "poorest" tenth per cent show a marked concentration

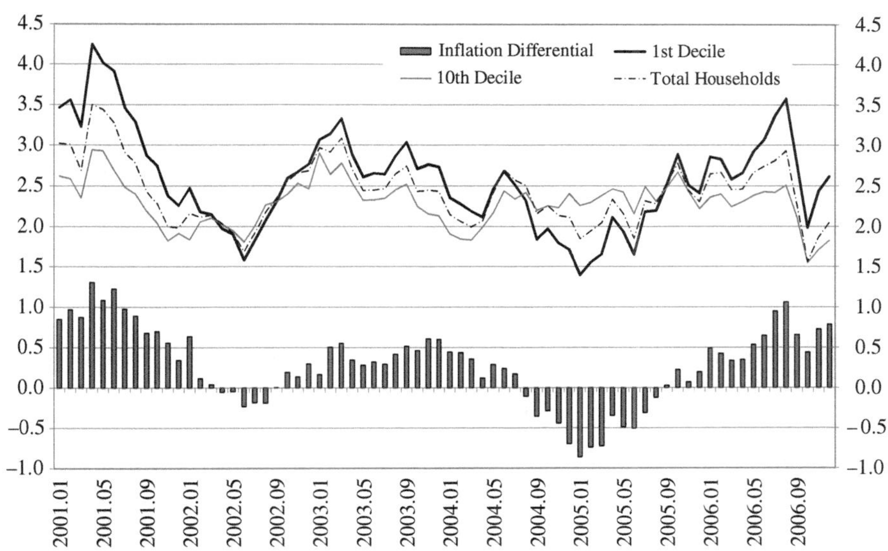

Fig. 1 Trend inflation by families' groups

Table 4 Plutocratic price indexes by families' types (chained indexes reference base, values at December of each year)

Year	Elderly households	Other households	Single person with children (age < 18)	Head of the family retired	Head of the family unemployed	Other households
2000	103.34	103.12	103.03	103.22	102.92	103.12
2001	105.55	105.15	105.00	105.35	104.90	105.14
2002	107.93	108.04	107.82	108.02	107.77	108.03
2003	110.78	110.62	110.58	110.73	110.57	110.60
2004	112.33	113.09	113.06	112.76	112.99	113.09
2005	114.91	115.70	115.55	115.41	115.63	115.67
2006	117.39	118.05	117.98	117.80	118.15	118.02

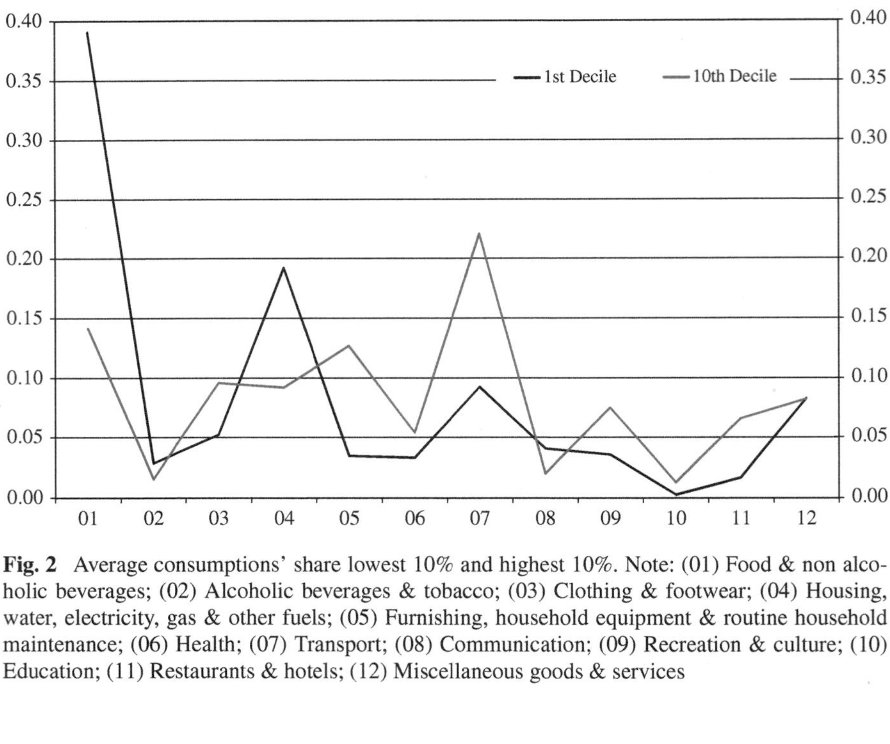

Fig. 2 Average consumptions' share lowest 10% and highest 10%. Note: (01) Food & non alcoholic beverages; (02) Alcoholic beverages & tobacco; (03) Clothing & footwear; (04) Housing, water, electricity, gas & other fuels; (05) Furnishing, household equipment & routine household maintenance; (06) Health; (07) Transport; (08) Communication; (09) Recreation & culture; (10) Education; (11) Restaurants & hotels; (12) Miscellaneous goods & services

in basic needs. Food (01), Housing (04) and Transport (07) account for more than two thirds of total expenditure, with the other shares ranging around 3%.

"Richest" households show a different consumption profile. Food and Housing' expenditures markedly decrease, covering together the same share of Transport alone (doubled than the poorer's one). As expected, these consumers allocate not negligible shares of their total allowances to a wider set of commodities: Clothing (03) and Furnishing (05), but also Recreation & culture (09) and Restaurants & hotels (11).

Even if at an aggregate (COICOP division) level, a significant temporal coincidence between greater-than-average trend inflation for Food,[15] higher consumer price indexes for poor, and divergence between perceived and official inflation, can be observed.

4.2 *Effects of Aggregation System*

As shown in Table 3, plutocratic and democratic indexes do not markedly differ, unless in the extreme (first two and last two tenths) of the distribution by equivalent expenditure. For the entire population, plutocratic bias was negative, i.e. changes in prices hurt the poor more than the rich, during the entire period.

Applying decile-specific weights, for the lowest tenth of households differences between the two approaches should decrease. However, while until May 2004 plutocratic bias ranged around zero, afterwards the gap began to grow (Fig. 3).

As repeatedly pointed out by Kokoski (2003), since statistical significance of these results is not know, quantitative conclusions cannot be drawn, but in our opinion qualitative suggestions from these data may be of some interest. Namely,

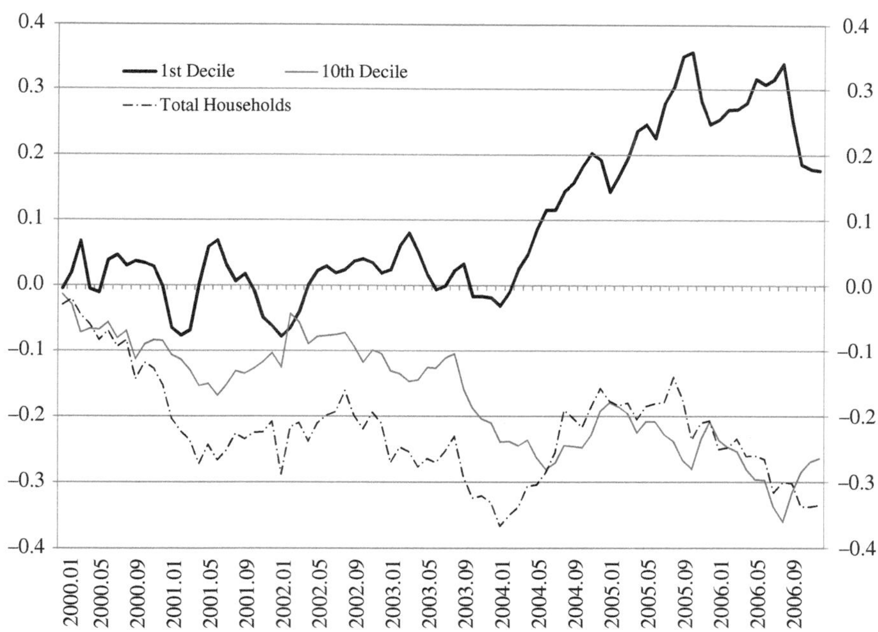

Fig. 3 Plutocratic bias lowest 10%, highest 10% and total households

---

[15]In 2003 yearly inflation rate according to data published by ISTAT were 4.32 for Food vs an average of 2.50.

positive gaps at the end of the study-period might indicate an increase in relative prices of the more expensive items in the poorest' basket. By the way, in 2005–2006 inflation for Housing grew at about 5 point per cent per year, thus inverting the ratio with Food inflation rates as observed in previous years. This result melted with the increased inequality in the expenditure profiles of households within the group (as shown by the increasing positive gap).

On the other side, from June 2000 to May 2004 negative plutocratic gaps for the highest tenth of households are smaller – in absolute value – than for the entire population, since then they became nearly coincident, suggesting a flattening of "representative" household's weights on the richest' ones.

4.3 Changing Basket

The last issue in our empirical analysis was to investigate the effect of changes in the choice of goods and services in the reference basket on consumer price indexes.

To find out if inflation rates measured with a basket limited to basic needs could be more indicative of perceived inflation, we used (Carlucci & Zelli, 1998) a basket of Necessities as expenditures for food and non-alcoholic beverages, electricity and household fuels; household cleaning products, communication charges.

Chained "Necessities" indexes are higher than CPI ones until the first half of 2004, with more pronounced gaps in 2003 (Fig. 4), for households in any decile[16] of equivalised expenditures. This result comes in line with previous ones, supporting

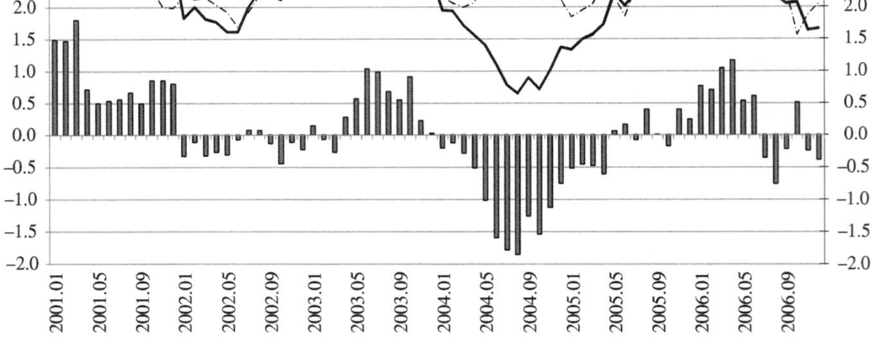

Fig. 4 Trend inflation with necessities and CPI basket

[16]Data available at request.

the idea that higher prices' increases in frequently purchased items could contribute to determining gaps between official and perceived inflation.

5 Final Remarks

One discussed issue in economic debate is the adequacy of CPI in measuring inflation rates actually experienced by different households.

Thus, aim of this study was an exploratory analysis of the empirical evidence on households' specific inflation rates.

To do so, we defined a one-to-one match between 155 HBS expenditure items and consumer price indexes, exhausting the entire reference domain of CPI listing.

We developed sub-indexes according to two criteria. Firstly, we broke down Italian families' by equivalised expenditure decile. Then, we selected households that were likely both to have specific consumption behavior and to be particularly sensitive to the loss of purchasing power: single person with dependent children; head of family workless through unemployment or economic inactivity; elderly.

Our results agree with previous studies in showing that the impact on inflation rates of differences in expenditure profiles across households grouped according to socio-demographic criteria is veiled by price heterogeneity within each group, whose extent could not be evaluated[17] with official data. Only income disposable for consumption has been found to affect so much expenditures' patterns as to determine a detectable grade of across-strata heterogeneity.[18]

Then, we have analyzed the gap between inflation measured according to CPI and according to a democratic index, i.e. the *plutocratic bias*, positive or negative depending on whether prices behave in an anti-rich or an anti-poor manner. Moreover, we checked the effect of changes in the sample of goods and services, developing a Laspeyres price index for a basket of frequently purchased goods and services, defined as common "Necessities" of any type of households, related to daily basic personal needs and housing maintenance.

Evidence supporting the hypothesis of an anti-poor dynamic of relative prices following euro changeover also come from the findings of negative plutocratic bias and of values of the price indexes based on the Necessities basket greater then CPI ones, from early 2002 to the first half of 2004.

These results strongly suggest the need, for further research on inflation differentials by households' groups, to combine official data on CPI and households expenditures with prices' data detailed by outlet and brand/quality, collected by private statistical institutes.

Detailed analyses of this type are crucial for social policies' aims, helping in correctly assessing changes in the standard of living of different socio-economic clusters of households, at different territorial areas.

---

[17] For an analysis of price variations with private scanner data sets see "Price Dispersion: the Case of "Pasta"" in this volume.

[18] With poorer households having experienced higher inflation than richer ones.

References

Adriani F., Marini G., & Scaramozzino, P. (2009). The inflationary consequence of a currency changeover: Evidence from the Michelin Red Guide. *Oxford Bulletin of Economics and Statistics, 71*(1), 111–133.

Álvarez González, L., Cuadrado Salinas, P., Jareño Morago, J., & Sánchez García, I. (2004). *El impacto de la puesta en circulation del euro sobre los precios de consumo. Banco de Espana, Servicio de Estudios*, Documentos ocasionales 0404. Retrieved 21 May, 2009, from http://www.bde.es

Antonides, G. (2008). How is perceived inflation related to actual price changes in the European Union? *Journal of Economic Psychology, 29*, 417–432.

Biggeri, L., & Leoni, L., (2003, December 4–5). *Families of consumer price indices for different purposes*. In Joint Unece-ILO meeting on consumer price indices, Geneva.

BLS. (1997). The consumer price index. In *BLS handbook of methods*, Bulletin 2490 (updated 06/2007). Retrieved 21 May, 2009, from http://www.bls.gov/opub/hom/pdf/homch17.pdf.

Boskin, M. J., Dulberger, E., Gordon, R., Griliches, Z., & Jorgenson, D. (1996, December 4). *Toward a more accurate measure of the cost of living* (Final Report to the Senate Finance Committee). Retrieved 21 May, 2009, form, http://www.ssa.gov/history/reports/boskinrpt.html

Brachinger, H. W. (2006). *Euro or "Teuro"? The Euro-induced perceived inflation in Germany* (Working Paper 5). Switzerland: Department of Quantitative Economics, University of Fribourg. Retrieved 21 May, 2009, from http://ideas.repec.org/p/fri/dqewps/wp0005.html

Carbonaro, G. (1985). Nota sulla scala di equivalenza. In *Commissione di Indagine sulla povertà Primo rapporto sulla povertà in Italia*. Roma: Presidenza del Consiglio dei Ministri.

Carlucci, M., & Zelli, R. (1998). *Expenditure patterns and equivalence scales*. Paper presented to the 25th IARIW General Conference, Cambridge.

Deaton, A. (1998). Getting prices right: What should be done? *Journal of Economic Perspectives, 12*, 37–46.

Del Giovane, P., & Sabbatini, R. (2006). Perceived and measured inflation after the launch of the euro: Explaining the gap in Italy. *Giornale degli Economisti e Annali di Economia, 65*(2), 155–192.

Del Giovane, P., Fabiani, S., & Sabbatini, R. (2008, January). What's behind "inflation perceptions"? A survey-based analysis of Italian consumers. TD No 655.

European Central Bank (2003, October). Recent developments in euro area inflation perceptions. *Monthly Bulletin*, 24–25.

European Central Bank (2007, May). Measured inflation and inflation perceptions in the euro area. *Monthly Bulletin*, 63–72.

Fluch, M., & Stix, H. (2005). Perceived inflation in Austria – extent, explanations, effects. *Monetary Policy & the Economy, 3*, 22–47.

Garner, T. I., Johnson, D. S., & Kokoski, M. F. (1996, September). An experimental consumer price index for the poor. *Monthly Labor Review, 119*, 32–42.

Goni, E., Lopez, H., & Servén, L. (2006, January). *Getting real about inequality evidence from Brazil, Colombia, Mexico, and Peru* (World Bank Policy Research Working Paper 3815). Retrieved 21 May, 2009, from http://econ.worldbank.org

Hill, R. J. (2004). Constructing price indexes across space and time: The case of the European Union. *American Economic Review, 94*(5), 1379–1410.

Istat (2007). Indicatori della dinamica dei prezzi al consumo per alcune tipologie di famiglie. Anni 2001–2006. Retrieved 21 May, 2009, from http://www.istat.it

Izquierdo, M., Ley, E., & Ruiz-Castillo, J. (2003). The plutocratic gap in the CPI: Evidence from Spain. *IMF Staff Papers 50*(1). Retrieved 21 May, 2009, from http://www.imf.org

Jungermann, H., Brachinger, H. W., Belting, J., Grinberg, K., Zacharias, E. (2007). The Euro changeover and the factors influencing perceived inflation. *Journal of Consumer Policy, 30*(4), 405–419.

Kokoski, M. F. (2003, December) *Alternative consumer price index aggregations: Plutocratic and democratic approaches* (BLS Working Paper 370). Retrieved 21 May, 2009, http://www.bls.gov

Kurri, S. (2006). Why does consumers' perceived inflation differ so much from actual inflation? *Bank of Finland Bulletin, 3*, 75–82. Retrieved 21 May, 2009, from http://www.bof.fi

Lyziak, T. (2009). Is inflation perceived by polish consumers driven by prices of frequently bought goods and services? *Comparative Economic Studies, 51*, 100–117.

Pollak, R. A. (1998). The consumer price index: A research agenda and three proposals. *Journal of Economic Perspectives, 12*(1), 69–78.

Ranyard, R., Missier, F. D., Bonini, N., Duxbury, D., & Summers, B. (2008). Perceptions and expectations of price change and inflation. *Journal of Economic Psychology, 29*(1), 378–400.

Rao, V. (2000). Price heterogeneity and "real" inequality: A case study of prices and poverty in rural south India. *Review of Income and Wealth, 46*(2), 201–211.

Schultze, C. L., & Mackie, C. (2002). *At what price? Conceptualizing and measuring cost-of-living and price indexes. Panel on conceptual, measurement, and other statistical issues in developing cost-of-living indexes*. Washington, DC: National Academy Press.

Silver, M., & Heravi, S. (2007). Why elementary price index number formulas differ: Evidence on price dispersion. *Journal of Econometrics, 140*, 874–883.

Son, H. H., & Kakwani, N. (2006). Measuring the impact of prices on inequality: With applications to Thailand and Korea. *Journal of Economic Inequality, 4*, 181–207.

Stewart, K. J. (2008, April). The experimental consumer price index for elderly Americans (CPI-E): 1982–2007. *Monthly Labor Review, 131*, 19–24.

Wunder, C., Schwarze, J., Krug, G., & Herzog, B. (2008). Welfare effects of the euro cash changeover. *European Journal of Political Economy, 24*, 571–586.

Price Dispersion: The Case of "Pasta"

Isabella Carbonaro, Raffaele Santioni, and Margherita Carlucci

1 Introduction and Problem Definition

The aim of our research is to explore the possibility of utilizing scanner data on pasta purchases to build bilateral and multilateral spatial price indexes, taking a binary approach in the latter.[1]

Pasta plays a major role in the Italian diet. Historically, pasta consumption was mainly concentrated in the Southern regions of the country but today pasta is perhaps the product most representative of the eating habits of the Italians. The range of pasta producers runs from firms of longstanding tradition (some of them mainly directed towards local markets, such as Mastromauro in Puglia) to well known international brands (such as Barilla and De Cecco).

The marked increase in pasta prices over the last two years has aroused great interest, but with little focus on spatial price diversity.

This study stems from the availability of an extremely detailed panel dataset (Nielsen data) on values and quantities of pasta purchased. This data was produced by the use of bar-code scanning at retail outlets and thus includes information which provides weights at an elementary level. The use of scanner data to construct price indexes is not new in literature and there is a widespread consensus on the advantages of this approach in achieving more representative indexes. Average prices (unit values) show a marked spatial price variability: even when only considering the five bestselling products, regional prices vary greatly.

The paper is set out as follows: Sect. 2 provides a description of the pasta scanner dataset and briefly looks for price variability; in Sect. 3 the requirements of comparability and representativity in the case of pasta are discussed; Sect. 4 deals with the methods and formulas chosen to obtain indexes for the regional comparisons of

I. Carbonaro (✉)
DET, University of Rome Tor Vergata, Rome, Italy
e-mail: isabella.carbonaro@uniroma2.it

[1] This research was funded by PRIN 2005. The Authors thank participants in the International Workshop on Price Index Numbers in Time and Space, held in Florence, September 29th 2008, for useful comments. The views expressed in this paper are those of the authors and do not involve the responsibility of the Bank of Italy.

L. Biggeri, G. Ferrari (eds.), *Price Indexes in Time and Space*, Contributions to Statistics, DOI 10.1007/978-3-7908-2140-6_8, © Springer-Verlag Berlin Heidelberg 2010

prices; Sect. 5 shows empirical results; in Sect. 6 a brief conclusion and suggestions for future work are given.

2 Pasta Scanner Dataset Description and Price Dispersion

In Italy, A.C. Nielsen has developed a database of scanner data at a regional level for the sales of major supermarket chains. Product groups included are those typically sold in supermarkets (food, beverages and a large number of other commodities for personal care, home cleaning, etc.). The data used results from the aggregation of the information from all the supermarkets examined. Individual product models (hereafter called "items") are identified by a unique product code based on the bar code. If bar codes correspond to a single product configuration, we can be sure that the product match based upon the code is exact.

In the case of pasta, the data covers the entire commodity group, including all types of pasta products: fresh filled pasta, fresh semola pasta, fresh egg pasta, dried filled pasta, dried semola pasta, dried egg pasta and other types. In this paper we concentrate on the sub-group "dried semola pasta/other" (category n. 499 in the Nielsen data). For this sub-group, spatial price variability is analyzed via the construction of spatial price indexes. For the selected category, scanner panel data provide information on 8,071 items of pasta identified by a product code (hereafter CPR). For each item, information on product characteristics (type of pasta as long or short, brand name, weight and variety of pasta e.g. *spaghetti*, *penne*, etc.) is also provided. In Italian, the latter characteristic is called *formato* or *trafila*. The timespan covers a period of 217 weeks (from June 2002 to July 2006). Spatial coverage is 17 regions (all Italian regions with the exception of Basilicata, Valle d'Aosta and Molise).

For each CPR at a temporal and regional level we have weekly sales, quantity and number of packages purchased. Prices are not included: monthly average revenue per unit sold has been used as a proxy. The whole data set consists of about 6 million items. Since the codes differ according to the different weights of packages, we only selected items weighting 500 grams. Some CPR lack a product characteristic description. These CPR were excluded and only 5,618 CPR (or items) were used. Weekly data were aggregated into 50 sets of monthly data. In the following tables, the data for the December of each year (2002, 2003, 2004 and 2005) is used. December 2006 is excluded because of the availability of data only until July 2006.

Scanner data provide highly detailed information which permits the investigation of price differences of the same basket of products sold in two neighboring or distant regions. Binary comparison may successively be inserted in a multilateral context of spatial price index construction. The use of scanner data is not new in the literature of spatial price indexes (Heravi, Heston, & Silver, 2003), but as far as we are aware, analysis of regional price dispersion of pasta at such a detailed level has never been carried out in Italy[2].

[2]Empirical evidence of regional PPPs at a more aggregated level can be found in De Carli (2008).

Tables 1 and 2 provide some database features at regional and aggregate (Italy) level. In Table 1 the number of CPR selected and quantities of purchases of pasta are shown. As can be observed the availability of the selected items varies greatly across regions. In Liguria, Sicilia and Trentino the number of available items is somewhat low, while in Campania, Emilia Romagna, Lazio, Lombardia, Piemonte and Veneto it exceeds 1,300 items. These differences depend mainly upon the number of inhabitants as well as on the tastes of the consumers. The different relative weights of traditional small outlets and modern distribution chains, as well as the selling policy of the latter, might in part account for the difference.

To extend the analysis of the spatial price variability to a comparable basket of items consumed in the various regions, further study is needed. Nielsen data give information on brands, types (long or short) and variety (spaghetti, penne, etc.) of pasta. As regards the construction of regional price indexes, brand appears to be the most relevant issue. Grouping the CPR by brand, two important results may be achieved. Firstly, the spatial indexes can be assigned to the types of pasta that correspond to the choices of Italians consumers, as in general these choices are influenced by the brand. The second result is that we may obtain a preliminary data set from which a subset can be derived which has the requisites of comparability[3] and representativity[4] required by spatial indexes.

Following these criteria, the CPR have been grouped according to brands: the results are shown in Table 2, where we see that the number of brands available to Italian consumers grew over the period from 2002 to 2006 (212 brands in December 2002 against 247 in the last month of July 2006). As can be seen in Table 2 there are also differences in the number of brands existing in each region, only partially dependent on the different sizes of the Italian regions.

Analysis of the five bestselling brands purchased at a regional level showed that on a monthly basis within the same region there are few changes, while there is much diversity between the brands purchased in the various regions.

At the aggregate level (Italy) the top five brands are the same in all the selected months, with a market share of approximately 71%. The consumption shares of pasta (quantity) goes on average from 67.6% in Campania to 90.5% in Trentino. Price dispersion is also high: a kilogram of pasta of the same brand has different average prices in Italian regions, but there is also a great deal of variability in the prices of the available brands of pasta within each region.

It is worth noting that regional coverage is achieved differently by the five bestselling brands. The lower results for Campania show that brands sold only in the local market attract higher preference in terms of quantity consumed. In the following we refer to this type of brand as local or typical brands. Typically, in each region

[3]Comparability depends on the way of defining each product, as 'product' can cover a large variety of types depending on various characteristics, such as raw materials used, weight, and packaging, all of which affect price. A product must have the same characteristics in order to be strictly comparable over the different areas (Biggeri, Brunetti, & Laureti, 2008).

[4]Representative products are defined here as products that are purchased in relatively large quantities in a country (Hill, 2008).

Table 1 Number of Cpr and quantity of pasta (thousand kilograms) sold in Italian regions

	December 2002		December 2003		December 2004		December 2005		July 2006	
Italian regions	N. CPR	Quantity	N. CPR	Quantity	N. CPR	Quantity	N. CPR	Quantity	N. CPR	Quantity
Abruzzo	1,195	839	1,335	964	1,465	942	1,296	796	1,446	886
Calabria	1,021	1,649	1,102	1,923	1,178	2,000	970	1,547	932	1,664
Campania	1,582	3,249	1,680	4,018	1,674	3,812	1,532	3,044	1,610	2,335
Emilia Romagna	1,371	2,203	1,582	2,629	1,709	2,451	1,415	2,094	1,526	2,053
Friuli V.G.	938	637	1,098	837	1,165	736	1,015	662	1,116	640
Lazio	1,554	3,783	1,700	4,469	1,805	4,051	1,598	3,389	1,773	2,986
Liguria	679	709	772	895	860	838	926	606	982	657
Lombardia	1,538	5,126	1,759	6,211	1,983	5,587	1,772	4,499	1,852	4,305
Marche	1,048	933	1,171	1,110	1,326	1,082	1,177	955	1,321	1,004
Piemonte	1,364	1,860	1,534	2,355	1,684	2,410	1,455	1,996	1,544	1,754
Puglia	1,413	2,397	1,498	2,692	1,497	2,307	1,456	2,007	1,496	2,073
Sardegna	913	1,250	1,093	1,414	1,051	1,457	718	929	802	919
Sicilia	582	491	620	558	648	505	665	418	847	388
Toscana	1,114	2,265	1,287	2,726	1,312	2,728	1,196	2,461	1,222	2,440
Trentino A.A.	423	412	590	592	645	545	573	416	611	406
Umbria	1,065	505	1,170	644	1,097	609	1,014	619	1,049	561
Veneto	1,329	2,424	1,525	2,993	1,655	2,667	1,450	2,340	1,593	2,251
Italia	3,806	30,728	4,031	37,031	4,218	34,726	4,009	28,779	4,163	27,323

Table 2 Number of brands sold in Italian regions

Italian regions	December 2002	December 2003	December 2004	December 2005	July 2006
Abruzzo	61	72	84	86	86
Calabria	52	54	67	57	52
Campania	70	77	78	71	76
Emilia Romagna	98	115	121	113	111
Friuli V.G.	66	85	89	82	79
Lazio	89	99	104	95	97
Liguria	51	61	54	65	72
Lombardia	109	123	138	133	134
Marche	68	80	91	75	82
Piemonte	88	100	113	110	116
Puglia	63	69	74	72	68
Sardegna	64	72	74	50	54
Sicilia	43	49	59	58	59
Toscana	73	85	93	85	90
Trentino A.A.	36	54	52	44	44
Umbria	60	70	69	69	65
Veneto	102	116	131	129	127
Italia	212	226	244	237	247

we see a large number of local brands, but only a few of these have considerable market shares[5].

3 Comparability and Representativity: List of Common Brands and List of Common Products

As is repeatedly emphasized in the literature, the most difficult steps in spatial price index construction are: (i) the preparation of the basket, i.e. a common list of CPR, (ii) compliance with the two important requirements of comparability and representativity which are usually in conflict with one another, (iii) avoiding a severe loss of characteristicity and (iv) the achievement of transitivity. The latter two refer to multilateral spatial price comparisons.

To achieve comparability a common list of types of pasta purchased in all regions must be compiled. Given that, as noted above, pasta consumption patterns differ between Italian regions, the risk of selecting marginally representative kinds of pasta for some regions must be avoided.

[5]If ten brands are involved, as in a study we subsequently developed, the consumption coverage grows on average to 92 percent, for example at December 2002, for all Italian regions. However, the increases show considerable differences between regions: the average increase in coverage passing from five to ten brands is about 11 percentage points, with a maximum of about 21 percentage points in Campania and a minimum in Trentino A.A. of about 5 percent.

As regards representativity, even if the constraint of an equally representative product list for all regions has been relaxed, the issue remains unresolved. Another difficulty stems from the varying meanings that the word "common" can assume in the case of pasta: it can refer to brands, items (CPR) or even to a particular variety of pasta (i.e. spaghetti, penne, etc.).

In our study we chose to select those brands, and the products (CPR) of these common brands, which are common to all regions.

Table 3 summarizes spatial common brands at an aggregate level. The number of common brands varies from 11 to 15. In every period, the bestselling brand is Barilla, followed by Agnesi, De Cecco, Divella, PL (Private Label) and Voiello. These brands, along with Colussi, are present in all the years considered. The numbers of CPR included in common brands are 1,253, 1,429, 1,507, 1,233 (respectively at December 2002, 2003, 2004 and 2005) and 1,251 at July 2006.

At regional level, in all periods common brands cover a substantial share of the aggregate sales of pasta. For example, in July 2006 this share exceeded 90% in 4 regions (Friuli V.G., Liguria, Piemonte and Trentino A.A.), 80% in 8 regions (Abruzzo, Calabria, Emilia Romagna, Lazio, Lombardia, Marche, Sicilia and Veneto) about 75% in Sardegna and 79% in Umbria, attaining minimum values in Puglia (62%) and Campania (63%). In these regions, particularly in Puglia and Campania, the importance of local brands reduces the market shares of common brands.

Table 4 shows the number of common CPR for each region.

Table 3 The common brands list

	Number of brands	Total expenditure (current euro)	Quantity purchased (kilograms)	Share of total quantity sold (%)
December 2002	12	27,236,346	23,700,733	77.1
December 2003	14	33,562,586	28,553,866	77.1
December 2004	15	31,252,208	27,546,265	79.3
December 2005	11	25,999,402	23,562,368	81.9
July 2006	11	24,863,682	22,416,679	82.0

Table 4 The common brands list. Number of CPR purchased

	Abr	Cal	Cam	Emi	Fri	Laz	Lig	Lom	Mar	Pie	Pug	Sar	Sic	Tos	Tre	Umb	Ven
December 2002	711	536	772	788	567	799	472	827	588	854	694	563	416	671	320	694	730
December 2003	879	588	863	940	699	918	541	1003	738	993	769	678	464	790	400	771	908
December 2004	930	673	828	960	679	980	611	1056	792	1019	769	657	458	770	416	743	899
December 2005	723	621	741	771	576	753	613	866	655	813	727	500	441	631	365	561	718
July 2006	770	592	767	839	643	801	638	911	711	860	734	549	578	655	396	605	812

A properly constructed spatial price index would require complete regional series of equal CPR within the common brands. This is not the case, because, even though the analysis is restricted only to products of common brands, observations are incomplete for many CPR, as there are many items that are not purchased (or do not exist) in a given region in the chosen month.

To construct a basket of comparable CPR, two lists of common products produced by common brands have been compiled.

The first list was compiled with the CPR present in the two regions being compared. These binary matches allow the measurement of the spatial variability between every pair of regions. The second list includes the CPR that are present in all regions in the selected months (December of every year from 2002 until 2005 and July 2006). The CPR included in these two lists ensure bilateral (first list) and multilateral (second list) comparability.

Table 5 shows the number of common CPR, in binary-matched cases, for every pair of regions. For the multilateral comparison, the number of selected CPR, common to all regions, is 128, 146, 145, 148 and 179 (at December 2002, 2003, 2004, 2005 and July 2006, respectively).

To comply with the representativity requirements, a slightly modified Sergeev approach (2003) was used. As in Sergeev, the CPR included in the lists are grouped in four subsets: the first includes CPR that are representative in both the regions involved in the comparison; the second contains the CPR representative for one region (the base region) but not for the other (the reference region); the third is formed by the CPR representative for the reference region but not for the base

Table 5 Common brands list. Binary-matched CPR at July 2006

Italian regions	Abr	Cal	Cam	Emi	Fri	Laz	Lig	Lom	Mar	Pie	Pug	Sard	Sic	Tos	Tre	Umb	Ven
Abr	–																
Cal	460	–															
Cam	602	540	–														
Emi	605	423	565	–													
Fri	515	349	478	614	–												
Laz	665	481	621	659	566	–											
Lig	514	321	461	591	497	533	–										
Lom	654	455	610	790	614	676	617	–									
Mar	624	439	571	617	534	642	500	646	–								
Pie	606	408	561	795	629	662	622	807	611	–							
Pug	599	485	628	578	506	625	476	597	589	576	–						
Sar	437	364	436	491	471	498	405	506	478	503	465	–					
Sic	485	413	521	482	438	516	404	518	491	489	510	407	–				
Tos	547	379	500	634	535	592	537	630	566	623	519	455	452	–			
Tre	370	228	306	385	370	376	367	396	354	379	329	275	285	371	–		
Umb	559	342	472	540	473	573	497	554	541	543	486	400	403	515	354	–	
Ven	604	411	551	718	642	654	571	758	615	731	562	515	490	622	397	533	–

region; the fourth subset includes the CPR which are not representative for either region and consequently, because irrelevant for the construction of the bilateral index, were deleted.

The CPR in each region have been considered representative on the basis of their share of total quantity of common CPR sales. Let us illustrate the procedure with an example: in July 2006 respectively 770 and 911 items were purchased in Abruzzo and Lombardia, but only 654 common CPR (see Table 5). Within this subset, a given CPR was considered representative if the quantity of it sold in each of the two regions is more than $1/654 = 0.001529$ (i.e. more than 0.1529%) of the aggregate quantity of common CPR sold in the same region. This threshold corresponds to the hypothesis of uniform distribution among the common CPR, in each region, of the purchased quantity of pasta.

On the basis of this procedure the 654 CPR that are common for Abruzzo and Lombardia have been grouped, as shown in Table 6, part (a), in the three subsets cited above: the first includes 72 representative CPR of both regions, that is the CPR whose "quantity bought" shares exceed the threshold in both regions; the second subset includes 39 representative CPR of Abruzzo (base region) but not for Lombardia; the third subset includes 62 representative CPR of Lombardia but not of Abruzzo. These three subsets (173 CPR) constitute the basket. The remaining 481 CPR are irrelevant for the construction of the spatial indexes, as they are not representative of the two regions considered in our example.

The procedure is the same for the two lists of common products of common brands: CPR with a quantity share above (below) the threshold are representative (unrepresentative) and define the four subsets.

It must be noted that in the first list the threshold changes for each pair of regions compared because the number of common CPR is different in the regions; in the second list the threshold changes only over time because the common list was based on the matched CPR for all regions (the number at July 2006 is always 179 and the threshold is 0.559%) as shown in Table 6, part (b).

Table 6 The representativity subsets. Coverage, number of CPR and threshold (%)

	Abruzzo (base)	Lombardia	No. of CPR	Abruzzo (base)	Lombardia	No. of CPR
**	64.48	64.30	72	65.80	57.48	31
–*	17.04	1.99	39	12.01	2.12	9
*–	4.26	17.38	62	8.77	20.38	26
Total	85.78	83.66	173	86.58	79.98	66
Threshold		0.1529			0.5587	
		Part (a)			Part (b)	

Note: ** representative items of both regions; –* items representative of base region; *– items representative for comparison region

4 Notation and Methodological Issues

The notation below refers only to the common brands in the selected month. Moreover, notation and methodological issues are mainly offered with respect to the first list, but with minor changes can also be adapted to the second, since the procedure is the same.

Let N_t denote the number of products that are included in the common brand in period t ($t = 1, 2, ..., T$). Let N_{tj} denote the number of products purchased in region $j(j= 1, 2, \ldots, m)$ in period t. Generally $N_{tj}<N_t$. Let N_{tjk} denote the number of products purchased both in region j and k on the same date. Generally $N_{tjk}<N_{tj}$ and $N_{tjk}<N_{tk}$. N_{tjk} is the number of products forming the first list in the comparison between region j and k. We denote with p^i_{tj} and q^i_{tj} respectively, the average price and quantity of i-th CPR in j-th region ($i = 1,2\ldots,N_{tj}; j = 1,2,\ldots, m$ and $t = 1,2,\ldots,$ T). Only with respect to the items belonging to the binary-matched N_{tjk} we do define the quantity share (%) for each product, $s^i_{tj} = \frac{q^i_{tj}}{\sum_{i=1}^{N_{tjk}} q^i_{tj}}$ in j-th region and $s^i_{tk} = \frac{q^i_{tk}}{\sum_{i=1}^{N_{tjk}} q^i_{tk}}$ k-th region. The threshold corresponding to the hypothesis of a uniform distribution of the purchased quantity of pasta among the common CPR is $thr_{tjk} = \frac{100}{N_{tjk}}$. Obviously, the threshold is the same for each pair of regions compared in the binary-matched case. Each item is representative of j-th region if $s^i_{tj} \geq thr_{tjk}$, the same is true for k-th region if $s^i_{tk} \geq thr_{tjk}$.

We will now define the following subsets of products:

- products that are representative in either region j and k. Their number is denoted by n^{**}_{tjk};
- products representative in base region j, but not in comparison region k: The number is: n^{-*}_{tjk};
- products representative in comparison region k, but not in base region j: number n^{*-}_{tjk}

There are also products that are not representative at all (neither in region j nor in region k) and denote the number of the products of this last subset with n^{--}_{tjk}. These are excluded from the analysis.

Let $w^i_{tj} = \frac{p^i_{tj}q^i_{tj}}{\sum_{i=1}^{N_{tjk}} p^i_{tj}q^i_{tj}}$ and $w^i_{tk} = \frac{p^i_{tk}q^i_{tk}}{\sum_{i=1}^{N_{tjk}} p^i_{tk}q^i_{tk}}$ be the expenditure share of the i-th item, respectively in the j-th and k-th region

The binary elementary indexes are calculated using unit value ratios of all the CPRs belonging to the three subsets identified among the matched CPR whose prices are available in both regions. Since expenditure share weights are available, the binary purchasing power parity between the same currency (euro) in the j-th region (base) vs k-th comparison of regions are calculated by the Törnqvist formula.

$$PPP^T_{j,k} = \prod_{i=1}^{n_{tjk}(**)} \left[\left(\frac{p^i_{tk}}{p^i_{tj}} \right)^{\frac{w^i_{tj}+w^i_{tk}}{2}} \right] \prod_{i=1}^{n_{tjk}(-*)} \left[\left(\frac{p^i_{tk}}{p^i_{tj}} \right)^{\frac{w^i_{tj}+w^i_{tk}}{2}} \right] \prod_{i=1}^{n_{tjk}(*-)} \left[\left(\frac{p^i_{tk}}{p^i_{tj}} \right)^{\frac{w^i_{tj}+w^i_{tk}}{2}} \right] \tag{1}$$

In the second list, N_{tjk} is equal for every comparison, because it includes those items which are purchased simultaneously in all regions. In this case we denote the set of matched items as ${}^{AR}N_t$where "AR" means "all regions". The same is also true for the threshold ${}^{AR}thr_t = \frac{100}{{}^{AR}N_t}$ because the denominator is constant for every pair of regions. With some modification it is possible to rearrange the notations and formula and derive Törnqvist-type binary PPPs, denoted with ${}^{AR}PPP^T$. A binary approach to multilateral comparison has subsequently been selected applying the modified GEKS[6] formula (Gini, 1924, 1931; Eltetö & Köves, 1964; Szulc, 1964) to binary Törnqvist PPPs to construct transitive multilateral PPPs (only for the second list):

$$GEKS_{jk} = \left[\prod_{l=1}^{m} {}^{AR}PPP^T_{j,l} * {}^{AR}PPP^T_{l,k} \right]^{\frac{1}{m}} \tag{2}$$

The GEKS formula is widely used in most countries and by international organizations (Balk, 1996; Hill, 1997; ILO et al., 2004). GEKS is recognized as an attractive method for various reasons, in particular because its property of transitivity[7] is imposed by construction. The GEKS method for multilateral comparisons also has its origins in the property of characteristicity[8]. The modified GEKS refers (i) to the use of Törnqvist binary indexes (Caves, Christensen and Diewert, 1982) instead of Fisher-type formula as originally proposed and (ii) to the choice of the Sergeev procedure to satisfy the requirements of representativity.

5 Results and Discussion

Tables 7, 8 and 9 show the results[9] regarding spatial price diversity of the CPR included in the baskets of the two common lists: bilateral PPPs for the first and modified GEKS for the second. The binary parities of Table 7 (the indexes are base

[6] Gini proposed this method long before the others, so EKS should in fact be termed GEKS and we therefore adopt this acronym.

[7] Transitivity requires that the application of a formula to make a direct comparison between j and k countries should result in the same numerical measure as an indirect comparison between j and k through a link country m (ILO et al., 2004).

[8] This property requires that any set of multilateral comparisons satisfying the transitivity property should retain the essential features of the binary comparisons constructed without the transitivity requirement (ILO et al., 2004).

[9] Due to space restraints, only results for July 2006 are given.

Table 7 The binary Törnqvist PPPs

Italian regions	Abruzzo	Calabria	Campania	Emilia Romagna	Friuli V.G.	Lazio	Liguria	Lombardia	Marche	Piemonte	Puglia	Sardegna	Sicilia	Toscana	Trentino A.A.	Umbria	Veneto
Abruzzo	100	97.22	99.87	98.07	97.89	97.98	101.86	99.17	100.85	100.80	96.40	104.61	109.94	92.36	98.63	92.63	97.67
Calabria	102.86	100	102.93	106.32	103.93	104.34	110.98	106.23	107.37	109.54	98.70	117.88	115.39	100.32	106.16	100.40	103.69
Campania	100.13	97.15	100	100.12	97.38	97.89	102.73	100.32	101.96	101.14	98.12	105.89	109.66	94.82	97.54	93.77	97.90
Emilia Romagna	101.97	94.06	99.88	100	99.86	99.20	102.74	100.79	102.36	101.34	96.45	103.09	111.16	95.23	100.49	94.16	98.67
Friuli V.G.	102.15	96.22	102.69	100.14	100	99.65	103.58	101.49	102.88	101.79	98.97	105.41	111.63	95.08	100.20	94.14	99.32
Lazio	102.06	95.84	102.16	100.80	100.35	100	104.97	101.84	103.20	102.52	98.15	107.51	111.95	95.30	100.58	95.06	99.88
Liguria	98.17	90.11	97.35	97.33	96.54	95.27	100	97.41	98.57	97.27	91.37	99.57	107.76	92.70	97.60	90.51	95.80
Lombardia	100.84	94.13	99.68	99.21	98.53	98.19	102.66	100	101.14	100.58	97.08	104.38	110.73	94.98	99.13	93.09	98.20
Marche	99.15	93.14	98.08	97.70	97.20	96.90	101.45	98.87	100	99.91	95.25	102.63	109.25	92.35	97.81	92.19	96.66
Piemonte	99.21	91.29	98.87	98.67	98.24	97.55	102.80	99.42	100.09	100	93.63	102.78	108.96	94.12	99.20	92.20	97.69
Puglia	103.74	101.31	101.91	103.68	101.04	101.88	109.44	103.01	104.99	106.81	100	112.72	115.23	99.90	101.65	100.29	102.53
Sardegna	95.59	84.83	94.44	97.00	94.87	93.01	100.44	95.80	97.43	97.30	88.72	100	105.68	89.96	93.25	88.84	93.86
Sicilia	90.96	86.66	91.19	89.96	89.58	89.32	92.80	90.31	91.54	91.78	86.79	94.62	100	84.06	90.61	84.71	88.86
Toscana	108.28	99.68	105.47	105.01	105.18	104.93	107.88	105.29	108.29	106.25	100.10	111.16	118.96	100	106.01	99.22	103.99
Trentino A.A.	101.39	94.20	102.52	99.52	99.80	99.42	102.46	100.87	102.24	100.80	98.38	107.24	110.37	94.33	100	92.92	99.30
Umbria	107.96	99.60	106.65	106.21	106.23	105.19	110.49	107.42	108.48	108.45	99.71	112.56	118.05	100.78	107.62	100	105.73
Veneto	102.38	96.44	102.15	101.35	100.69	100.12	104.38	101.83	103.46	102.36	97.53	106.55	112.54	96.16	100.71	94.58	100

Table 8 Ranking of Italian regions due to GEKS PPPs

(1) Toscana	100	(10) Lombardia	107.42
(2) Umbria	100.25	(11) Abruzzo	107.44
(3) Calabria	100.25	(12) Campania	108.7
(4) Puglia	103.11	(13) Piemonte	108.98
(5) Lazio	105.15	(14) Marche	109.09
(6) Veneto	105.53	(15) Liguria	110.8
(7) Friuli Venezia Giulia	105.83	(16) Sardegna	114.61
(8) Trentino	106.11	(17) Sicilia	119.28
(9) Emilia Romagna	106.81		

reversible, so the figures above the principal diagonal are the inverse of the corresponding figures below the same diagonal) show considerable price differences between the pairs of regions. Reading the table by row, the main results are the following:

- Calabria, Puglia, Umbria and Toscana are the regions where prices are lower relative to other regions (it must be stressed that the comparison of Table 7 regards pairs of regions; consequently the figures of the rows cannot be used to build rankings, as can be done with the figures of Table 9). For Calabria and Puglia, 15 indexes of the respective rows are above 100. In Toscana and Umbria more than 14 indexes are over 100.
- As regards the scale of the differences between Calabria and other regions, in 7 of these (Abruzzo, Campania, Friuli-Venezia Giulia, Lazio, Toscana, Umbria and Veneto) prices are higher by less than 5%. In 8 other regions (Emilia-Romagna, Liguria, Lombardia, Marche, Piemonte, Sardegna, Sicilia and Trentino) the difference exceeds 5%, with Piemonte (+9.54%), Liguria (+10.98%), Sicilia (+15.39%) and Sardegna (+17.88%) at the top of the list.
- The results are different for Toscana and Umbria, In many of the other regions, prices are higher than in Toscana by less than 6%; in Abruzzo, Liguria, Marche and Sardegna the difference is between 6 and 11%; in Sicilia 19%. Price differences are greater in Umbria. In 13 regions common basket prices exceed those of Umbria by more than 5% and in 5 of these (Liguria, Marche, Piemonte, Sardegna and Sicilia) by more than 8% (in Sicilia +18%).
- The results for Sicilia and Sardegna must be interpreted with caution, as they may be due in part to two factors: transport costs for the common brands and the exclusion from our analysis, as not being common, of local brands (for instance Tomasello, Piatti and Gallo in Sicilia, Pastificio Cellino Sardegna, Casa del Grano, Pastificio Cellino Sant'Alberto and Pastifici Cagliaritani in Sardegna), which offer the same products as the common brands at lower prices. In July 2006 these local brands accounted for substantial consumption shares (about 13% in Sicilia and 20% in Sardegna).The problem of products that are typical in one region and are not purchased or do not exist in the other regions is common in binary

Table 9 The modified GEKS PPPs

Italian regions	Abruzzo	Calabria	Campania	Emilia Romagna	Friuli V.G.	Lazio	Liguria	Lombardia	Marche	Piemonte	Puglia	Sardegna	Sicilia	Toscana	Trentino A.A.	Umbria	Veneto
Abruzzo	100	93.31	101.17	99.42	98.50	97.87	103.13	99.98	101.53	101.44	95.97	106.67	111.02	93.07	98.76	93.30	98.22
Calabria	107.17	100	108.43	106.55	105.57	104.89	110.53	107.15	108.82	108.71	102.86	114.33	118.99	99.75	105.85	100.00	105.27
Campania	98.84	92.23	100	98.27	97.37	96.73	101.94	98.82	100.36	100.26	94.86	105.44	109.74	92.00	97.62	92.23	97.09
Emilia Romagna	100.59	93.85	101.76	100	99.08	98.44	103.73	100.56	102.13	102.03	96.53	107.30	111.67	93.62	99.34	93.85	98.80
Friuli V.G.	101.52	94.72	102.71	100.93	100	99.35	104.69	101.49	103.07	102.98	97.43	108.29	112.71	94.49	100.26	94.72	99.71
Lazio	102.18	95.34	103.38	101.59	100.65	100	105.38	102.16	103.75	103.65	98.07	109.00	113.44	95.10	100.91	95.34	100.36
Liguria	96.97	90.48	98.10	96.40	95.52	94.90	100	96.95	98.45	98.36	93.06	103.44	107.66	90.25	95.77	90.47	95.24
Lombardia	100.02	93.33	101.19	99.44	98.53	97.89	103.15	100	101.56	101.46	95.99	106.70	111.05	93.10	98.78	93.32	98.24
Marche	98.49	91.90	99.64	97.92	97.02	96.39	101.57	98.47	100	99.90	94.52	105.06	109.35	91.67	97.27	91.89	96.74
Piemonte	98.59	91.99	99.74	98.01	97.11	96.48	101.67	98.56	100.10	100	94.61	105.17	109.45	91.76	97.36	91.98	96.83
Puglia	104.20	97.22	105.42	103.59	102.64	101.97	107.46	104.17	105.79	105.69	100	111.15	115.68	96.98	102.90	97.22	102.34
Sardegna	93.74	87.47	94.84	93.20	92.34	91.74	96.67	93.72	95.18	95.09	89.97	100	104.08	87.25	92.58	87.47	92.07
Sicilia	90.07	84.04	91.12	89.55	88.72	88.15	92.89	90.05	91.45	91.36	86.44	96.08	100	83.83	88.95	84.04	88.47
Toscana	107.44	100.25	108.70	106.81	105.83	105.15	110.80	107.42	109.09	108.98	103.11	114.61	119.28	100	106.11	100.25	105.53
Trentino A.A.	101.26	94.48	102.44	100.67	99.74	99.09	104.42	101.23	102.81	102.71	97.18	108.01	112.42	94.24	100	94.47	99.45
Umbria	107.18	100.00	108.43	106.55	105.57	104.89	110.53	107.15	108.82	108.72	102.86	114.33	118.99	99.75	105.85	100	105.27
Veneto	101.81	95.00	103.00	101.22	100.29	99.64	105.00	101.79	103.37	103.27	97.71	108.61	113.03	94.76	100.55	94.99	100

comparisons. The debate on how to include products that are characteristic in a single region in the analysis (or area) continues (Biggeri et al., 2008).
- The differences of prices in Puglia with other regions are small, being generally less than 5%. Only in Liguria, Piemonte, Sardegna and Sicilia are prices higher than in Puglia by more than 5%, being from 7 to 15%.
- Liguria, Piemonte, Marche, Sardegna and Sicilia are the regions where prices, in binary comparisons, are highest relative to other regions.

Turning to multilateral comparisons, Table 8 shows the non-decreasing ranking of the 17 regions according to GEKS PPPs (Table 9).

The differences are notable. For Sicilian consumers prices are 19% higher than for the inhabitants of Toscana, while for Sardinians they are 14% higher. In 11 regions prices exceed prices in Toscana by more than 5%; in Campania, Liguria, Marche and Piemonte the difference is greater than 8%.

As regards Sardegna and Sicilia, we refer to the considerations above. From the above ranking it appears that in the Northern and Central regions, pasta prices are lower than in the Southern regions.

It may be observed that, within the baskets utilized to calculate GEKS, we have implicitly defined a minimum basket. This is formed by those items which are perfectly comparable (matched in all the regions) and also representative in all the regions (after the applied procedure to construct binary PPPs). This subset is composed of 13 varieties of pasta belonging to 2 brands.

6 Conclusion and Future Work

Pasta is perhaps the product most representative of the eating habits of Italians. Using the extremely rich information provided by AC Nielsen (about 6 million pieces of elementary data) we have calculated binary and multilateral PPPs for 17 Italian regions.

One result emerges. There are vast differences in regional prices: living in Toscana but also in Umbria, Calabria and Puglia entails a remarkable saving in the expenditure for pasta. This is confirmed by both Törnqvist binary parities and GEKS indexes.

In our analysis it was necessary to make many choices, in some cases necessitated by the data available and in others inevitably discretionary.

Due to the characteristics of the available data, we had to use unit values such as average prices and unit value ratios as elementary binary indexes. Other choices relate to (i) the brands, given the preference of the Italian consumers for certain brands, (ii) the products of each brand, (iii) the procedure for representativity, (iv) the thresholds and (v) the index numbers methodology.

The choice regarding the brands (only those present in all the regions have been selected) was suggested by the need to ensure comparability of results and by the high percentage of purchases of pasta produced by the brands present in all regions.

A major problem is the exclusion of local brands. Including them would have extended the control for comparability, now limited to some features of the product (in our case brand, type, variety, weight etc.), to the percentage of purchases. We are currently at work on developing solutions to this problem.

Products were selected according to the various aims of the binary and the multilateral indexes. With the former, the aim was to analyze price differences between pairs of regions in order to explore diversities between contiguous regions or regions of different macro areas. The aim of multilateral indexes is to rank the regions on the basis of prices: this requires that the products considered for the comparison are the same and are present in all regions. Consequently, the basket used is smaller than the one used for binary comparisons.

The procedure for evaluating product representativity follows the Sergeev approach. We modified the procedure both by using the expenditure shares of each CPR, information available from the Nielsen data set, and by determining the thresholds for representativity.

The thresholds were set on the basis of the number of the products and of the quantities purchased in the regions matched. Obviously, these could have been defined on the basis of various criteria (the first 10 or more products, or those representing a given percentage of overall sales, for instance) and more complex methods. In any case, some arbitrariness is unavoidable. The soundness of our choice was confirmed by the fact that the baskets developed by our method proved to be well balanced: the number of representative products for one region and not-representative for the other, and vice versa, do not differ greatly from the total expenditure shares achieved in most cases.

For the construction of multilateral indexes we selected the binary approach, which is widely used in many countries and by international organizations. The use of the Törnqvist index for the calculation of binary parities was suggested, in addition to its characteristic of ideal index, by the Sergeev procedure and by the availability of the expenditure shares. Moreover, the use of the Törnqvist index instead of the Fisher formula for the construction of GEKS indexes is widely accepted in the relevant literature. The need to weight GEKS indexes has often been discussed in the literature (Ferrari & Riani, 1998; Rao, 2001), chiefly when consumption habits or when the general economic conditions of the countries or areas compared are very different. We chose not to use weighting, as we are analysing regions in the same country.

In conclusion, a methodological suggestion derives from the consideration that the selected baskets can be viewed as "nested baskets": (i) CPR perfectly comparable but not always representative (the CPRs inserted in the second list); (ii) representative but no longer perfectly comparable baskets (first list) and iii) a final basket (minimum basket) where the products are both fully comparable and representative. Comparing all these baskets (even with less regions, products or periods) might be useful in order to better evaluate the trade–off between comparability and representativity.

Acknowledgments Data have been kindly provided by CEIS of Tor Vergata University of Rome, whose Chief manager, Prof. Giovanni Tria, we would like to warmly thank.

References

Balk, B. M. (1996). A comparison of ten methods for multilateral international price comparisons. *Journal of Official Statistics, 12*(2), 199–222.

Biggeri, L., Brunetti, A., & Laureti, T. (2008). *The Interpretation of the Divergences between CPIs at Territorial Level: Evidence from Italy*. Paper presented at the Joint UNECE/ILO meeting on Consumer Price Indexes, May 8–9, Geneva.

Caves, D. W., Christensen, L. R., & Diewert, W. E. (1982). Multilateral Comparisons of Output, Input, and Productivity Using Superlative Index Numbers. *Economic Journal, 92*, 73–86.

De Carli, R. (2008). An Experiment to Calculate PPPs at Regional Level in Italy: Procedure Adopted and Analyses of the Results. Paper presented at the Joint UNECE/ILO meeting on Consumer Price Indexes, May 8–9, Geneva.

Eltetö, O., & Köves, P. (1964). On a problem of index number computation relating to international comparison.*Statisztikai Szemle, 42*, 507–518.

Ferrari, G., & Riani, M. (1998). On purchasing power prities calculation at the basic heading level. *Statistica, 58*, 91–108.

Gini, C. (1924). Quelques Considérations au Sujet de la Construction des Nombres Indexes des Prix et des Questions Analogues. *Metron, 4*, 3–162.

Gini, C. (1931). On the circular test of index numbers. *International Review of Statistics, 9*(2), 3–25.

Heravi, S., Heston, A., & Silver, M. (2003). Using scanner data to estimate country price parities: A hedonic regression approach. *Review of Income and Wealth, 49*(1), 1–21.

Hill, R. (1997). A Taxonomy of multilateral methods for making international comparisons of prices and quantities. *Review of Income and Wealth, 43*(1), 49–69.

Hill, R. (2008). *Elementary Indexes for Purchasing Power Parities (PPPs)*. Paper presented at the Joint UNECE/ILO meeting on Consumer Price Indexes, May 8–9, Geneva.

ILO et al. (2004). Consumer price index manual: Theory and practice, Geneva. An electronic updated version of the manual can be found at the web site of ILO.

Rao, D. S. Prasada (2001). *Weighted EKS and Generalized Country Product Dummy Methods for Aggregation at Basic Heading Level and Above Basic Heading Level*. Seminar on PPPs, World Bank Washington, Jan/Feb 2001.

Sergeev, S. (2003). *Equi-representativity and Some Modifications of the EKS Method at the Basic Heading Level*. Paper presented at the Joint Consultation on the European Comparison Programme, ECE, Geneva 31 March-2 April 2003.

Szulc, B. J. (1964). Indexes for multiregional comparisons. *Przeglad Statystyczny(Statistical Review), 3*, 239–254.

Measuring the Production of Non-Market Services

Paul Schreyer

1 Introduction

Although much effort is spent on measuring the value of GDP at current prices, an even more important objective of the National Accounts is to derive a measure of the growth of GDP and its components *in volume*. This implies decomposing current price measures into a price and a volume component. For complex services such as education and health, and moreover, in a context where there are no economically significant prices, this is a difficult task. Traditional methodologies have relied on measuring the volumes or the prices of inputs to obtain a measure of the volume or price of outputs. Quantity and quality of outputs are not identified consequently it is not possible to capture any productivity change in the production of health and education services. Productivity growth exists when more or better output can be produced with the same resources. As there is much evidence of changing quality of output in health and education services, ignoring productivity changes means foregoing important information for analysts and policy makers and measuring volume growth inaccurately.

The discussion about the deficiencies of input-based methods is by no means new but has resurfaced in the recent past. Eurostat (2001) stated in principle the desirability of applying output-based measures to non-market services. In the United Kingdom, the topic was taken up by the Atkinson (2005) report that was widely discussed inside and outside the UK. The measurement of services output and productivity has been a longstanding topic of interest in the United States, with several new publications recently (Triplett & Bosworth, 2004; Abraham & Mackie, 2006).

P. Schreyer (✉)
OECD Statistics Directorate, Paris Cedex 16, France
e-mail: Paul.Schreyer@OECD.org

Opinions voiced in this document reflect the views of the author and not necessarily those of the Organisation for Economic Co-operation and Development or its Member countries.

L. Biggeri, G. Ferrari (eds.), *Price Indexes in Time and Space*, Contributions to Statistics, DOI 10.1007/978-3-7908-2140-6_9, © Springer-Verlag Berlin Heidelberg 2010

Over the past two years, the OECD has also looked at the topic and organised several workshops.[1] One of the objectives of the work at the OECD is to come up with a handbook that provides guidance on the measurement of volume output of health and education services[2] and the present paper draws on this work. Views about the feasibility to implement different measures in the official national accounts are not unanimous but it is clear that there is significant demand for research and a need to get a better handle on the measurement of outputs, inputs and productivity of major economic activities such as health and education that are largely or partly characterised by non-market production.

It will be argued later in this paper that the distinction between market and non-market producers is not very relevant for the many of the statistical issues at hand. While it is clear that market production provides a basis for the measurement of production at current prices whereas the value of non-market production is typically estimated by summing up costs, most of the tricky issues associated with the measurement of volume output or with the measurement of price changes apply to both market and non-market producers. In particular, the difficulties to keep track of quality change and of entering and exiting products are present independently of the institutional affiliation of the producing unit. These problems are associated with the increasingly complex nature of modern services, for example in health care, and not with the question whether these services are provided under competitive conditions or not. Moreover, the quality of health and education services is difficult to observe and measure independently of whether they are provided by market or non-market producers. In consequence, and despite the title of this paper, many issues of output measurement are relevant for the activity as a whole, and not only under conditions of non-market provision.

A good deal of this paper will therefore be devoted to the question of quality adjustment because it is mainly in this context that the notion of outcome enters the picture and has to be distinguished from the notion of output. However, a second conclusion of this paper is that the story does not end here – in concept as in practice, it is very difficult to conceive a notion of output that is independent from the notion of outcome – hence we have two distinct, but related concepts.

The principle objective of our efforts here is to shed light on the measurement of volume measures of output for complex and economically important service activities and we shall frame our discussion with examples from health and education, although they may apply more broadly. Terminology is important in this context and we therefore start by stating our definition of "output" with a view to distinguishing it from "outcome". For completeness, we also define "inputs" and "activities".

---

[1] For documentation see
http://www.oecd.org/document/47/0,3343,en_2825_495684_37733615_1_ 1_1_1,00.html.

[2] A first and yet incomplete draft of this handbook can be found on the website for the 2007 meeting of the OECD National Accounts Working Party:
http://www.oecd.org/document/42/ 0,3343,en_2649_34245_38677418_1_1_1_1,00.html.

2 Terminology

Our point of reference is a production framework, i.e., an economic unit that transforms volumes of inputs into volumes of outputs. *Inputs*, the goods and services to be transformed comprise labour services, capital services and intermediate inputs. Inputs are combined and transformed by way of a production technology. *Outputs* are suitably differentiated counts of actions or activities (in the case of services), and counts of physical units (in the case of goods).

Outcome is a state that is valued by consumers – a functioning car, the state of health, the level of knowledge etc. Outcomes are influenced by many factors, and one of them may be the level of outputs. For example, the state of health (an outcome) is a function of medical care (output of the health industry), peoples' lifestyles and the natural environment. Often, outcomes manifest themselves with a considerable lag to the provision of output as would be the case of long-term effects on human health. Outcomes are therefore different from outputs. In principle, the production boundary as defined in the national accounts encompasses outputs, but not outcome. A superficial conclusion would therefore be that the national accounts statistician does not have to worry about outcomes, only outputs.

Different Meanings of Outcome

"Outcome" has been used in different ways in the relevant literature on non-market services. Two usages are common: in the health care literature, 'outcome' is typically defined as the resulting change in health status that is directly attributable to the health care received. Triplett (1998) indicates this usage in the cost-effectiveness literature and quotes Gold, Siegel, Russel, and Weinstein (1996) who define a health outcome as the end result of a medical intervention, or the change in health status associated with the intervention over some evaluation period or over the patient's lifetime. Employed in this sense, some authors suggest that the 'output' of the health care industry be measured by 'outcome'. Among national accountants, 'outcome' is typically used to describe a state that consumers value, for example the health status without necessarily relating the change in this state to the medical intervention. For example, Eurostat (2001) gives as examples of "outcome indicators" the level of education of the population, life expectancy, or the level of crime. Atkinson (2005) has the same usage of the word. Understood in this sense, outcome in itself cannot be a useful way to measure output or the effectiveness of the health or education system. In terms of national accounts semantics, the 'marginal contribution of the health care industry to outcome' is the equivalent to the notion of 'outcome' as used in the health care literature. As long as a particular definition is used consistently, the substance of the argument is of course unaffected and the only question is how useful a particular definition is for the purpose at hand. The present paper follows in the line of Eurostat (2001) and the Atkinson Review (2005), and employs the term 'outcome' in the sense of the national accounts literature.

However, things are more complicated. While outcomes are different from outputs, they are not independent. One of the conclusions of the present paper will in fact be that it is virtually impossible – in particular for health and education services – to define quality-adjusted outputs without invoking outcomes one way or the other.

For the forthcoming discussion, it is useful to refine the definition of outputs and outcomes in two ways. First, outputs are broken down into two components: *activities or processes* and the *quality adjustment* applied to them. Processes are observable and countable actions by which services are delivered although their characteristics may change over time. For education, a typical process measure is the number of pupils or the number of pupil hours taught in a particular grade. For health, a typical process measure is the number of treatments of a particular disease such as hip replacements. A second refinement consists in distinguishing between

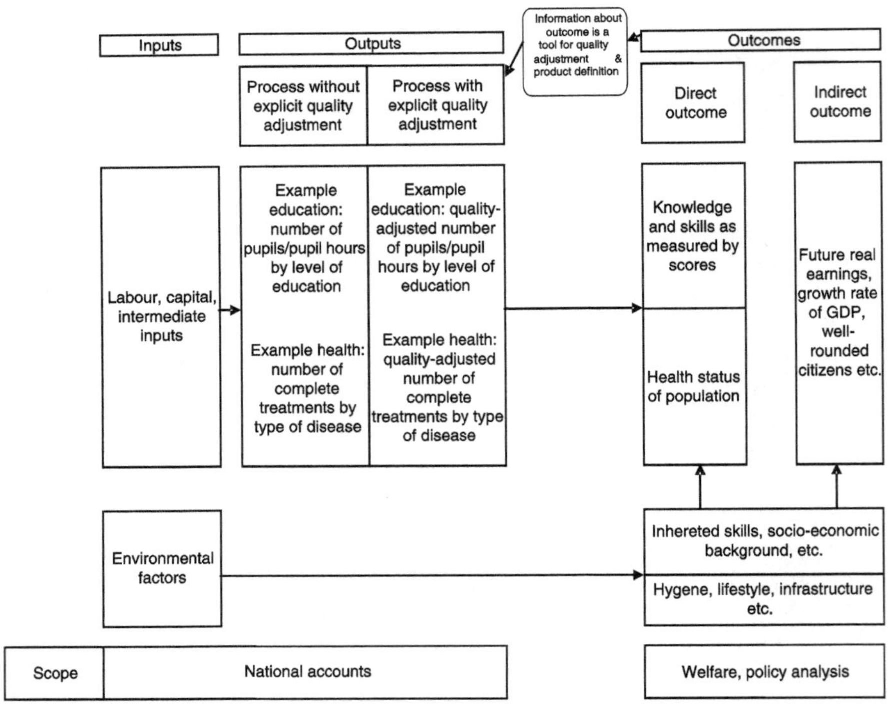

Fig. 1 Inputs, outputs and outcomes

direct and *indirect outcomes*. Direct outcomes are closer to the act of service provision than indirect outcomes although neither is a measure of service nor output itself. For example, in the case of education, a direct outcome is the state of knowledge of pupils, estimated by scores or degrees. The indirect outcome associated with education is employment possibilities and enhanced real earnings due to better human capital. Indirect outcomes associated with health services are fewer working days lost due to diseases, or individual well-being. These distinctions between inputs, activities/processes, quality adjustments and direct and indirect outcomes are shown in Fig. 1 below. Figure 1 also depicts the scope of national accounts measures which are defined via the production boundary. However, as has been mentioned above, there may be a need to bring considerations of outcome into the measurement of output to capture the quality of services. This link is also indicated in the figure and will be a central topic in the discussion in the sections to follow.

3 Competitive Markets, No Quality Change in Products

3.1 The Consumer Side

Our discussion starts with a simple market model of producers and consumers. For the moment, we assume that products are well-defined and transactions on the market are observable. On the demand side, consumers purchase the goods and services supplied by producers. Standard economic theory attributes a utility function to consumers where *utility* depends on the quantity of goods and services consumed. The utility function indicates how the consumer appreciates (in unobserved "utils") the quantity of products purchased.

To fix ideas, let households' utility function be $U^t = U(H^t)$, where U^t stands for the level of utility in period t which depends positively on H^t, a state of the world that consumers value. H^t corresponds to our notion of direct outcome. In general, H^t will be vector-valued because there are many different states valued by consumers – for example, different aspect of the status of health, the level of knowledge or the state of the natural environment. Consumers attach utility to a good or to a service because it affects *outcome*, i.e., a particular state that they value and which can be measured. We could also say that outcome is an intermediate step between consumption and utility and this is indeed the way it has been treated in the literature. In an application to health care, Berndt et al., (1998) distinguish between medical care ("output" in our terminology), the state of health ("direct outcome" in our terminology) and utility. They envisage a relationship whereby utility depends, among other variables, on the state of health and where the state of health is itself dependent on health care services, on the environment, lifestyle etc.

If we follow this idea, outcomes depend on *N* different services consumed by households, and we shall label quantities of these services in period t as $\mathbf{y}^t = [y_1{}^t, y_2{}^t, y_3{}^t, \ldots y_N{}^t]$. Importantly, the outcome variable H^t depends not only on services $\mathbf{y}^t$ that are purchased or obtained from producers but also on a host of other factors, Ω^t. Examples include households' behaviour with regard to smoking, alcohol consumption or physical exercise when H^t stands for the health status. Or Ω^t could stand for student's efforts or natural giftedness when H^t stands for educational attainment and the state of knowledge of the population. We shall come back to these "environmental" or "conditional" factors Ω^t in the discussion on quality change. For the moment, we formulate our "outcome" function and insert it into the utility function defined earlier:

$$H^t = H(y^t, \Omega^t) \text{ and therefore}$$
$$U^t = U[H(y^t, \Omega^t)] = V[y^t, \Omega^t]. \tag{1}$$

Households demand services to the point where prices equal the marginal utility generated by these services. In money terms, this gives $p_i^t = [\delta U / \delta y_i^t]\lambda^t$. Here, λ^t is the marginal utility of income that is needed to convert "utils", the unobserved units of utility, into currency units. It is not difficult to see that the marginal contribution of a good or a service to utility is in fact a composite term, namely the product of the marginal contribution of the product to outcome, $\delta H/\delta y_i^t$, and the marginal contribution of outcome to utility, $\delta U/\delta y_i^t = [\delta U/\delta H][\delta H/\delta y_i^t]$.

Although the above discussion does not give us any immediate insights into measurement, there is one point of importance for our discussion on output and outcome that emerges from the simple consumer demand relation above: the marginal contribution to outcome for a particular product i, $\delta H/\delta y_i^t$, relative to the marginal contribution of another product j, equals the price ratio of the two products:

$$[\delta H / \delta y_i^t] / [\delta H / \delta y_j^t] = p_t^t / p_j^i. \tag{2}$$

The implication of this relation is that outcomes are linked with product classifications, and products should be grouped into a category i if they generate similar contributions to outcome – then consumers will attach the same value, i.e., the same price to them and they should be grouped into a different product category if their contribution to outcome is different. Note that in our simple presentation here we make the implicit assumption that there is exactly one homogenous unit y_i^t and one price p_i^t. In practice, this is not the case and similar types of products have to be grouped. In the context of an elementary consumer price index, for example, the p_i^t's would be the un-weighted averages of individual items within product group i. Similarly, the y_i^t's for a volume index could be thought of as un-weighted average number of individual items for product type *i*. How individual products ("items") are grouped is a question that has to be answered with respect to the purpose of the price or volume index. Above, it was shown that from a consumer perspective, the criterion for grouping individual items is that they potentially offer the same contribution to outcomes, i.e., they satisfy the same or similar consumer needs. Put differently, they are substitutes from a consumer perspective. Conversely, if different items are not interchangeable from a consumer perspective, they should be treated as different products. In the presence of quality change or new and disappearing items, the question of grouping items becomes important. But the point to retain is that the organisation of price or quantity measurement, in particular how products are classified and stratified, cannot proceed without some reference to outcome, if one wants to bring in a consumer perspective.

3.2 The Producer Side

We now turn to producers and take it that their production technology can be represented by a cost function[3] that shows the minimum costs required during a given period to produce a quantity of N products $\mathbf{y}^t=[y_1{}^t, y_2{}^t, y_3{}^t, \ldots y_N{}^t]$, for a given set of input prices and for a given technology. In the case of health service producers, a particular product could consist of a (complete) treatment of a particular disease, in the case of education it could be the hours of teaching provided for a particular grade. To keep the exposition tractable, it will be assumed that there is a technology set $S_i{}^t$ for each product that links output $y_i{}^t$ to a set of M inputs, $\mathbf{x}_i{}^t=[x_{i1}{}^t, x_{i2}{}^t, \ldots x_{iM}{}^t]$. Inputs are purchased at the prices $\mathbf{w}^t=[w_1{}^t, w_2{}^t, \ldots w_M{}^t]$ on factor markets where producers are price takers. A cost function for product i can then be written as

$$C_i^t = C_i^t(y_i,\mathbf{w}) = min\{\mathbf{w}_i \cdot x{:}(y_i,\mathbf{w}) \text{ is contained in } S_i^t(y_i,\mathbf{w})\}; i = 1,\ldots N. \quad (3)$$

A unit or average cost function for output type i is defined as $c_i{}^t(y_i,\mathbf{w}_i) \equiv C_i{}^t(y_i,\mathbf{w})/y_i$ and measures the costs per unit of output y_i during period t. With constant returns to scale which we shall assume for convenience here, unit costs are independent of the level of output, and average costs equal marginal costs. When producers minimize costs, it follows that total costs for a particular product are equal to minimum costs:

$$C_i^t(y_i^t,\mathbf{w}^t) = c_i^t(\mathbf{w}^t)y_i^t = \mathbf{w}^t \cdot x_i^t. \quad (4)$$

We follow Diewert (2008) and define the producer's or sector's total cost function C^t as the sum of cost functions for different products. This will be helpful in the definition of index numbers below:

$$C^t(\mathbf{y}^t,\mathbf{w}^t) = \Sigma_i^N C_i^t(y_i^t,\mathbf{w}^t) = \Sigma_i^N c_i^t(\mathbf{w}^t)y_i^t = \mathbf{w}^t \cdot x_i^t. \quad (5)$$

Under constant returns to scale, and in a competitive market, producers provide services at the point where prices equal marginal costs. More generally, under market conditions, excess profits are competed away and prices will equal average costs so that:

$$p_i^t = c_i^t(\mathbf{w}_i^t). \quad (6)$$

---

[3]For more information about cost functions see Shephard (1970) or Diewert (1974).

When the consumer and the producer side are combined, market equilibrium is characterized by

$$[\delta U/\delta H][\delta H/\delta y_i^t]\lambda^t = p_i^t = c_i^t(w_i^t). \tag{7}$$

Thus, marginal utility from consumption and marginal costs of production are equal in equilibrium. There is of course nothing new about this statement which can be found in any micro-economics textbook. It is nonetheless useful because we see that in equilibrium, consumer valuation and producer valuation of a good or of a service coincide at the margin when there is a market-clearing price. The implication is that when weights are needed to aggregate across products, there is no need to invoke either a consumer or a producer perspective – the value of market transactions is all that is needed and it combines the two sides of the market.[4] Also, in the absence of quality change and as long as there are no new or exiting products, it would appear that there is little to worry about the distinction between output and outcome – market prices jointly value the marginal contribution of consumption to outcome and also reflect marginal production costs.

4 Non-Market Production, no Quality Change

Having established the competitive case with a well-defined set of products that does not change over time, we shall now direct attention to the case of non-market production. When goods or services are provided by non-market producers, they are provided at a price that does not cover costs and which may even be zero. In this case, the price at which products are transacted loses its significance as an indicator of marginal or average costs. Nor is the price necessarily linked to a utility-maximising quantity of consumer demand, as was the case in the competitive environment. Therefore, our convenient link between the producer and the consumer side, established in the first section, breaks down:

$$[\delta U/\delta y_i^t]\lambda^t \neq c_i^t(w^t) \tag{8}$$

[4]This is a simplification. In practice, and in the presence of transport costs or taxes, there is no unique market price and a distinction has to be made between different valuations. For example, from a consumer perspective, a valuation at purchasers prices is appropriate, which is inclusive of taxes and transportation margins. From a producer perspective, a valuation at basic prices would be more appropriate, which excludes for example taxes payable and subsidies receivable in conjunction with production or sale. The statement in the text is also a simplification in the sense that output price indices for producers and input price indices for consumers are typically compiled at different levels of aggregation and on the basis of different classifications. This may also lead to differences in weights between output price indices for producers and input price indices for consumers. However, even when aggregation happens differently in the producer and in the consumer case, the price at the lowest level of aggregation at which the transaction takes place, is a market price and reflects the joint influences of producers and consumers.

An immediate consequence from this situation is that the well-established body of literature on the theory of producer price indices, notably Fisher and Shell (1972); Archibald (1977); IMF et al. (2004) no longer applies. The theory of the output price index relies on revenue functions for producers and stipulates revenue-maximising behaviour, given a set of market prices for producers' outputs. In a non-market environment,[5] this is not a useful assumption, and consequently, revenue functions cannot be used as a conceptual basis for output price indices.

However, measurement can be based on *cost-based or quasi prices*. When transaction prices are significantly below cost or zero, it is customary in the national accounts to measure the money value of output as the sum of costs. One could also say that output is valued at quasi prices. They are those (unobserved) prices that emulate a competitive situation where prices equal average costs per product. With unit costs at hand, they can be treated *as if they were prices*:

$$p_i^t \equiv c_i^t(\mathbf{w}^t). \tag{9}$$

If for the moment we maintain the (courageous) assumption that non-market producers are cost-minimising units, then minimum costs equal actual costs or $c_i^t(\mathbf{w}^t)y_i^t = \mathbf{w}^t \cdot \mathbf{x}_i^t$, and it follows that

$$p_i^t y_i^t \equiv c_i^t(\mathbf{w}^t).y_i^t = \mathbf{w}^t \cdot x_i^t. \tag{10}$$

Expression (10) states the obvious, namely that with cost-based prices, the value output of product i equals the value of inputs used in production of product i. This is the way non-market output is valued in the *System of National Accounts*.[6] What is important for the purpose at hand is the fact that this equality of inputs and outputs in value does *not* imply equality of inputs and outputs in volume or quantity. If this were the case, our efforts to derive volume measures of output that are separate from volume measures of inputs would be put in question and with it any attempt to measure productivity change in non-market production.

The main difference between cost-based prices of outputs and prices of inputs is that the former correspond to costs per unit of output (such as the costs for one treatment of a heart attack or the costs for one year of schooling) whereas the latter correspond to the costs per unit of input (such as wages per hour of a nurse or the salary of a teacher).

Diewert (2008) shows formally how a cost-based volume index of output can be defined. He proposes a family of cost-based output quantity indices and focuses on

[5] Note that despite the fact that our discussion has been couched in terms of non-market producers, it carries over to the more general case of regulated industries. For example, Lawrence and Diewert (2006), measure the quantity index of output for New Zealand electricity utilities with a cost-based index because there are no meaningful revenue shares or prices for the three types of outputs identified for utilities: throughput of electricity, system line capacity and connections.

[6] For a genesis of the treatment of non-market production in the national accounts and the many issues associated with it, see Vanoli (2002).

the Laspeyres (Q_L), on the Paasche (Q_P) and on the Fisher (Q_F) case. In line with the economic approach towards index numbers, Diewert defines the Laspeyres version of a cost-based output quantity index as the (hypothetical) total cost $C^0(y^1, w^0)$ of producing the output vector y^1 of period 1 under the conditions of period *0* technology and input prices, divided by the actual costs of period 0, $C^0(y^0, w^0)$. Similarly, he defines a Paasche type index as the actual costs of period 1, $C^1(y^1, w^1)$ divided by the hypothetical costs $C^1(y^0, w^1)$ that would have been incurred, had the products of period *0* been produced in period *1*, under the technological constraints of period *1* and given period *1* input prices:

$$\begin{aligned} Q_L &= C^0(y^1, w^0)/C^0(y^0, w^0) = \Sigma_i^N c_i^0 y_i^1 / \Sigma_i^N c_i^0 y_i^0 \\ Q_P &= C^1(y^1, w^1)/C^1(y^0, w^1) = \Sigma_i^N c_i^1 y_i^1 / \Sigma_i^N c_i^1 y_i^0 \\ Q_F &= [Q_L Q_P]^{1/2}. \end{aligned} \tag{11}$$

For a number of practical reasons, we prefer working with a "price" rather than a quantity index à la Diewert for non-market producers and then deflate total costs by a price index. To this end, we construct an *indirect index of quasi prices* by dividing total costs by the volume index of output:

$$\begin{aligned} P_L &= [C^1(y^1, w^1)/C^0(y^0, w^0)]/Q_P = \Sigma_i^N c_i^1 y_i^0 / \Sigma_i^N c_i^0 y_i^0 \\ P_P &= [C^1(y^1, w^1)/C^0(y^0, w^0)]/Q_L = \Sigma_i^N c_i^1 y_i^1 / \Sigma_i^N c_i^0 y_i^1 \\ P_F &= [P_L P_P]^{1/2}. \end{aligned} \tag{12}$$

Note a useful interpretation of this quasi-price index that is obtained by re-writing the Laspeyres or Paasche version in expression (12). For example, after inserting the theoretical expression for Q_P into the first line of (12), P_L can be presented as the product of two terms:

$$\begin{aligned} P_L &= [C^1(y^1, w^1)/C^0(y^0, w^0)]/Q_P \\ &= [C^1(y^1, w^1)/C^0(y^0, w^0)]/[C^1(y^1, w^1)/C^1(y^0, w^1)] \\ &= [C^1(y^0, w^1)/C^0(y^0, w^0)] \\ &= [C^1(y^0, w^1)/C^1(y^0, w^0)][C^1(y^0, w^0)/C^0(y^0, w^0)] \end{aligned} \tag{13}$$

The first term in the last line of (13) is an index of input prices: costs are compared between two situations, with technology and the level of output held fixed but input prices are allowed to vary. The second term in the same line is an inverted productivity index: for a given reference output and input prices, changes in minimum costs between the periods are compared. Similar transformations could be applied to P_P and P_F, but there is no need to present them here. The main point can easily be explained with the decomposition of P_L only: we recall that in a market situation, and under competition, a productivity index equals an input price index divided by an (output) price index: if output prices rise less rapidly than input prices, this implies productivity improvements. In the present non-market case, the quasi-price index for outputs plays the same role as the market price index for outputs in a market situation. If unit costs rise less rapidly than input prices, there has been productivity change.

But (13) also shows that despite that fact that much of the discussion about non-market producers has been by way of costs, we *are* dealing with an output perspective: unit costs or quasi prices are productivity-adjusted input prices and the productivity adjustment marks the movement from an input perspective towards an output perspective in measuring non-market activity. This is not always well understood, because costs are rightly seen as input-related variables. But the above makes it clear that considering costs per unit *of output* differentiates an output perspective from considering costs per unit *of input*, i.e., the input perspective.

5 Non-Market Production and Quality Change

5.1 Direct Outcome, Stratification and Implicit Quality Adjustment

There is an extensive literature on how to deal with quality change in existing products, with the exit of old products and with the entry of new products if one wants to compile price or quantity indices. Quality change counts among the most serious measurement issues in estimating price indices. Early references include Stone (1956); Griliches (1971), and more recent ones IMF et al. (2004), as well as Triplett (2006). The reader is referred to these volumes for a complete discussion. The task in this section is twofold: discuss how the measurement of quality change relates to outputs and outcomes, and propose a method for quality adjustment for the non-market case. To start, recall some key principles and methods that are followed in measuring quality change.

> Agencies that estimate price indexes employ, near universally, one fundamental methodological principle. The agency chooses a sample of sellers [...] and of products. It collects a price in the initial period for each of the products selected. Then, at some second period, it collects the price for exactly the same product, from the same seller, that was selected in the initial period. The price index is computed by matching the price for the second period with the initial price, observation by observation or 'model by model' as it is often somewhat inaccurately called. (Triplett, 2006, p.15)

One technique to deal with quality change in products is thus to group them such that only products of the same specification are compared over time or in space. Such grouping or matching ensures that only prices or quantities of products of the same or very similar quality are compared. The idea is that products of different quality are treated as different products. Examples for such grouping in education are establishments that provide different services in addition to education, such as boarding schools as opposed to day-time schools or hospitals with different levels of non-medical services.

Note, however, that grouping also relies on an important assumption: to show a price or quantity movement that is representative of a product group, the price or quantity movements of those products that *are* matched have to be a good indicator of the price or quantity movements of those products that are *not* matched – in particular products that are newly entering the market. Price or quantity changes that arise outside the sample of matched products are ignored.

The non-market case on which our attention is focused here shows also the importance of choosing the right level of aggregation where matching takes place. And it is again considerations of direct outcome that govern this choice. Take the case that is described in Box 1. Two medical procedures are considered, of different unit costs, but equally interesting from the consumer's viewpoint – both procedures treat the same medical problem equally well. In other words, the contribution of each treatment to outcome, from a consumer perspective, is the same. Treating each procedure as a separate product, i.e., setting the elementary level for the construction of the price or volume index below each procedure can lead to counter-intuitive results: more of the cheaper but equally helpful treatment translates into a decline in the volume of output because, in a non-market context, the new procedure only gradually replaces the old procedure. When both procedures are treated as providing the same service and at the elementary level both procedures are treated as the same product, some of the bias is eliminated. But no judgement about substitutability can be made without invoking at least implicitly, some judgement about outcomes. By definition, (complete) substitutability of services implies that they are equally appreciated by consumers, in other words, they generate the same outcome.

We conclude that even in a situation where matching between products is perfect, and no explicit quality adjustment may be needed, *some* reasoning about outcome is in place, if only to group substitutable products together in one stratum.

But matching is rarely perfect. Matching of the quasi prices or volumes of non-market services such as health and education services is unlikely to control completely for particular characteristics associated with the provision of these services. Or a representative service may change its characteristics, akin to a product that price collectors are no more able to find in a particular outlet and that has to be replaced with a new product. These are the instances where explicit quality adjustment comes into play, of which more later.

5.2 Indirect Outcome and Quality Adjustment Through Re-Definition of Products

Jorgenson and Fraumeni (1989, 1992) were the first to apply a human capital approach towards the measurement of output of the education sector of the United States. Their approach constitutes a particular way of accounting for quality change and despite the fact that the human capital approach cannot easily be generalised to other services, it is worth putting the work by Fraumeni and Jorgenson into the context of our present discussion. The approach has a clear theoretical foundation and has stood the test of empirical implementation.

At the core of the human capital approach towards measuring education output lies the idea that educational services are investment flows that add to human capital. Private households, it is assumed, demand educational services to the point where the marginal costs of an extra year of education (in the form of fees or income foregone, for example) equal the marginal benefits from education. The latter are measured as the discounted differentials in lifetime income due to the additional year of schooling. This supposes in turn that wages on labour markets correspond to the marginal productivity of workers – in this way, the relative level of human capital and consequently the relative level of worker's marginal productivity can be gauged via relative additions to discounted life income of workers. The quantity measures that enter the calculation are a set of student enrolment numbers, stratified by various criteria such as level of education, type of studies, gender and so forth. At this, lowest level of aggregation, quantity measures for the education services provided are not different from the unadjusted quantity measures that formed the body of our discussion earlier and which would be matched over time to control for quality change. However, under the human capital approach, the same quantities are now quantities of investment goods and need to be matched by investment goods prices for aggregation, not by unit costs as was the case earlier.

Let there be I different education services and let the price of the educational investment for each particular type of education in period t be $p_{edu,i}^{t}$ $(i=1,2...I)$. The value of $p_{edu,i}^{t}$ is not a market value that is readily observed but has to be imputed. Jorgenson and Fraumeni (1989) show how to implement such computations empirically. Let the quantity of the educational investment be y_i^t, so that the value of education services is $\Sigma_i^I\, p_i^t y_i^t$. The Laspeyres, Paasche and Fisher-type volume index of education services between periods 1 and 0 are then:

$$\begin{aligned} Q_{L,edu} &= \Sigma_i^I p_{edu,i}^0 y_i^1 / \Sigma_i^I p_{edu,i}^0 y_i^0 \\ Q_{P,edu} &= \Sigma_i^I p_{edu,i}^1 y_i^1 / \Sigma_i^I p_{edu,i}^1 y_i^0 \\ Q_{F,edu} &= [Q_{L,edu} Q_{P,edu}]^{1/2}. \end{aligned} \tag{14}$$

The different y_i's are thus valued with the corresponding returns on extra education. When the relative valuation of the different types of education changes, the volume index of education services changes. For example, if the returns to one extra year of tertiary education rise quicker than the returns to an extra year of secondary education, and if more students undergo the extra training, the resulting change in the volume index would not only capture an increase in years of schooling but also reflect the fact that the average quality of human capital has risen. In this sense, the human capital approach adjusts volume measures of education output for changes in the quality of human capital. This is achieved by re-defining education output as the production of an investment good and by imputing the price of this investment good.

Note, however, that the human capital approach assumes that at the lowest level of aggregation, the quantities y_i^t are not subject to quality change – one hour of a particular course taught this semester equals one hour of the same course taught the following semester.

How does the human capital approach relate to outcome? There is no formal link here except that the quantity of human capital should in some way be related to the stock of knowledge of humans having undergone education. But in terms of the underlying economic theory, the human capital approach does not really need a variable H as in our formulation earlier. Human capital is only a means to insure wage income on the labour market. Households' utility depends on the goods and services that can be purchased with these labour incomes, but there is no need for an explicit introduction of the stock of knowledge as an argument in the utility function. Outcome only enters in a monetary sense, as the reward on the labour market. In terms of the terminology introduced at the beginning of this paper, we are dealing with indirect outcome.

In summary, the human capital approach is an example of how certain (important) aspects of quality change can be captured by re-defining education output as an investment good and by relying on labour market outcomes for its valuation.[7] But it is also a fact that not all aspects of quality can be captured. Further, the human capital approach has to rely on some strong assumptions. In particular, it has to be true that observed differences in remuneration on labour markets are attributable to differences in educational attainment. Finally, there is no immediate way of applying a similar approach to other areas of non-market services such as health or general government. Further, future incomes are not the only motivation for education – learning per se and the nature of future work can give satisfaction. This aspect cannot be captured with discounted lifetime incomes.

---

[7] We note several accounting issues that would arise if the human capital approach were implemented in the national accounts. First, there is a question about the scope of inputs to and outputs from the production of educational investment goods. For example, Jorgenson and Fraumeni (1989) consider students' time as one of the inputs into the production process. On the output side, they value not only future income from labour market activities but they also value non-market labour. Other authors, for example Ervik, Holmoy and Haegeland (2003) have argued that in a national accounts context, students' time should not be part of inputs. They also exclude non-market labour from the output computations. This can lead to significant differences in the measured value of production of the education sector. A second issue is that the value of the education output at current prices under the human capital approach (which could be seen as a social gross production) is not necessarily equal to the value of inputs. This implies a kind of gross operating surplus for non-market producers that is hard to deal with in an established accounting framework. Ervik et al. (2003) estimate the human capital-based output for the Norvegian education sector to be around NOK 77 billion in 1995, compared to the cost-based national accounts estimate of about NOK 14 billion.

5.3 Outcome and Explicit Quality Adjustment

If matching is insufficient to control for key characteristics of service provision, other, explicit, techniques have to be invoked to account for quality change. In general, the quality of a product can be expressed by the quantity of its characteristics. Quality change can then be captured by the change in characteristics. Similarly, price changes in products can be attributed to pure price changes and to those price changes that reflect changes in product characteristics. This is the approach followed by hedonic price indices[8] that are now well established among statistical agencies.

Quality adjustments require the identification of a set of characteristics such as the speed, engine size or equipment for a car or the processor speed for a computer. Berndt and Cutler (2001) use patient characteristics, information on different types of depression, variables on medication and the like to estimate a hedonic price model for the treatment of depression; the idea being to isolate those price changes that are due to changes in characteristics from those price changes that constitute "inflation". An important result of the hedonic model is that it allows the identification of characteristics and provides a market valuation of each one.[9] Market valuation, in turn, is a convenient way of aggregating across characteristics because everything is expressed in a single monetary unit.

At first glance, hedonic regressions would appear to be badly suited for a non-market environment, given that hedonic methods are all about extracting "market" information from examining the link between market prices and product characteristics. At a second glance, however, the basic principles of hedonic regressions *would* appear to be adaptable for a non-market environment, albeit with a different interpretation. Schreyer (2009) shows how, in the case of health, the regression of unit costs of treatments against variables that are indicative either of the severity of the disease or of treatment attributes, can be used for an explicit quality adjustment. What is important is that the attributes of the treatment are those relevant from the perspective of the patient. Coefficients in such a regression provide an indication about the technology of supply of characteristics and its cost structure. Supply in a non-market context does not interact with demand. In such a situation, regression coefficients are indicative of costs or producer valuation. However, despite the fact that the *valuation* of characteristics is a cost valuation, the *choice* of characteristics should largely reflect a consumer perspective as was explained earlier.

---

[8] See Triplett (2006) for a comprehensive discussion.

[9] Rosen (1974) demonstrated that in general, those characteristics of a product will show up in the function that are valued by consumers *and* that have cost implications for producers. Triplett (2006) writes on this: "It is well-established – but still not sufficiently understood – that the functional form of [the hedonic equation] cannot be derived from the form of the utility function or of the production function. Neither does [the hedonic equation] represent a 'reduced form' of supply and demand functions derived from [the utility and the production functions] as the term is conventionally used. Establishing these results requires consideration of buyer and seller behaviour towards characteristics]" (page 231).

Box 1. Why Narrow Specifications of Products may not always be Sufficient to Capture Quality Change

A straightforward technique of dealing with quality change in a price or in a volume index is to match models, i.e., to compare only prices or quantities of products that are tightly specified. In other words, products are treated as different products whenever their characteristics are different. The more specific the characteristics of a particular product, the less likely it is that a modification of the product goes unnoticed and that a change in quality is not recognised as such. Such implicit quality-adjustment is well adapted when the set of products observed is stable and when it is representative for the universe of products. It may, however, be insufficient, when products change, when there are substitution processes between new and old products and when there are no markets or when existing markets operate imperfectly. This is best illustrated by way of an example. We use a quantity index here but the same points could also be made by way of a price index that is subsequently used to deflate values.

Suppose there are two treatments for a disease, traditional surgery and laser treatment, and assume that laser treatment is introduced in period 1. In addition, as may well be the case, the unit cost of laser treatment is lower than the unit cost of traditional surgery. The total number of interventions remains the same.

	Traditional surgery			Laser surgery		
	Period 0	Period 1	Period 2	Period 0	Period 1	Period 2
Unit cost	100	100	100	-	90	90
Number of interventions	50	40	5	0	10	45
Total cost	5000	4000	500	0	900	4050

Now consider a *matched-model approach* towards calculating a volume change from period 0 to period 1. In the simplest case, the volume index is given by the quantity changes in the two treatments, each weighted by the cost share it occupies in period 0. As laser surgery does not yet exist in period 0, it receives a zero weight so that the volume index of treatments is simply the change in the number of traditional surgery, or (40/50–1) = –20%. Between periods 1 and 2, the corresponding volume index equals $[s_T(5/40)+s_L(45/10)]-1 = -7.1\%$ where $s_T = 82\%$ and $s_L = 18\%$ are the period 1 cost shares of the traditional and of the laser treatment respectively. This approach treats the two treatment as different products and the sharp drop in the total volume index in period 1 reflects the "new goods" problem that arises when new products enter the sample that cannot be compared with quantities in the base period. Note also the assumption implicit in this model: consumer valuation of the two products is captured by the relative unit costs, so if laser surgery is cheaper than traditional surgery, this method implicitly quality-adjusts *downward* the quantity of laser surgery when it is combined with traditional surgery. In a perfectly operating market, the price of the traditional treatment would see an instantaneous downward adjustment, bringing consumer valuation of the two processes in line but in the practice of health services provision, in particular in a non-market context, this would appear an unrealistic assumption.

A different result arises when it is considered that the two treatments are perfect substitutes, i.e., that they are in fact the same product. In this case, no cost weighting is applied between the two treatments – and the number of treatments is simply added up. As there are 50 interventions in every period, the result is a volume index that shows zero growth and a declining price index, reflecting the drop in average unit costs of treatment.

The previous method is justified if it is plausible that consumers are indifferent about the two treatments. If this is not the case, and they prefer laser over traditional surgery because the former is less intrusive or requires fewer days of recovery, an explicit quality-adjustment is needed. Such an adjustment can be applied to the quantity measures, either by scaling up the quantity of laser treatments or by scaling down the quantity of traditional treatments. Whichever way this is done, the implication is always that one treatment is expressed in equivalents of the other treatment, and the ratio should in some way reflect consumer preferences. Alternatively, prices or unit costs could be rescaled before constructing a price index. Suppose the adjustment factor is 1.1 – each laser treatment is the equivalent of 1.1 traditional treatments. Then, expressed in "traditional surgery-equivalents", the number of treatments is 50 in period 0, 40+10*1.1=51 in period 1 and 5+45*1.1=54.5 in period 2. The resulting volume index is +2% in period 1 and +6.9% in period 2. Obviously, the difficulty lies in determining the adjustment factor which should (i) reflect consumer preferences; (ii) be uni-dimentional.

5.4 Output as the Marginal Contribution to Direct Outcome

One consequence of (8) was the inapplicability of the theory of output price indices. We solved this by reverting to cost-based or quasi prices. In so doing, we implicitly signed up to another decision on how to deal with the inequality of marginal utility and marginal costs: cost-based output prices (or cost-based output volumes à la Diewert (2008)) imply that weights are cost shares of goods and services, and not utility shares. When it came to quality change, we stuck with a cost-based approach but applied explicit quality adjustment to unit costs, and implicitly, to volume measures of output. Outcomes played an indirect role only in the sense that they helped stratifying activity counts and choosing the set of characteristics for quality adjustment.

An alternative way of constructing volume and price indices of output consists of directly invoking the consumer perspective and constructing measures of output that are based on a product's marginal contribution to outcome. Atkinson (2005) puts this forward as one of the methods of measuring the output of non-market producers. In our notation, this would mean tracking movements of output through observation of movements of outcomes caused by the change in output, i.e., $\delta H/\delta y_i^t$. One obvious advantage of this procedure is that – at least in concept – it provides a solution to the problem of quality adjustment. If all that consumers care about is direct outcome, and we are able to trace outcome and measure its change then there is no need to explicitly quality-adjust measures of production.

Outcome-based Measures of Health Services: A UK Example

The UK Department of Health (DH) estimated outcome-based measures of production in two areas where there is evidence of health gain being significantly higher than cost; one of them is prescription of statins.

"Statins reduce cholesterol, which can block arteries, and so reduce the risk of heart disease and stroke. DH assesses the health gain from statin therapy in terms of the risk of coronary heart disease (CHD) for different groups, based on analysis of the Health Survey for England (HSE) in 1998 and 2003, which contained questions and clinical measures relevant to CHD. The estimates of health gain take account of HSE evidence on preexisting CHD and stroke, diabetes (which increases risk of CHD), total cholesterol, HDL cholesterol, blood pressure, smoking status, age and sex. This approach makes it possible to estimate the health gain from statins as used in the population, making use of evidence from clinical trials on effectiveness of statins for different groups. DH estimates that statin therapy in 2003 added 77,000 life years, compared with no therapy, for the 1.9 million patients who took the drug. It estimates that each prescription has a marginal benefit of 0.0038 life years. If each life year is valued at £30,000, the value of each prescription is £115. This compares with £14 as the unit cost in 2005/06. Further Developments updates previous work and estimates that growth in health care output would be higher by 0.7–0.9% points a year, on different assumptions if a value weight replaced the cost weight." (Lee, 2008)

There are many practical considerations before any of these approaches can be implemented and output measures that rely on the marginal contribution to outcome are a long way from being regular features in statistical programmes. Note the following empirical issues.

It is not obvious how to identify the marginal impact of output to outcome. As was mentioned earlier in this paper, outcomes (for example the state of health) are affected by many factors, not only by the good or service under consideration. Thus, the effect of health care on the state of health should not be affected by any other factors that influence consumer outcome such as the lifestyle of patients. It is empirically difficult to control for these other influences on outcome.

A more subtle issue is the following: a measure of the contribution to outcomes should reflect the normal, or expected effect of the activity whose output is to be measured. "Normal", "average" or "expected" effects should be considered rather than ex-post, individual effects. This distinction arises from in the context of service provision where the consumer is typically actively involved in the provision of the product. It would seem that a measure of service production should not be influenced by the *individual* capacity of the consumer to make use of these services. For example, the same teaching activity performed on two different students, should be measured as the same quantity of teaching services towards each student, even if it turns out that one of them benefits less from teaching than the other. Or the same medical treatment, applied to two different persons with the same disease (and similar patient characteristics) should be measured as equal quantities of medical services. Unless, that is, the two persons come from different groups of patients where "groups" is understood as sets of patients with characteristics that require a-priori different services, for example young and old persons for particular treatments. It is easy to see that making this difference in practice will be difficult and sometimes impossible. This limits the applicability of output methods that rely on the direct contribution to outcomes.

6 Conclusions

This paper looked at the notions of output and outcome in the context of service activities with market and non-market producers. We started by defining the various notions, in particular inputs, processes/activities, outputs, and outcomes. The following main conclusions were derived:

- Even in the simplest market model without any quality change in products, output measurement may require *some*, often implicit, reference to outcome when products are grouped and classified. Thus, output and outcome are different, but they are not independent of each other.
- With non-market producers, the equality of marginal utility and unit cost ratios breaks down. The money value of output is determined as the sum of costs of inputs, but it is possible to construct meaningful indices of cost-based volumes and quasi prices along with indices of volumes and prices of inputs. Akin to the simple market case, cost-based indices are not void of implicit information about

utility and outcome – typically, this enters via the product classification on the basis of which aggregation takes place.

- In the presence of quality change, all existing methods require implicit or explicit information or reasoning about outcome. The simplest and most widely-used way of controlling for quality change – matching or stratification – requires implicit statements about the similarity of product in their capacity to generate direct outcome when products are grouped together. A second method uses explicit information on product characteristics to adjust unit costs for differences in characteristics. The choice of characteristics cannot proceed without considering outcome.
- Problems of quality adjustment arise whether services are provided by market or non-market producers. The fact that there is an observable market transaction in one period and another market transaction in the next does not imply that they are comparable – otherwise, price statistician would be in the convenient situation of only having to deal with quality adjustment for non-market production. This is of course not the case. Although the distinction between market and non-market producers is useful and has consequences for measuring the current price value of output, it loses most of its significance when we come the thorny part of measuring output – the treatment of quality change and the treatment of exiting and new products.
- A pragmatic approach will be called for to proceed with services measurement. In particular, there is no reason to approach every type of service with the same method for quality adjustment – methodologies should be robust but will also have to be reflective of data availability and transparency for users.
- Measuring output for complex services is difficult but the conclusion should not be that it is simply too difficult to do anything. Health and education account for a too large and growing part of the economy to ignore output measurement for them. It may take a while before consensual and internationally comparable methods are agreed upon but active research and data development is vital to achieve this objective.

References

Abraham, K. G., & Mackie C. (2006). A framework for nonmarket accounting. In D. Jorgenson, J. S. Landefeld, & W. Nordhaus (Eds.), *A new architecture for the U.S. National Account,* NBER, Studies in Income and Wealth. Chicago: University of Chicago Press.

Archibald, R. B. (1977). On the theory of industrial price measurement: Output price indexes. *Annals of Economic and Social Measurement, 6*, 57–72.

Atkinson Review (2005). *Final Report: Measurement of Government Output and Productivity for the National Accounts*. Basingstoke: Palgrave McMillan.

Berndt, E. R., Cutler, D. M., Frank, R. G., Griliches, Z., Newhouse, J. P., & Triplett, J. E. (1998). *Price indexes for medical care goods and services: An overview of measurement issues*. (NBER Working Paper Series, No 6817).

Berndt, E. R., & Cutler D. M. (Eds.). (2001). *Medical care output and productivity* (Vol. 62) NBER Studies in Income and Wealth. Chicago: University of Chicago Press.

Diewert, W. E. (2008, January 17). *The measurement of nonmarket sector outputs and inputs using cost weights* (Discussion Paper 08-03). Department of Economics, University of British Columbia. Accessed from http://www.econ.ubc.ca/diewert/dp0803.pdf.

Diewert, W. E. (1974). Application of duality theory. In M. D. Intriligator & D. A. Kendrick (Eds.), *Frontiers of quantitative economics* (Vol. II). Amsterdam: North Holland.

Ervik, A. O., Holmøy, E., & Hægeland T. (2003, July). *A theory-based measure of the output of the education sector* (Statistics Norway Research Department Discussion Paper No. 353).

Eurostat. (2001). *Handbook on price and volume measures in national accounts.* Luxembourg: European Communities.

Fisher, F. M., & Shell, K. (1972). *The economic theory of price indices.* New York and London: Academic Press.

Gold, M. R., Siegel, J. E., Russel, L. B., & Weinstein, M. C. (Eds.). (1996). *Cost effectiveness in health and medicine.* New York: Oxford University Press.

Griliches, Z. (Ed.). (1971). *Price indexes and quality change: Studies in new methods of measurement.* Cambridge MA: Harvard University Press.

International Monetary Fund, International Labour Organisation, Organisation for Economic Co-operation and Development, United Nations Commission for Europe, The World Bank. (2004). *Producer price index manual, theory and practice.* Washington, DC: International Monetary Fund.

Jorgenson, D. W., & Fraumeni, B. (1989). The accumulation of human and nonhuman capital, 1948–1984. In R. E. Lipsey & H. S. Tice (Eds.). The measurement of saving, investment and wealth; national bureau of economic research studies in income and wealth (Vol. 52) pp. 227–282. Chicago: University of Chicago Press.

Jorgenson, D. W. & Fraumeni, B. (1992). Investment in education and U.S. economic growth. *Scandinavian Journal of Economics, 94*(Suppl), 51–70.

Lawrence, D. A., & Diewert, W. E. (2006). Regulating electricity networks: The ABC of setting X in New Zealand. In T. Coelli & D. Lawrence (Eds.), *Performance Measurement and Regulation of Network Utilities* (pp. 207–241). Cheltenham, UK: Edward Elgar.

Lee, P. (2008). *Public service productivity: Health care; UK office for national statistics.* http://www.statistics.gov.uk/cci/article.asp?ID=1922.

Rosen, S. (1974). Hedonic prices and implicit markets: Product differentiation in pure competition. *Journal of Political Economy, 82*(1), 34–55.

Schreyer, P. (2009). Output and outcome in health and education. Unpublished manuscript.

Shephard, R. W. (1970). *Theory of cost and production functions*, Princeton: Princeton University Press.

Stone, R. (1956). *Quantity and price indexes in national accounts.* Paris: Organisation for European Economic Cooperation.

Triplett, J. E. (1998). What's different about health? Human repair and car repair in national accounts and in national health accounts. In D. Cutler & E. Berndt (Eds.), *Medical Care Output and Productivity; NBER Studies in Income and Wealth* (Vol. 62). Chicago: The University of Chicago Press.

Triplett, J. E. (2006). *Handbook on hedonic indexes and quality adjustments in price indexes: Special application to information technology products.* Paris: OECD.

Triplett, J. E., & Bosworth, B. P. (2004). *Productivity in the U.S. services sector: New sources of economic growth.* Washington, DC: The Brookings Institution.

Vanoli, A. (2002). *Une Histoire de la Comptabilité Nationale.* Paris: La Découverte.

Part IV
Price Indexes in National Accounts

Total Factor Productivity Surpluses and Purchasing Power Transfers: An Application to the Italian Economy

Giorgio Garau, Patrizio Lecca, and Lucia Schirru

1 Introduction

At the present, the national statistic institutes prefer to pursue the manageable double deflation approach producing as a result an equilibrating system of accounts at constant prices. The main intention of this paper is, instead, to provide a comprehensive framework to support the idea according to which a constant price system of account is in nature unbalanced.

Whilst double deflation is consistent with the balancing rule, some objections arise if we consider that relative price change might reflect change in productivity. Indeed, when the purpose is a system of accounts at constant prices the application of the usual methods might hide some important processes that the analysis of constant prices figures would allow, as e.g. the effect of technical progress. In the main body of this paper, we attempt to justify our thesis using several contributions that implicitly or explicitly sustain this view. In doing so, we also pay attention to the complementarities between constant prices measures and productivity transfers among agents. As far as we know, the economic literature has often treated separately the technical change generation from the distribution of productivity gains among economic agents. We begin with the seminal Fontela's (1989) work that set up the distributional rule of productivity gains in the input output context, ending up with an extension that identify a measure of surplus called Purchasing Power Transfer (PPT) originally developed by Garau (1996 and 2002). This measure is given by the productivity gains and the market surplus generated by extra-profit conditions derived from rental position detained by agents. Such a decomposition is very useful from our point of view since it would provide information about the degree of non-competitiveness in different markets.

The rest of the paper is organized as follows. In the next section we deal with some basic concepts of constant price measures in the National Accounts and in Sect. 3, we show the drawbacks of using the double deflation method. In Sect. 4 we

G. Garau (✉)
DEIR, University of Sassari, Sassari, Italy
e-mail: giorgio@uniss.it

L. Biggeri, G. Ferrari (eds.), *Price Indexes in Time and Space*, Contributions to Statistics, DOI 10.1007/978-3-7908-2140-6_10, © Springer-Verlag Berlin Heidelberg 2010

compute and explain Fontela's model of Total Factor Productivity Surplus (TFPS) with an application to the case of Italy, while in Sect. 5, the theoretical model of PPT comes with the explanation of the results we have obtained. Finally, in Sect. 6 remarks and conclusions are drawn.

2 Basic Concepts of Constant Price Measures in the National Account

The Handbook on Price and Volume Measures in National Account (Eurostat, 2001) defines a constant price system of accounts as an economic situation for a particular year, expressed in the price of another year. A system of account, either in constant or current prices, should respect the accounting constraint, that is to say, total supply and total use should be equal for each product and for each industry. Such an accounting constraint is considered by the Handbook as an important advantage because it allows comparison and consistency among different estimates.

Three methods are available to produce a system of account at constant prices: the *revaluation method*, the *quantity extrapolation method*, and the *deflation method*. Whilst the first two are based on quantity indexes, the deflation method is based instead on price indexes.

The *revaluation method* consists in the direct collection of physical quantities and values using base year prices. This approach, albeit very powerful and meaningful, is very demanding because it requires direct observation of physical quantities. Actually, it is especially adopted for agricultural products and for goods produced for own final use. In the *quantity extrapolation method* a quantity indicator obtained by a quantity index is used to update the base year value. The *deflation approach* is recommended by the SNA93 and ESA95 to develop constant-price transactions for a homogenous good or service. This implies the division of each period's current value by an appropriate price index. This approach, unlike the others, is very straightforward and furthermore it allows for an easier adjustment to account for quality changes. Indeed, the two methods based on quantity indexes might improperly allocate quality changes to changes in price and as it has been pointed out in ESA95 the changes in characteristics have to be recorded as changes in quantities and not as changes in prices.

The SNA93 requires the use of two separated approaches to produce constant price measures for the value of final goods and services: the *expenditure approach* and the *output* or *double deflation* approach. According to the first one, for each component of the final demand, constant price measures are separately produced. The second one, instead, consists in determining constant price measures separately for gross output and intermediate inputs for each industry in the economy. So the GDP can be seen from the expenditure side as the aggregated constant price measures of the final demand, whilst from the output side it can be seen as the difference between gross output and intermediate inputs[1].

Although the SNA93 recommend deriving constant price estimates from both sides (output and expenditure) official statistics usually report only the result obtained from the production side. This is because the GDP calculated from the output side is often different from the one calculated from the expenditure side.

The SNA93 advice the use of alternative methods when the data required for constant price estimates of both output and intermediate consumption are not available. Such methods might be single deflation or single extrapolation.

Single deflation means to deflate the value added at current prices directly using the producer price index which measures the price-changes of the output. The extrapolation method is based on indicators that reflect the movements in the volume changes of the industries inputs. Such inputs might be employment, labour-inputs, intermediate consumption or total inputs (weighted average of labour and intermediate inputs). Basically, the value added of the previous period is multiplied by an extrapolator indicator. This technique is mainly used to deflate value added in the financial sector and in the non-tradable sector.

An alternative method to the ones proposed by the SNA93 has been formalized in Durand (1994). Starting from the property that the real value-added of commodities delivered to final demand must be equal to the sum of the real values added by industries, he produces a value added matrix, industries x commodities, whose sum in row gives the value added of the final demand while the sum in column gives the value added by industries. This allows for consistency between the value added seen from the side of expenditure (final demand) and the value added seen from the side of output (or production). Firstly, Durand obtains the value added matrix in current prices and then by deflating each column of the matrix by the corresponding commodity price he obtains the value added matrix at constant prices. So, each cell of the value-added matrix represents the contribution of a specific industry to the real value-added of a given commodity.

In recent literature, a quite interesting method to derive constant price measures has appeared. This method developed by Rampa (2008) begins with the assumption that the constant price estimates reported in the official statistics, such as the chained real value added time series and price and quantity indexes are not very accurate. Quantity and price indexes are not consistent with each other for certain periods or sectors and the real value added series is not as smooth as expected. Thus, Rampa (2008) proposes a Subjective Weighted Least Square estimation (SWLS) for the deflation of a yearly series of current price IO table for Italy.

---

[1] In addition to the expenditure and output approach the GDP at constant prices can also be obtained from an income side approach. However, when we move into a constant prices context only output and expenditure measures can be used since the income measure of GDP would require a direct observation of its components, both labour income and operative surplus. While the former is, to some extent, directly observable, the latter is usually determined as a residual.

3 Double Deflation and Productivity Analysis

The double deflation method albeit widely used has been strongly criticized in the economic literature as it provides a measure of the net output of industries only under extremely restrictive assumptions. This method is in fact feasible only for constant price estimates which are additive, such as those calculated using a fixed-base price index. Furthermore, the use of the double deflation could hide some important processes behind economic growth such as technical progress.

Although the double deflation allows one to produce an equilibrating system of accounts even at constant prices, we argue that such a constraint might produce some important disadvantages. While the use of such a method can be accepted in current prices, some doubts arise as to whether a constant price measure has to be constructed. Essentially, we disagree on the requirement to produce an equilibrating system of account when we move into a constant prices framework since we believe that we might lose some important effects concerning economic growth such as efficiency, rent spillovers and all those elements that may concern disembodied technical change.

A single deflation procedure, instead, would allow one to determine a measure of productivity gains (Flexner, 1959; Fontela, 1989; Babeau, 1978; Garau, 1996) and to understand the generating process of economic growth.

Furthermore, the use of a constant price method not only gives us the opportunity to obtain information about the internal generation process of productivity but also the external determinant of growth that are behind the effect of change in the terms of trade, if, of course, proper index prices are used to deflate imports and exports. This would yield quite interesting results since the literature on economic growth has now recognized the role of knowledge spillovers as the most important driving forces behind economic growth. As knowledge is incorporated in commodities, trade with high technological countries means high quality and sophisticated inputs (either intermediate or capital goods), that improve efficiency and, in turn, the competition among regions. Such a potential finding has been identified in Flexner's (1959) original paper and estimates of external rent spillovers may be found for the Swiss economy in Antille and Fontela, (2003).

Let us suppose to have for a given period t: $\mathbf{X}$ and $\overline{\mathbf{X}}$, $\mathbf{l}$ and $\bar{\mathbf{l}}$, $\mathbf{k}$ and $\overline{\mathbf{k}}$, $\mathbf{m}$ and $\overline{\mathbf{m}}$, $\mathbf{f}$ and $\bar{\mathbf{f}}$, $\mathbf{e}$ and $\bar{\mathbf{e}}$; the matrix of intermediates flows, a vector of labour income, the capital return, the import flows, the final demand and export demand respectively in current and constant prices. According to the accounting constraint both the following equations must hold:

$$\mathbf{X}'\iota + \mathbf{l} + \mathbf{k} + \mathbf{m} = \mathbf{X}\iota + \mathbf{f} + \mathbf{e} \tag{1}$$

$$\overline{\mathbf{X}}'\iota + \bar{\mathbf{l}} + \overline{\mathbf{k}} + \overline{\mathbf{m}} = \overline{\mathbf{X}}\iota + \bar{\mathbf{f}} + \bar{\mathbf{e}} \tag{2}$$

where ι is a unit vector. Now, as we cannot observe $\overline{\mathbf{k}}$, the value added $(\bar{\mathbf{l}} + \overline{\mathbf{k}})$ must be of course obtained as a residual. However if we were able to deflate every single item of Eq. (1) including $\mathbf{k}$ or at least to find a proper deflator for the value added as a whole, it would be quite plausible that the equilibrating relationship represented in Eq. (2) does not hold.

As pointed out by Flexner (1959) even though we were able to remove all the statistical discrepancies due to calculation and statistical approximations, Eq. (2) would be inadequate to represent constant price relationship since whenever productivity change arises between base year values and current values, this must be reflected in a balancing item in Eq. (2).

Accordingly, we may argue that a well defined system of account may provide a measure of productivity resulting from the difference between the amount of goods produced and the amount of inputs of production used. Such a measure will take positive value only if the quantity variation of the output is greater than the variation of all inputs. Therefore, the relationship in Eq. (2) must not hold and the balancing term has a precise economic meaning, which is called by Fontela (1989), Total Factor Productivity Surplus (TFPS):

$$\left[\overline{\mathbf{X}}\iota + \bar{\mathbf{l}} + \bar{\mathbf{k}} + \overline{\mathbf{m}}\right] + \mathbf{TFPS} = \overline{\mathbf{X}} + \bar{\mathbf{f}} + \bar{\mathbf{e}} \qquad (3)$$

4 The Transfer of Productivity Gains

Fontela (1989) calls the differences between output and inputs, both measured at constant prices, TFPS:

$$\mathrm{TFPS}_{i,t} = \sum_j p_{i,j,0} \cdot q_{i,j,t} - \sum_j p_{j,i,0} \cdot q_{j,i,t} \qquad (4)$$

where $TFPS_{i,t}$ corresponds to the amount of real resource flows between time t and time 0, $q_{i,j,t}$ is the flow of output of sector i towards sector j and $p_{i,j,0}$ is the market price to its base year value. Since $\sum_j p_{i,j,t} \cdot q_{i,j,t} = \sum_j p_{j,i,t} \cdot q_{j,i,t}$, the expression (4) can be re-written in terms of price variations as follows (Fontela, 1989):

$$\mathrm{TFPS}_{i,t} = -\sum_j q_{i,j,t} \cdot (p_{i,j,t} - p_{i,j,0}) + \sum_j q_{j,i,t} \cdot (p_{j,i,t} - p_{j,i,0}) \qquad (5)$$

Whereas Eq. (4) measures the creation of TFPS using the index number approach, Eq. (5) can be interpreted as a distributional rule of TFPS. As it is self-evident such a distribution depends on the price variations of outputs (first element on the right hand-side) and inputs (second element on the right hand-side) and can be transposed into the traditional IO context if the entire accounting system in current and constant price is available. For a given period, t, the following definition of TFPS can be considered:

$$\mathbf{TFPS} = \left(\mathbf{S}'\iota + \mathbf{s_k} + \mathbf{s_l} + \mathbf{s_m}\right) - \left(\mathbf{S}\iota + \mathbf{s_f} + \mathbf{s_e}\right)$$

where ι is a unit vector, $\mathbf{S}\left[s_{i,j}\right] = \mathbf{X} - \overline{\mathbf{X}}$, $\mathbf{s_k}\left[sk_i\right] = \mathbf{k} - \overline{\mathbf{k}}$, $\mathbf{s_l}\left[sl_i\right] = \mathbf{l} - \bar{\mathbf{l}}$, $\mathbf{s_m}\left[sm_i\right] = \mathbf{m} - \overline{\mathbf{m}}$, $\mathbf{s_f}\left[sf_i\right] = \mathbf{f} - \bar{\mathbf{f}}$ and $\mathbf{s_e}\left[se_i\right] = \mathbf{e} - \bar{\mathbf{e}}$.

By considering a given year *t*:

- $s_{i,j} > 0$, it means that industry *j* is transferring surplus to the industry *i*,; and the reverse applies when $s_{i,j} < 0$, that is to say, industry *j* is paying relatively less for the inputs provided by industry *i*. Particularly interesting is the net industry contribution: $s_{n,i} = \sum_j s_{j,i} - \sum_j s_{i,j}$. When $s_{n,i} > 0$, industry *i* is transferring surplus to the rest of the economy more than it is gaining from all other sectors.
- Industry *i* is transferring surplus to its primary inputs when, sl_i and sk_i are positive.
- When the price of some commodity falls, industries transfer additional surplus to consumers making $sf_i < 0$
- From the trade side, we have an inflow of productivity gains from the Rest of the World when $se_i > 0$ and $sm_i < 0$. And, the reverse applies when $se_i < 0$ and $sm_i > 0$. Then we can compute, as in Antille and Fontela (2003), the net outflow $sm_i - se_i > 0$ or net inflow in the opposite situation, $sm_i - se_i < 0$.

4.1 The Distribution Process of the TFP Surpluses: The Case of Italy for the Period 1995–2002

The analysis of the TFPS is carried out for the period 1995–2002 for the Italian economy. The index prices are shown in Table 1. Index prices for consumption, production, imports and exports are supplied by ISTAT (2008a) whilst index price for labour, capital and investment are deflators obtained from the Italian System of National Accounts (ISTAT, 2008b). The symmetric Italian Input-Output, is obtained from the Make and Use Table for the year 2002, published by ISTAT (2008c).

The results are shown in Table 2. From the last column of this table, we see that the total amount of TFPS is negative, meaning that in the period 1995–2002, the Italian economy is not able to generate TFP gains and also to create an available surplus to be transferred through a reduction in sales prices to the consumers and investors.

The figures also show that the rate of return of one factor of production, paradoxically increases albeit inefficiency in production (negative innovation gains). Indeed, the rate of return to capital raises whilst the real wage rate fall, reflecting that production activities are transferring TFPS to capital and absorbing TFPS from labour. These results, we believe, are the consequences of the national labour market reform undertaken in 1993 (Income Policy Agreement) which has generated a high labour-capital conflict ending up giving advantage to the capital side and increased labour

Table 1 Price index 2002 base 100 = 1995

	Production	Consumption	Export	Import	Investment	Capital	Labour
Agriculture, forestry and fishing	109.60	109.60	119.67	88.62	1.18	1.06	1.13
Mining and Quarrying	111.90	111.90	104.93	169.94	0.96	1.34	1.27
Manufacture of food products, beverages and tobacco	109.30	117.28	108.02	97.28	1.05	1.20	1.18
Manufacture of textiles and wearing appareal	109.60	120.80	115.80	113.00	0.96	1.07	1.25
Manufacture of leather and related products	116.40	126.80	141.34	118.43	1.43	1.09	1.29
Manufacture of wood and wood products	104.10	104.10	95.64	95.87	1.32	1.03	1.36
Manufacture of paper and paper products	95.30	95.30	93.48	75.99	1.26	1.38	1.22
Manufacture of printing and reproduction of recorded media	118.20	118.20	93.48	75.99	1.26	1.38	1.16
Manufacture of coke and refined petroleum products	117.00	117.00	198.75	166.24	1.13	1.30	1.27
Manufacture of chemicals and pharmaceutical	107.60	107.60	90.47	93.58	1.20	1.10	1.27
Manufacture of rubber and plastic products	105.60	105.60	96.51	95.44	1.17	1.18	1.25
Manufacture of other non-metallic mineral products	119.80	119.80	107.07	106.35	1.45	1.19	1.19
Manufacture of fabricated metal products, except machinery and equipment	103.40	103.40	92.14	88.47	1.42	1.13	1.22
Manufacture of machinery and equipment	113.30	113.30	116.14	120.07	0.99	1.13	1.25
Manufacture of electrical equipment	103.30	103.30	106.98	106.54	1.14	1.28	1.21
Manufacture of transport equipment	113.00	113.00	111.69	117.61	0.84	1.09	1.20
Other manufacturing	115.30	115.30	113.45	111.48	1.16	1.10	1.32
Electricity, Gas and water supply	126.10	126.10	126.10	82.72	1.06	1.16	0.99
Construction	116.48	116.48	116.48	97.28	2.15	1.26	1.17
Wholesale and Retail trade; Repair of Motor vehicles and motorcycles	118.80	118.80	118.80	97.28	1.52	1.30	1.38
Accommodation and food service activities	126.80	126.80	126.80	97.28	1.64	1.35	1.38
Transportation and storage	113.30	113.30	113.30	97.28	1.29	1.33	1.16
Financial and Insurance activities	158.94	158.94	158.94	97.28	1.06	1.07	1.09
Real estate activity	118.80	118.80	118.80	97.28	1.12	1.14	0.90
Scientific research and development	118.80	118.80	118.80	97.28	1.12	1.14	1.41
Legal, accounting, management and other professional activities	118.80	118.80	118.80	97.28	1.12	1.14	1.17
Public administration and defence; Compulsory social security	118.80	118.80	118.80	97.28	1.36	1.13	1.03
Education	119.50	119.50	119.50	97.28	1.42	1.28	1.00
Human health services	121.30	121.30	121.30	97.28	1.66	1.22	1.10
Other service activities	115.34	115.34	115.34	97.28	1.91	1.33	1.20

Table 2 The distribution of TFP surpluses in Italy, 1995–2002; values in millions of Euros

	Agricolture, forestry and fishing	Mining and Quarrying	Manufacture of food products, beverages and tabacco	Manufacture of textiles and wearing appareal	Manufacture of leather and related products	Manufacture of wood and wood products	Manufacture of paper and paper products	Manufacture of printing and reproduction of recorded media	Manufacture of coke and refined petroelum products	Manufacture of chemicals and pharmaceutical	Manufacture of rubber and plastic products	Manufacture of other non-metallic mineral products	Manufacture of fabricated metal products, except machinery and equipment	Manufacture of machinery and equipment	Manufacture of eletrical equipment	Manufacture of transport equipment	Other manufacturing	Electricity, Gas and water supply	Construction	Wholesale and Retail trade; Repair of Motor vehicles and motorcycles	Accomodation and food sevice activities	Transporation and storage	Financial and Insurance activities	Real estate activity	Scientific research and development	Legal, accounting, managment and other professional activities	Public administration and defence; Compulsory social security	Education	Human healt services	Other service activities	Total
Total TFPS[a]	-2618	10512	-1703	-696	-3070	983	3020	-374	-168	3310	3355	-295	6620	5780	5482	10393	-673	-6382	49380	4317	-3195	9183	25502	-7011	1554	-4407	-9178	-8950	-5406	1987	-62513
Lowering(-)/increasingr(+) the cost of primary inputs {b= [b1+b2+b3]}	-2908	479	271	-494	-69	-79	667	758	237	83	309	132	172	498	1197	22	109	-1428	2018	14232	5658	4805	-4757	-5703	2276	-1921	-8457	-8290	-2342	1364	-1205
Labour [b1]	-388	108	-9	724	331	357	91	-88	78	628	325	51	720	1015	374	174	640	-1187	-127	6614	2830	-611	-2473	-693	2656	-329	-7649	-8994	-3041	458	-7414
Capital [b2]	-2523	371	282	-1223	-402	-437	574	844	148	-553	-18	74	556	-524	818	198	-533	-263	2134	7590	2825	5384	-2313	-5019	-383	-1609	-844	696	671	894	5910
Net Taxes [b3]	3	1	-2	5	1	1	2	1	10	8	2	7	9	6	4	3	2	22	12	28	4	32	29	9	3	16	36	7	29	11	299
		-0.77																													
Positive/Negative spillovers from Import [c]	3031	8084	3989	624	-23	783	3194	493	-1554	9381	1324	329	8218	-364	3801	136	266	820	80	1451	861	2489	947	1228	573	1815	0	22	0	306	36138
Total available for distribution {a+[-b]+c}	3321	1949	2016	422	-3024	1846	5547	-639	-1959	12608	4369	-98	14665	4919	8086	10550	-517	-4134	51319	-8464	-7992	6867	19798	-79	-149	-671	-721	-637	-3063	929	-25171
Distribution Process																															
Lowering (+) or increasing (-) prices for intermediate inputs [d]	2547	1798	540	497	-383	1470	3769	-1025	-844	2899	1919	-910	9832	-3005	3258	1324	-1542	-3024	-1682	-2817	-3118	4276	-11909	287	-606	-777	-53	142	-677	461	0
Lowering(+) or increasing(-) prices for consumers [e]	834	0	331	-700	-603	131	688	-12	117	1769	304	-29	180	186	1367	889	270	-1093	85	-1044	-4873	2344	-6945	-691	-51	-52	-668	-778	-2386	812	-9620
Lowering (+) or increasing (-) prices for investors [f]	-1	90	-124	142	-2	-68	-1	-16	18	-3	5	-78	-1262	6904	1037	9415	131	0	-49726	-4454	0	-291	0	346	518	215	0	0	0	-364	-37568
Lowering (+) or increasing(-) the price of export [g]	-59	59	1269	483	-2036	313	1091	414	-1250	7943	2141	919	5914	835	2424	1570	624	-17	5	-149	-1	539	-944	-21	-11	-57	0	0	0	19	22017
Total distribution {d+e+f+g}	3321	1949	2016	422	-3024	1846	5547	-639	-1959	12608	4369	-98	14665	4919	8086	10550	-517	-4134	-51319	-8464	-7992	6867	-19798	-79	-149	-671	-721	-637	-3063	929	-25171
Net outflow(-)/inflow(+) {c-g}	3090	-8144	2720	141	2013	470	2103	79	-304	1438	-817	-590	2303	-1199	1378	1434	-358	837	75	1599	862	1951	1890	1250	584	1872	0	22	1	287	14121

market flexibility[2] leading to the reduction of the bargaining power of workers and the purchasing power of wages.

As far as the net foreign flows are concerned, overall, during the period in analysis Italy experienced positive terms of trade effects. The net foreign inflow (14121) reflects the high capacity of the Italian economy to gains innovation spillovers from the Rest of the World (ROW). However, the capacity of the Italian economy to gain from terms of trade improvement is not able to offset the negative domestic productivity performance. This gives us a picture of an economic system where consumers (Government and Households) and investors have to give up part (or all) of their TFPS to pay a high price for consumption and capital goods, respectively.

From a sectoral investigation we understand that *Mining and Quarrying* and *Manufacture of Transport Equipment* are those sectors able to generate the highest TFP gains. However, these sectors act a different distribution process of the TFPS.

Mining and Quarrying is able to distribute just a minimum part of its TFP gains since in this sector the rate of return to capital and the real wages increase, and its position with the rest of the world is weak. Indeed, we see that about 77% of the TFPS is absorbed by the rest of the world through an increase in the cost of imports (144, is the price index). Furthermore, the overall position with respect to the ROW is negative. Substantially, under the period in analysis, this sector has suffered of negative terms of trade with the results of a net outflow of TFP surpluses (8084).

With regards to *Manufacture of Transport Equipment*, the total surplus available for distribution (10550) is greater than the innovation gains generated (10393). This is happening because *Manufacture of Transport Equipment*, is not only able to generate TFP surpluses but, at the same time, is also able to lower its costs through a reduction in the overall cost of primary inputs (22) and take advantage of positive spillovers from imports. On the other hand, the distribution process gives advantages to more investors (9415) and the ROW (1570) by a fall in the price of capital goods and exports, respectively.

The worst performance in term of innovation gains is coming from *Construction* and *Financial and Insurance Activities.*In the former sector, not only the TFPS is negative (49380) but also there is a net transfer of purchasing power toward primary inputs which is going to advantage only capital. Albeit the net benefit from the ROW is positive, this is not able to cover the rise in the price of value added and the negative TFPS, meaning that *Construction* is a sector that has a strong rental position in the market, absorbing as a consequence TFP surpluses from the final demand with important drawbacks especially for investors. In the latter sector,

[2]The work of Devicienti, Maida, and Pacelli (2007) shows that, after the national labour market reform, wages became more flexible since they are now more responsive to local unemployment. Before the reform wages were set within a centralized bargaining with automatic indexation of wages to the real inflation. The reform has, instead, introduced a new bargaining system. The centralized bargaining process still remains in order to set the industry wide national wage, but with indexation to the Government's target inflation (which is always lower than the real inflation). The additional wage distributed to the workers (or the top up component) is now set according to the firm and regional conditions.

instead, the negative TFPS (25502) is partially offset by a reduction in the cost of primary inputs and positive external spillover from imports. Nevertheless, the total surplus available for distribution still remain negative (19798) with the consequences that consumers and investors have to pay a high price for consumption and capital goods, respectively.

If we just consider the net outflow or inflow of TFP surpluses, the best performance is experienced by *Agricultural, Forestry and Fishing*, for which we have a total net inflow of TFP gains (3090). Indeed, the related price index for imports is equal to 75.10 while the price index for export is 101.41 with an improvement of the terms of trade. On the contrary, the worst external performance is for *Mining and Quarrying* that experience a deterioration of the terms of trade (import price index is equal to 144.08 while the export price index is 88.92).

5 The Purchasing Power Transfer (PPT)

In Eq. (5), the computation of the TFPS is obtained through market prices. We may decompose the market price $p_{i,j}$, as follows:

$$p_{i,j,0} = p^{*}_{i,j,0} + p^{**}_{i,j,0}$$

where $p^{*}_{i,j,0}$, is the price we would have if the agents present in the market were not able to gain from rental positions. It means a situation where the prices of all sectors adjust to their productivity and the rate of productivity growth is the same in all sectors. That is to say, a situation in which extra profits are zero and there is no modification of relative prices. $p^{**}_{i,j,0}$, is instead an index of market bias that identifies the presence of extra-profits and those agents that are in the position to gain from market imperfections.

Substituting the definition of market price in Eq. (5), we have that:

$$TFPS_{i,t} = -\sum_{j} q_{i,j,t}(p_{i,j,t} - p^{*}_{i,j,0} - p^{**}_{i,j,0}) + \sum_{j} q_{i,j,t}(p_{j,i,t} - p^{*}_{j,i,0} - p^{**}_{j,i,0}) \quad (6)$$

Furthermore, with some simple adjustment, the Eq. (6) can be re-written ascribing to the TFPS and to the following decomposition the meaning of a measure of Purchasing Power Transfer (PPT):

$$\begin{aligned} \mathrm{PPT}_{i,t} = & \underbrace{-\sum_{j} q_{i,j,t} \cdot (p_{i,j,t} - p^{**}_{i,j,0})}_{1} + \underbrace{\sum_{j} q_{i,j,t} \cdot p^{*}_{i,j,0}}_{2} + \\ & + \underbrace{\sum_{j} q_{j,i,t}(p_{j,i,t} - p^{**}_{j,i,0})}_{3} - \underbrace{\sum_{j} q_{j,i,t} \cdot p^{*}_{j,i,0}}_{4} \end{aligned} \quad (7)$$

The market price indexes, $p_{i,j}$, allow us, through Eq. (5), to compute the overall purchasing power transfers (PPT). With the decomposition process, in Eq. (7), we can distinguish the technological performance of each sector (TFP), that is to say, the difference between the terms 4 and 2 on the right side of Eq. (7), and a measure of the alteration of the natural market mechanisms, given by difference between terms 3 and 1 on the right side of Eq. (7).

With regards to the TFP component, the difference between real outputs and real inputs should be computed using ideal price indexes or prices consistent with the neoclassical framework (Wolf, 1985, 1989). This measure reflects the welfare gains of innovations that allows sector i, to increase its outputs faster than its inputs between time *0* and t, if $\left[\sum_j q_{i,j,t} \cdot p^*_{i,j,0} - \sum_j q_{j,i,t} \cdot p^*_{j,i,0}\right] > 0$.

The market component of the PPT decomposition is given by (1) and (3):

$$\text{Market Surplus} = \underbrace{-\sum_j q_{i,j,t}(p_{i,j,t} - p^{**}_{i,j,0})}_{1} + \underbrace{\sum_j q_{j,i,t}(p_{j,i,t} - p^{**}_{j,i,0})}_{3}$$

The difference between these two terms can be interpreted as redistribution among the different economic agents of the market power generated through changes of the prices of inputs and outputs. If $p_{i,j,t} < p^{**}_{i,j,0}$ then the industry is losing bargaining power and it is transferring part of its purchasing power to its customers (intermediate producers or final users) by supplying its products at a lower relative price. Accordingly, the relative price of output of an industry decreases and the term (1) becomes positive. If $p_{j,i,t} > p^{**}_{j,i,0}$, the industry is transferring part of its purchasing power to its suppliers since is paying relatively more for its intermediate and primary inputs. This means that the term (3) becomes negative.

With regards to the computation of the measure of TFP in Eq. (7), we adopt a discrete approximation of the Divisia index given by the Törnqvist chain index[3]:

$$p^*_j = \prod_{i=1}^{n+2} \left(\frac{r^1_{i,j}}{r^0_{i,j}}\right)^{\frac{1}{2}\left(w^0_{i,j} + w^1_{i,j}\right)}$$

where r_i are the inputs prices, (n intermediate inputs and 2 primary inputs) and w^0_i (w^1_i) are input shares, at constant prices, per unit of output for the two periods (the first, 1 and the last, 0). If current price values are deflated by their Törnqvist price index, the relative transfers reflect the effects of technical change. Indeed, in the neoclassical perfect competition context, it is recommended for consistency to use Törnqvist-Divisia indexes for the measurement of TFP (Wolff, 1989; Fontela, 1994).

---

[3]The use of Tornqvist's index for productivity measures has been strongly recommended by Wolf (1985, 1989) and Fontela (1994). For the limitations of the Törnqvist Divisia index see Martini (1992).

In order to compute the total measure of PPT, we should also calculate the Market Surplus (MS) through the distortionary price $p^{**}_{j,i,0}$. However, we do not have such price index. Then, in order to compute the MS component of Eq. (7), we can use the definition of the market price seen above, $p_{i,j,0} - p^{*}_{i,j,0} = p^{**}_{i,j,0}$. This also means that MS can be easily obtained as residual.

According to Eq. (7), if $PPT_{i,t} > 0$, the sector i is transferring purchasing power to the rest of the economy, through a reduction in the market power and an increase in TFP. While if $PPT_{i,t} < 0$ the sector i is absorbing resources from the rest of the economy. We may distinguish different situations:

$$
\mathrm{PPT}_{i,t} > 0, \begin{cases}
\left[-\sum_j q_{i,j,t}\left(p_{i,j,t} - p^{**}_{i,j,0}\right) + \sum_j q_{j,i,t}\left(p_{j,i,t} - p^{**}_{j,i,0}\right)\right] > 0; \quad \left[\sum_j q_{i,j,t}\cdot p^{*}_{i,j,0} + \sum_j q_{j,i,t}\cdot p^{*}_{j,i,0}\right] > 0 & \langle 1\rangle \\
\left[-\sum_j q_{i,j,t}\left(p_{i,j,t} - p^{**}_{i,j,0}\right) + \sum_j q_{j,i,t}\left(p_{j,i,t} - p^{**}_{j,i,0}\right)\right] > 0; \quad \left[\sum_j q_{i,j,t}\cdot p^{*}_{i,j,0} + \sum_j q_{j,i,t}\cdot p^{*}_{j,i,0}\right] < 0 & \langle 2\rangle \\
\left[-\sum_j q_{i,j,t}\left(p_{i,j,t} - p^{**}_{i,j,0}\right) + \sum_j q_{j,i,t}\left(p_{j,i,t} - p^{**}_{j,i,0}\right)\right] < 0; \quad \left[\sum_j q_{i,j,t}\cdot p^{*}_{i,j,0} + \sum_j q_{j,i,t}\cdot p^{*}_{j,i,0}\right] > 0 & \langle 3\rangle
\end{cases}
$$

In $\langle 1\rangle$ sector i is distributing purchasing power through an increase in efficiency (TFP > 0) and through a reduction in their market power, MS > 0. Basically, the market is imposing to sell their goods to a lower relative price. For case $\langle 2\rangle$, the PPT is positive although the generic sector i is experiencing loss of efficiency (TFP<0). Indeed, the negative productivity impact is totally offset by an increase in the transfer of purchasing power to the rest of the economy through a loss of market power. In $\langle 3\rangle$, the capacity to transfer purchasing power of sector i is partially offset by negative market imperfections meaning that the market conditions allow this sector to absorb resources from the rest of the economy. However, the net transfer of purchasing power is positive.

For the case of PPT<0, we can have:

$$
\mathrm{PPT}_{i,t} < 0, \begin{cases}
\left[-\sum_j q_{i,j,t}\left(p_{i,j,t} - p^{*}_{i,j,0}\right) + \sum_j q_{j,i,t}\left(p_{j,i,t} - p^{*}_{j,i,0}\right)\right] < 0; \quad \left[\sum_j q_{i,j,t}\cdot p^{**}_{i,j,0} + \sum_j q_{j,i,t}\cdot p^{**}_{j,i,0}\right] < 0 & \langle 1'\rangle \\
\left[-\sum_j q_{i,j,t}\left(p_{i,j,t} - p^{*}_{i,j,0}\right) + \sum_j q_{j,i,t}\left(p_{j,i,t} - p^{*}_{j,i,0}\right)\right] > 0; \quad \left[\sum_j q_{i,j,t}\cdot p^{**}_{i,j,0} + \sum_j q_{j,i,t}\cdot p^{**}_{j,i,0}\right] < 0 & \langle 2'\rangle \\
\left[-\sum_j q_{i,j,t}\left(p_{i,j,t} - p^{*}_{i,j,0}\right) + \sum_j q_{j,i,t}\left(p_{j,i,t} - p^{*}_{j,i,0}\right)\right] < 0; \quad \left[\sum_j q_{i,j,t}\cdot p^{**}_{i,j,0} + \sum_j q_{j,i,t}\cdot p^{**}_{j,i,0}\right] > 0 & \langle 3'\rangle
\end{cases}
$$

Considering the situation $\langle 1'\rangle$, the sector absorbs purchasing power from the rest of the economy because of negative productivity and favourable market distortion. For $\langle 2'\rangle$, the reduction of the market power is not enough to cover the negative impact in term of productivity. In the last situation $\langle 3'\rangle$, the sector i, overall, absorb resources from the other sectors. Here, essentially, the appropriations of purchasing power through the exploitation of their rental position overwhelm the capacity to generate TFP.

5.1 TFP and Market Surpluses in Italy, for the Period 1995–2002

The results of the operations are presented in Table 3. The first column is the difference between terms (2) and (4) of Eq. (7), whilst the second one is the difference between the terms (1) and (3) of the same equation. The last column is the total effect.

With regards to the manufacturing sectors, the best performance in terms of TFP is are in the *Manufacture of Fabricated Metal Products* and in the *Manufacture of Chemicals and Pharmaceutical. Manufacture of Fabricated Metal Products*, has

Table 3 Total factor productivity and market surpluses in Italy, 1995–2002, in millions of Euros

	TFP	MS	PPT=TFPS
Agricolture, forestry and fishing	4182	−6800	−2618
Mining and quarrying	−11495	22007	10512
Manufacture of food products, beverages and tabacco	547	−2249	−1703
Manufacture of textiles and wearing appareal	−153	−543	−696
Manufacture of leather and related products	−665	−2405	−3070
Manufacture of wood and wood products	717	266	983
Manufacture of paper and paper products	2239	781	3020
Manufacture of printing and reproduction of recorded media	−135	−239	−374
Manufacture of coke and refined petroelum products	2776	−2944	−168
Manufacture of chemicals and pharmaceutical	4727	−1417	3310
Manufacture of rubber and plastic products	432	2922	3355
Manufacture of other non-metallic mineral products	287	−582	−295
Manufacture of fabricated metal products, except machinery and equipment	5933	687	6620
Manufacture of machinery and equipment	−2468	8249	5780
Manufacture of eletrical equipment	−507	5989	5482
Manufacture of transport equipment	−3827	14220	10393
Other manufacturing	22	−695	−673
Electricity, gas and water supply	3699	−10081	−6382
Construction	−2425	−46955	−49380
Wholesale and retail trade; repair of motor vehicles and motorcycles	−16472	20789	4317
Accomodation and food sevice activities	−6010	2815	−3195
Transporation and storage	−4399	13581	9183
Financial and insurance activities	−1034	−24467	−25502
Real estate activity	2808	−9819	−7011
Scientific research and development	−2402	3955	1554
Legal, accounting, managment and other professional activities	658	−5066	−4407
Public administration and defence; compulsory social security	7172	−16350	−9178
Education	6858	−15808	−8950
Human healt services	905	−6311	−5406
Other service activities	−1747	3734	1987
Total	−9779	−52734	−62513

positive productivity (TFP > 0) and also positive market surplus (MS > 0). This implies that this sector is giving up part of its purchasing power to the rest of the system. On the contrary, for *Manufacture of Chemicals and Pharmaceutical*, the capacity to generate productivity is partially offset by an increase of its market power.

What is interesting is the position of the *Manufacture of Machinery and Equipment,* and *Manufacture of Transport Equipment* that lose productivity but, at the same time, give up part of their market power allowing the system to regain PPT.

The manufacturing sectors that, more than other, experience an increase in its market power, is the *Manufacture of Coke and Refined Petroleum Products*. Its strong market position overwhelms the capacity to transfer purchasing power in terms of TFP.

From the side of services it is worth noting that some public services such as *Public Administration and Defense*, *Education* and *Human Health Services*, albeit the real outputs is greater than the real inputs, the distortionary market conditions prevent a redistribution through a favorable change in prices.

Also for *Electricity, Gas and Water supply*, we have the same kind of situation. By and large, it follows that an increase in TFP in a given sector of the economy is not necessarily leading to a decrease of its relative market prices, nor a decrease in its market power. Since our analysis takes in consideration the period 1995–2002, we are not able to capture the effect of the liberalization of the electricity supply that started in Italy at the beginning of 2004[4]. Two years (2000–2002) may not be enough to produce positive effect. From a liberalization policy we would expect not only an increase in TFP, but also a transfer of purchasing power to the rest of the system, given that a more competitive market in energy supply should lead to a decrease in its relative price change.

Construction and *Financial activities* not only have negative performance in terms of productivity but they also increase their purchasing power by increasing their relative price.

In conclusion, we can say that the capacity to generate productivity does not automatically produce downward pressure on relative prices. There is not a mechanical process according to which positive innovation gains corresponds to an increase in the purchasing power of workers or capitalists, nor to a transfer of resources to consumers. Indeed, the total PPT available for distribution depends on the rule of distribution that in turn is the result of the structure of different markets. Prices in the market might be different from the ones we would expect in a perfect competitive market. So, an industry may adjust its selling price increasing its purchasing power that is to say, enlarging the gap between the actual market price and the ideal price. The same might occur for instance in the market of labour or capital. Specifically, it is the combination of the degree of distortion in the market that determines the rule of distribution of the PPT.

---

[4]We thank an anonymous referee for making this point.

6 Conclusion

In this paper we have highlighted the importance of adopting a single deflation method as the necessary approach to produce a system of economic accounts at constant prices that is in equilibrium only if it accounts for productivity gains. Furthermore, with a well-defined system of account at constant price, we are able to produce a productivity model that allows one to understand the distribution of purchasing power among agents.

Such a model could be very helpful for policy makers, since it gives a picture of the inter-industry diffusion and distribution of the welfare gains of innovations, that might be used to reorient economic priorities and managing the process of price adjustment when, for instance, the industrial policy takes the form of selective subsidies. Indeed, public investment in a give sector might not produce the expected positive outcome if this industry does not transfer part of its purchasing power to the rest of the system. So, selective subsidies can be oriented to correct distortions or imperfections in the market mechanism or addressed towards those progressive sectors that have a sufficiently high rate of innovation and operate, transferring massive welfare gains to the rest of the economy.

In our point of view, there is also scope for further development. Specifically, the analysis of the distributional rule of TFPS can also be integrated in the Leontief multiplier in order to capture the impact of a policy in terms of distributions of innovation gains. Yet a cost-linkage function can be constructed in order to improve our understanding of the mechanism of price adjustment.

References

Antille, G., & Fontela, E. (2003). The terms of trade and the international transfers of productivity gains. *Economic System Research, 15*(1), 3–20.

Babeau, A. (1978). The application of the constant price method for evaluating the transfer related to inflation: The case of french households. *Review of Income and Wealth, 24*(4), 391–414.

Devicienti, F., Maida, A., & Pacelli, L. (2008). The resurrection of the Italian wage curve. *Economics Letters, 98*(3), 335–341.

Durand, R. (1994). An alternative to double deflation for measuring real industry value added. *Review of Income and Wealth, 40*(3), 303–316.

Eurostat. (2001). Handbook on price and volume measures in national accounts.

Flexner, W. (1959). An analysis of the nature of aggregates at constant price. *Review of Economic and Statistics, 41*(4).

Fontela, E. (1989). Industrial structure and economic growth: An input output perspective. *Economic System Research, 1*(1), 45–53.

Fontela, E. (1994). Inter-industry distribution of productivity gains. *Economic System Research, 6*(3), 227–236.

Garau, G. (1996). La distribution des Gains de la Croissance: une analyse entrees sorties, ed. Lang, Berna.

Garau, G. (2002). Total factor productivity surplus in a sam context. *I International Conference on Economic and Social Statistics*, China: Canton.

ISTAT. (2008a). Prezzi alla produzione e prezzi al consume. http://www.istat.it

ISTAT. (2008b). Contabilità Nazionale (anni 1995–2006). http://www.istat.it

ISTAT. (2008c). Make and Use Tables. http://www.istat.it.

Martini, M. (1992). I Numeri Indice in un approccio assiomatico. Giuffré Editore.

Rampa, G. (2008). Using weighted least squares to deflate input output tables. *Economic Systems Research, 40*(4).

Wolff, E. N. (1985). Industrial composition, interindustry effects and the US productivity slowdown. *Review of Economics and Statistics, 67*(2), 268–277.

Wolff, E. N. (1989). *Dynamics of Growth in Input-Output Analysis*. Paper presented at the OECD International Seminar on Science Technology and Economic Growth, June 1989, Paris.

Jointly Consistent Price and Quantity Comparisons and the Geo-Logarithmic Family of Price Indexes

Marco Fattore

1 Introduction

In the axiomatic approach to composite index numbers, price and quantity indexes are dually linked since their product has to decompose the value index in a multiplicative way (Balk, 1995). In practice, the price index is often given some prominence, so that its formula is selected first and its cofactor is *de facto* chosen as the quantity index assuring for the value index decomposition to hold. This fact breaks the symmetry existing between price and quantity indexes and may have unexpected consequences on the consistency of the comparisons. In fact, usually no requirement is given connecting the choice of the price index to the properties satisfied by its cofactor, so that the former is very often selected irrespective of the axiomatic features of the latter. Unfortunately, few axiomatic properties of the price index are automatically inheritated by its cofactor, so that even an apparently "good" price index may have a cofactor which is not acceptable from an axiomatic point of view. As a result, both the implicit quantity comparison and the value index decomposition may result axiomatically inconsistent. It is important to note that this issue is not only theoretical. For example, when comparing real GDPs across countries or across time, an implicit quantity index is in fact computed, derived from some price index. The lacking of properties like proportionality (for instance) in such quantity index may lead to scarcely consistent GDP comparisons and arguable economic evaluations.

Trying to find out conditions assuring a price index to share good axiomatic properties with its cofactor, in the early '90s the Italian statistician Marco Martini (Martini 1992a, b) proposed a new parametric class of price indexes, known as *geo-logarithmic* family (or $\mathbf{P}_{xy}$ family). Its peculiarity is that all of its members satisfy the essential axioms of proportionality, commensurability and homogeneity together with their cofactors, so that both prices and quantities can be jointly compared in a consistent way. Indeed, there are price indexes sharing these properties with their

M. Fattore (✉)
Dipartimento di Metodi Quantitativi per le Scienze Economiche ed Aziendali, Università degli Studi di Milano - Bicocca, Piazza dell' Ateng Nuoro 1, 20126, Milano, Italy
e-mail: marco.fattore@unimib.it

L. Biggeri, G. Ferrari (eds.), *Price Indexes in Time and Space*, Contributions to Statistics, DOI 10.1007/978-3-7908-2140-6_11,

cofactors and not belonging to $\mathbf{P}_{xy}$, but on an empirical ground, it appears that all of them are linked to the geo-logarithmic family and particularly to a larger family $\overline{\mathbf{P}}_{xy}$, generated from $\mathbf{P}_{xy}$ by some simple transformations. So, when the joint consistency of price and quantity comparisons is of interest, the geo-logarithmic family should be considered.

In the following, we first analyze the relationship between the properties of a price index and the properties of its cofactor, showing that the latter ones are largely independent from the former ones. Next we introduce the geo-logarithmic family, discussing the axiomatic properties satisfied by its members and their cofactors. Some original results are proved, mainly pertaining monotonicity. Finally, we show how the geo-logarithmic family can be extended to the larger family $\overline{\mathbf{P}}_{xy}$. Some conclusions end comments end the chapter.

Most of the contents presented here are due to Prof. Marco Martini, even if they are exposed in an original way, with the addition of some new results. This chapter is in fact meant to be also a tribute to him and to his contribution to index number theory.

2 Notation

Let $\boldsymbol{p}_a, \boldsymbol{p}_b, \boldsymbol{q}_a$ and $\boldsymbol{q}_b$ be four n-dimensional vectors ($n \geq 2$) of strictly positive components, representing the prices and the quantities of the same n goods in situation a and in situation b (b will be always considered as the *basis*, i.e. the reference situation, for the comparison). We indicate with p_{ai}, p_{bi}, q_{ai} and q_{bi} ($i = 1, \ldots, n$) the i-th component of $\boldsymbol{p}_a, \boldsymbol{p}_b, \boldsymbol{q}_a$ and $\boldsymbol{q}_b$. The ratio

$$V = \frac{\sum_{i=1}^{n} p_{ai} q_{ai}}{\sum_{i=1}^{n} p_{bi} q_{bi}} \tag{1}$$

is called the *value index* between situations a and b. The aim of price and quantity index theory is to decompose the value index as the product of two strictly positive functions

$$V = P(\boldsymbol{p}_a, \boldsymbol{p}_b, \boldsymbol{q}_a, \boldsymbol{q}_b) \cdot Q(\boldsymbol{p}_a, \boldsymbol{p}_b, \boldsymbol{q}_a, \boldsymbol{q}_b) \tag{2}$$

where P (the *price index*) accounts for the variation of the prices and Q (the *quantity index*) accounts for the variation of the quantities between situations a and b. Given a price index $\boldsymbol{P}(\boldsymbol{p}_a, \boldsymbol{p}_b, \boldsymbol{q}_a, \boldsymbol{q}_b)$, the *implicit quantity index*

$$CofP(\boldsymbol{p}_a, \boldsymbol{p}_b, \boldsymbol{q}_a, \boldsymbol{q}_b) = \frac{V}{P(\boldsymbol{p}_a, \boldsymbol{p}_b, \boldsymbol{q}_a, \boldsymbol{q}_b)} \tag{3}$$

is called the *cofactor* of P, while the quantity index defined by

$$CorP(\boldsymbol{p}_a, \boldsymbol{p}_b, \boldsymbol{q}_a, \boldsymbol{q}_b) = P(\boldsymbol{q}_a, \boldsymbol{q}_b, \boldsymbol{p}_a, \boldsymbol{p}_b) \tag{4}$$

obtained exchanging the roles of prices and quantities within the formula for P is called the *correspondent* of P. For future reference, it is useful to consider *Cof* and *Cor* as two operators acting on a price index P and whose images are respectively the cofactor and the correspondent of P.

In the following, the notion of *basis antithesis* and *factor antithesis* of a price index will be repeatedly used. To ease the discussion, it is useful to introduce two operators $\boldsymbol{B}$ and $\boldsymbol{F}$ whose actions are defined as:

$$\boldsymbol{B}P(\boldsymbol{p}_a,\boldsymbol{p}_b,\boldsymbol{q}_a,\boldsymbol{q}_b) = \frac{1}{P(\boldsymbol{p}_b,\boldsymbol{p}_a,\boldsymbol{q}_b,\boldsymbol{q}_a)}; \tag{5}$$

$$\boldsymbol{F}P(\boldsymbol{p}_a,\boldsymbol{p}_b,\boldsymbol{q}_a,\boldsymbol{q}_b) = \frac{V}{P(\boldsymbol{q}_a,\boldsymbol{q}_b,\boldsymbol{p}_a,\boldsymbol{p}_b)}. \tag{6}$$

$\boldsymbol{B}P$ and $\boldsymbol{F}P$ are called the *basis antithesis* and the *factor antithesis* of P and the operators $\boldsymbol{B}$ and $\boldsymbol{F}$ can be called the *basis antithesis operator* and the *factor antithesis operator* respectively (Fattore & Quatto, 2004; Fattore, 2006b). A direct computation shows that $\boldsymbol{B}$ and $\boldsymbol{F}$ commute (Vogt & Barta 1997, Fattore & Quatto 2004, Fattore 2006b) and that both of them are idempotent. Thus the idempotent operator $\boldsymbol{D} = \boldsymbol{B} \circ \boldsymbol{F} = \boldsymbol{F} \circ \boldsymbol{B}$ can be defined. It is easily checked that $\boldsymbol{B}, \boldsymbol{F}, \boldsymbol{D}$ and the identity operator $\boldsymbol{I}$ form a commutative group (here called the *Antithesis group*) isomorphic to the Klein group (Miller, 1972; Tung, 1985; Bosch 2003). For future reference, it is worth noting that $\boldsymbol{F} = Cof \circ Cor = Cor \circ Cof$, i.e. that the factor antithesis of a price index is just the correspondent of its cofactor (or the cofactor of its correspondent, since *Cof* and *Cor* commute).

Finally, we define two *crossing* operators $\otimes^B$ and $\otimes_F$, whose actions on a price index are given by:

$$\otimes^B P(\boldsymbol{p}_a,\boldsymbol{p}_b,\boldsymbol{q}_a,\boldsymbol{q}_b) = \sqrt{P(\boldsymbol{p}_a,\boldsymbol{p}_b,\boldsymbol{q}_a,\boldsymbol{q}_b) \cdot \boldsymbol{B}P(\boldsymbol{p}_a,\boldsymbol{p}_b,\boldsymbol{q}_a,\boldsymbol{q}_b)}; \tag{7}$$

$$\otimes_F P(\boldsymbol{p}_a,\boldsymbol{p}_b,\boldsymbol{q}_a,\boldsymbol{q}_b) = \sqrt{P(\boldsymbol{p}_a,\boldsymbol{p}_b,\boldsymbol{q}_a,\boldsymbol{q}_b) \cdot \boldsymbol{F}P(\boldsymbol{p}_a,\boldsymbol{p}_b,\boldsymbol{q}_a,\boldsymbol{q}_b)}. \tag{8}$$

By construction, $\otimes^B P$ and $\otimes_F P$ are invariant under the actions of $\boldsymbol{B}$ and $\boldsymbol{F}$ respectively. Since the crossing operators $\otimes^B$ and $\otimes_F$ commute, the operator $\otimes_F^B = \otimes^B \circ \otimes_F = \otimes_F \circ \otimes^B$ can be defined, whose action is given by:

$$\otimes_F^B P = \sqrt[4]{P \cdot \boldsymbol{B}P \cdot \boldsymbol{F}P \cdot \boldsymbol{D}P}. \tag{9}$$

$\otimes_F^B P$ is invariant under the simultaneous action of $\boldsymbol{B}$ and $\boldsymbol{F}$. A direct computation shows that $\otimes^B \circ \otimes_F^B = \otimes_F \circ \otimes_F^B = \otimes_F^B \circ \otimes_F^B = \otimes_F^B$, so that $\otimes^B$, $\otimes_F$, $\otimes_F^B$ and the identity operator $\boldsymbol{I}$ form a commutative monoid, here called the *Crossing monoid*.

3 The Axioms for Price and Quantity Indexes

In the axiomatic setting, the basic idea is to identify a set of axioms that a function of prices and quantities has to satisfy, in order to be accepted as a price or a quantity index and then to derive or look for formulas satisfying them. The literature about Axiomatic Index Number Theory is very wide (Eichhorn & Voeller 1976, Eichhorn & Voeller 1990, Balk 2008, Krtscha 1988) but there is no universal agreement on the axiomatic properties needed for a formula to be considered as an index (IMF 2004). For this reason, here we focus on a set of six axioms that are generally accepted, even if some authors do criticize some of them, particularly, axioms 4 and 6 (Diewert & Nakamura 1993, Reinsdorf & Dorfman 1999, IMF 2004). The list is given below.

1. **Proportionality axiom**. Let α be a strictly positive real number, then

$$P(\alpha \boldsymbol{p}_b, \boldsymbol{p}_b, \boldsymbol{q}_a, \boldsymbol{q}_b) = \alpha.$$

2. **Commensurability axiom**. Let U be a $n \times n$ diagonal matrix of strictly positive weights, then

$$P(U\boldsymbol{p}_a, U\boldsymbol{p}_b, U^{-1}\boldsymbol{q}_a, U^{-1}\boldsymbol{q}_b) = P(\boldsymbol{p}_a, \boldsymbol{p}_b, \boldsymbol{q}_a, \boldsymbol{q}_b).$$

3. **Homogeneity axiom**. Let α and β be two strictly positive real numbers, then

$$P(\alpha \boldsymbol{p}_a, \beta \boldsymbol{p}_b, \boldsymbol{q}_a, \boldsymbol{q}_b) = \frac{\alpha}{\beta} P(\boldsymbol{p}_a, \boldsymbol{p}_b, \boldsymbol{q}_a, \boldsymbol{q}_b).$$

4. **Monotonicity axiom**. Let $\hat{\boldsymbol{p}}_a > \boldsymbol{p}_a$, then

$$P(\hat{\boldsymbol{p}}_a, \boldsymbol{p}_b, \boldsymbol{q}_a, \boldsymbol{q}_b) \geq P(\boldsymbol{p}_a, \boldsymbol{p}_b, \boldsymbol{q}_a, \boldsymbol{q}_b);$$

   similarly, let $\hat{\boldsymbol{p}}_b > \boldsymbol{p}_b$, then

$$P(\boldsymbol{p}_a, \hat{\boldsymbol{p}}_b, \boldsymbol{q}_a, \boldsymbol{q}_b) \leq P(\boldsymbol{p}_a, \boldsymbol{p}_b, \boldsymbol{q}_a, \boldsymbol{q}_b)$$

   where $\hat{\boldsymbol{p}}_s > \boldsymbol{p}_s$ if $\hat{p}_{si} \geq p_{si}$ for all $i = 1, \ldots, n$ and j exists such that $\hat{p}_{sj} > p_{sj}$, with $s = a,b$.
5. **Basis reversibility axiom**. Let the situations a and b be exchanged and let a be taken as the basis for the comparison, then

$$P(\boldsymbol{p}_b, \boldsymbol{p}_a, \boldsymbol{q}_b, \boldsymbol{q}_a) = \frac{1}{P(\boldsymbol{p}_a, \boldsymbol{p}_b, \boldsymbol{q}_a, \boldsymbol{q}_b)},$$

   or, in terms of the basis antithesis operator, $\boldsymbol{B}P = P$.

6. **Factor reversibility axiom**. Let the vectors $\boldsymbol{p}_a$ and $\boldsymbol{p}_b$ be exchanged with the vectors $\boldsymbol{q}_a$ and $\boldsymbol{q}_b$ respectively, then

$$P(\boldsymbol{q}_a, \boldsymbol{q}_b, \boldsymbol{p}_a, \boldsymbol{p}_b) = \frac{V}{P(\boldsymbol{p}_a, \boldsymbol{p}_b, \boldsymbol{q}_a, \boldsymbol{q}_b)}.$$

In terms of the factor antithesis operator, this property can be simply stated as $\boldsymbol{F}P = P$.

An analogous axiomatic system holds for quantity indexes. It is obtained simply exchanging the role of price and quantity vectors in the above list of axioms 1–6. In general, price index formulas used for concrete price comparisons do not fulfill all the axioms listed above (Fisher 1922, Swamy 1965, Balk 1995). As a matter of fact, not all the axioms are given the same relevance: axioms 1–3 are retained as fundamental, while axioms 4–6 are treated as less relevant. Hence proportionality, commensurability and homogeneity can be regarded as a minimal subset of axiomatic properties that a candidate price index should satisfy to be adopted in price comparisons. Similar considerations hold for quantity indexes.

3.1 *Properties of the Cofactor of a Price Index*

By virtue of (2), when a price index is chosen, its cofactor is *de facto* selected as the associated quantity index. As a consequence, a function $P(\boldsymbol{p}_a, \boldsymbol{p}_b, \boldsymbol{q}_a, \boldsymbol{q}_b)$ should not be accepted as a price index, irrespective of the axiomatic properties satisfied by its cofactor. As already mentioned, this issue is not of purely theoretical relevance, since usually cofactors of axiomatically compliant price indexes do not fulfill the axioms for a quantity index.

Trivially, if a price index is factor reversible then it and its cofactor share the same set of axiomatic properties, since in this case the two indexes have the same functional form. Nevertheless, the factor reversibility axiom is a very strong requirement and some argue that it is not an essential property (IMF 2004); moreover, most of the indexes currently used or studied does not fulfill it. In full generality, only few axiomatic properties of a price index are automatically inherited by its cofactor, namely commensurability and basis reversibility (Martini 1992a):

Proposition 1 *(i) CofP is a commensurable quantity index if and only if P is a commensurable price index; (ii) CofP is a basis reversible quantity index if and only if P is a basis reversible price index.*

Proof. (i) The value index V is clearly commensurable, so that by hypothesis *CofP* is the ratio of two commensurable functions and thus it is commensurable too. (ii) It is straightforward to show that $\boldsymbol{B}(V/P) = \boldsymbol{B}V/\boldsymbol{B}P$, so that

$$\boldsymbol{B}(CofP) = \boldsymbol{B}(V/P) = \frac{\boldsymbol{B}V}{\boldsymbol{B}P} = CofP \tag{10}$$

since V and P are basis reversible. □

On the contrary, proportionality, homogeneity and monotonicity of the cofactor are independent from the analogous properties of the price index, as we show by means of the following counterexamples:

1. The Törnqvist price index is proportional, but its cofactor is not (Martini 2003);
2. The Edgeworth-Marshall-Bowley price index is homogeneous, but its cofactor is not (Martini 2003);
3. The Walsh price index is monotonic, but its cofactor is not (see Proposition (7) below).

Thus, selecting a "good" price index formula does not assure the implicit quantity index to be similarly "good". In order to assure the joint consistency of both price and quantity comparisons, a possible alternative to restricting to factor reversible price indexes, is to search for a class of price index formulas satisfying at least a subset of fundamental axiomatic properties *together with* their cofactors. In the axiomatic setting, the most relevant axioms are those of proportionality, commensurability and homogeneity and we will refer to these three properties when jointly satisfied by the price index and its cofactor using the acronym PCH. We are thus led to search for PCH price indexes and this, in turn, leads to the geo-logarithmic family of price indexes.

4 The Geo-Logarithmic Family

The starting point for the definition of the geo-logarithmic family is the geo-logarithmic decomposition of the value index (Martini 2003), that we state without proof.

Proposition 2 *Let* $\boldsymbol{p}_a$, $\boldsymbol{p}_b$, $\boldsymbol{q}_a$ *and* $\boldsymbol{q}_b$ *be the price and quantity vectors of situations a and b and let*

$$V = \frac{\sum_{i=1}^{n} p_{ai} q_{ai}}{\sum_{i=1}^{n} p_{bi} q_{bi}} \tag{11}$$

be the value index. Then the following geo-logarithmic decomposition holds true:

$$V = \prod_{i=1}^{n} \left(\frac{p_{ai}}{p_{bi}} \right)^{\frac{\tau(w_{ai}, w_{bi})}{\sum_{j=1}^{n} \tau(w_{aj}, w_{bj})}} \cdot \prod_{i=1}^{n} \left(\frac{q_{ai}}{q_{bi}} \right)^{\frac{\tau(w_{ai}, w_{bi})}{\sum_{j=1}^{n} \tau(w_{aj}, w_{bj})}} \tag{12}$$

where

$$w_{ai} = \frac{p_{ai}q_{ai}}{\sum_{i=1}^{n} p_{ai}q_{ai}}, \quad w_{bi} = \frac{p_{bi}q_{bi}}{\sum_{i=1}^{n} p_{bi}q_{bi}} \quad i = 1,\ldots,n \tag{13}$$

and $\tau(w_{ai},w_{bi})$ is the logarithmic mean of w_{ai} and w_{bi}, defined for $w_{ai},w_{bi} > 0$ as

$$\tau(w_{ai},w_{bi}) = \frac{w_{ai} - w_{bi}}{\ln w_{ai} - \ln w_{bi}}, \tag{14}$$

if $w_{ai} \neq w_{bi}$ and as $\tau(w_{ai},w_{bi}) = w_{ai}$, if $w_{ai} = w_{bi}$ (Carlson 1972).
Note that

$$\prod_{i=1}^{n} \left(\frac{p_{ai}}{p_{bi}}\right)^{\frac{\tau(w_{ai},w_{bi})}{\sum_{j=1}^{n} \tau(w_{aj},w_{bj})}} \tag{15}$$

and

$$\prod_{i=1}^{n} \left(\frac{q_{ai}}{q_{bi}}\right)^{\frac{\tau(w_{ai},w_{bi})}{\sum_{j=1}^{n} \tau(w_{aj},w_{bj})}} \tag{16}$$

are, respectively, the Sato-Vartia price index and the Sato-Vartia quantity index (Sato 1976, Vartia 1976).

For $x,y \in [0,1]$, let $\boldsymbol{q}_x$ and $\boldsymbol{q}_y$ be two vectors, whose components are defined by

$$q_{xi} = q_{ai}^{x} q_{bi}^{1-x}, \quad q_{yi} = q_{ai}^{y} q_{bi}^{1-y} \quad i = 1,\ldots,n \tag{17}$$

and let

$$w_{xi} = \frac{p_{ai}q_{xi}}{\sum_{i=1}^{n} p_{ai}q_{xi}}, \quad w_{yi} = \frac{p_{bi}q_{yi}}{\sum_{i=1}^{n} p_{bi}q_{yi}} \quad i = 1,\ldots,n. \tag{18}$$

The geo-logarithmic family (or the $\mathbf{P}_{xy}$ family) is the class of price indexes defined by:

$$P_{xy}(\boldsymbol{p}_a,\boldsymbol{p}_b,\boldsymbol{q}_a,\boldsymbol{q}_b) = \prod_{i=1}^{n} \left(\frac{p_{ai}}{p_{bi}}\right)^{\frac{\tau(w_{xi},w_{yi})}{\sum_{j=1}^{n} \tau(w_{xj},w_{yj})}}. \tag{19}$$

In the following, to ease the notation we will write

$$P_{xy}(\boldsymbol{p}_a,\boldsymbol{p}_b,\boldsymbol{q}_a,\boldsymbol{q}_b) = \prod_{i=1}^{n} \left(\frac{p_{ai}}{p_{bi}}\right)^{\tau_{xyi} / \sum_{j=1}^{n} \tau_{xyj}} \tag{20}$$

where $\tau_{xyi} = \tau(w_{xi},w_{yi})$.

The geo-logarithmic family can be introduced in a slightly different way. Let us indicate with $P_{10}(\boldsymbol{p}_a, \boldsymbol{p}_b, \boldsymbol{q}_a, \boldsymbol{q}_b)$ the Sato-Vartia price index. Then, we can define a P_{xy} index by

$$P_{xy}(\boldsymbol{p}_a, \boldsymbol{p}_b, \boldsymbol{q}_a, \boldsymbol{q}_b) = P_{10}(\boldsymbol{p}_a, \boldsymbol{p}_b, \boldsymbol{q}_x, \boldsymbol{q}_y) \quad x,y \in [0,1] \tag{21}$$

where $\boldsymbol{q}_x$ and $\boldsymbol{q}_y$ are defined by means of (17). In other words, the members of the $\mathbf{P}_{xy}$ family are just the Sato-Vartia index computed on the virtual quantity vectors $\boldsymbol{q}_x$ and $\boldsymbol{q}_y$.

For future reference, note that the maps

$$\begin{aligned} \boldsymbol{T}_{xy} &: \mathbb{R}^{+n} \times \mathbb{R}^{+n} \mapsto \mathbb{R}^{+n} \times \mathbb{R}^{+n} \\ &:(\boldsymbol{q}_a, \boldsymbol{q}_b) \to (\boldsymbol{q}_x, \boldsymbol{q}_y) \end{aligned}$$

defined for $x,y \in [0,1]$ are invertible if and only if $x \neq y$, the inverse maps being (Fattore 2006a)

$$\boldsymbol{T}_{xy}^{-1}(\boldsymbol{q}_x, \boldsymbol{q}_y) = (\hat{\boldsymbol{q}}_a, \hat{\boldsymbol{q}}_b) \tag{22}$$

where

$$\hat{q}_{ai} = q_{xi}^{\frac{y-1}{y-x}} q_{yi}^{\frac{x-1}{x-y}} \quad i = 1, \ldots, n \tag{23}$$

and

$$\hat{q}_{bi} = q_{xi}^{\frac{y}{y-x}} q_{yi}^{\frac{x}{x-y}} \quad i = 1, \ldots, n. \tag{24}$$

The geo-logarithmic family contains some well known price indexes. Trivially, P_{10} is the Sato-Vartia price index. Moreover, if $x = y$ a few computations show that the formula of a geo-logarithmic price index reduces to (Martini 1992a)

$$P_{xx} = \frac{\sum_{i=1}^{n} p_{ai} q_{xi}}{\sum_{i=1}^{n} p_{bi} q_{xi}}. \tag{25}$$

Thus the $\mathbf{P}_{xx}$ subfamily is composed of expenditure ratios and it is easily verified that P_{00} is the Laspeyres price index, P_{11} is the Paasche price index and $P_{0.5\,0.5}$ is the Walsh price index.

4.1 Axiomatic Properties of Geo-Logarithmic Price Indexes

In this section the main axiomatic properties fulfilled by geo-logarithmic price indexes are discussed. It is first proved that the elements of the $\mathbf{P}_{xy}$ family fulfill the proportionality, commensurability and homogeneity axioms and that they are basis reversible if and only if $y = 1 - x$ (Martini 1992a). Then, it is shown that monotonic geo-logarithmic price indexes are characterized by the condition $x = y$ and that, among them, only the Laspeyres and the Paasche price indexes have monotonic factor antithesis (this last result will be particularly relevant when discussing axiomatic properties of geo-logarithmic cofactors).

Proposition 3 *Geo-logarithmic price indexes satisfy (i) the proportionality axiom, (ii) the commensurability axiom (iii) the homogeneity axiom. Moreover, (iv) they are basis reversible if and only if $y = 1 - x$.*

Proof. (i) Let α be a strictly positive real number, then:

$$P_{xy}(\alpha \boldsymbol{p}_b,\boldsymbol{p}_b,\boldsymbol{q}_a,\boldsymbol{q}_b) = \prod_{i=1}^{n} \alpha^{\tau_{xyi} / \sum_{j=1}^{n} \tau_{xyj}} = \alpha \quad x,y \in [0,1] \tag{26}$$

thus the proportionality axiom is fulfilled by each element of the $\mathbf{P}_{xy}$ family; (ii) let U be a $n \times n$ diagonal matrix of strictly positive weights. For every $x,y \in [0,1]$ and every $i = 1,\ldots,n$, the ratios p_{ai}/p_{bi} and the coefficients w_{xi} and w_{yi} are invariant under the simultaneous transformations $\boldsymbol{p}_a \to U\boldsymbol{p}_a$, $\boldsymbol{p}_b \to U\boldsymbol{p}_b$, $\boldsymbol{q}_a \to U^{-1}\boldsymbol{q}_a$ and $\boldsymbol{q}_b \to U^{-1}\boldsymbol{q}_b$, thus any geo-logarithmic index fulfills the commensurability axiom; (iii) let α and β be two strictly positive real numbers. For every $i = 1,\ldots,n$ and every $x,y \in [0,1]$, both w_{xi} and w_{yi} are invariant under the transformations $\boldsymbol{p}_a \to \alpha\boldsymbol{p}_a$ and $\boldsymbol{p}_b \to \beta\boldsymbol{p}_b$, thus

$$P_{xy}(\alpha\boldsymbol{p}_a,\beta\boldsymbol{p}_b,\boldsymbol{q}_a,\boldsymbol{q}_b) = \prod_{i=1}^{n} \left(\frac{\alpha p_{ai}}{\beta p_{bi}}\right)^{\frac{\tau(w_{xi},w_{yi})}{\sum_{j=1}^{n} \tau(w_{xj},w_{yj})}} = \frac{\alpha}{\beta} P_{xy}(\boldsymbol{p}_a,\boldsymbol{p}_b,\boldsymbol{q}_a,\boldsymbol{q}_b) \tag{27}$$

so that the homogeneity axiom is fulfilled by geo-logarithmic price indexes; (iv) requiring $\boldsymbol{B}P_{xy} = P_{xy}$ is the same as requiring $P_{1-y\,1-x} = P_{xy}$ or $y = 1 - x$. □

When first introduced, geo-logarithmic price indexes were supposed to satisfy the monotonicity axiom (Martini 1992a). Later, the Sato-Vartia price index was proved not to be monotonic (Reinsdorf & Dorfman 1999). Using this result, the following proposition shows that a geo-logarithmic index is monotonic if and only if it belongs to the subfamily $\mathbf{P}_{xx}$ of the expenditure ratios (Fattore 2006a).

Proposition 4 *An element of the $\mathbf{P}_{xy}$ family is monotonic if and only if $x = y$.*

Proof. Let

$$P_{xy}(\boldsymbol{p}_a,\boldsymbol{p}_b,\boldsymbol{q}_a,\boldsymbol{q}_b) = P_{10}(\boldsymbol{p}_a,\boldsymbol{p}_b,\boldsymbol{q}_x,\boldsymbol{q}_y) \quad x,y \in [0,1] \tag{28}$$

be an element of the geo-logarithmic family. As previously shown, if $x \neq y$, the map $\boldsymbol{T}_{xy}$ is invertible, thus P_{xy} is not monotonic since P_{10} is not. If $x = y$, (28) simplifies in

$$P_{xx} = \frac{\sum_{i=1}^{n} p_{ai} q_{xi}}{\sum_{i=1}^{n} p_{bi} q_{xi}} \tag{29}$$

which is clearly monotonic for every $x \in [0,1]$. □

Due to its properties, the $\mathbf{P}_{xx}$ subfamily is the most important subclass of the geo-logarithmic family. In general, its elements are not basis reversible (the intersection of $\mathbf{P}_{xx}$ and $\mathbf{P}_{x\,1-x}$ reduce to the Walsh index only), but it is closed under the action of the basis antithesis operator, since $\boldsymbol{B}P_{xx} = P_{1-x\,1-x}$. On the contrary, neither a factor reversible element of $\mathbf{P}_{xx}$ exists, nor is this class closed under the action of $\boldsymbol{F}$, since only P_{00} and P_{11} (respectively the Laspeyres and the Paasche price indexes) have monotonic factor antithesis, as proved in the following proposition:

Proposition 5 *Let $P_{xx} \in \mathbf{P}_{xx}$. $\boldsymbol{F}P_{xx}$ satisfies the axiom of monotonicity if and only if $x = 0$ or $x = 1$. In other words, the only monotonic elements of the $\mathbf{P}_{xy}$ family having monotonic factor antithesis are the Laspeyres and the Paasche indexes.*

Proof. A direct computation shows that

$$\boldsymbol{F}P_{xx} = \frac{\sum_{i=1}^{n} p_{ai} q_{ai}}{\sum_{i=1}^{n} p_{bi} q_{bi}} \cdot \frac{\sum_{i=1}^{n} q_{bi} (p_{bi})^{x} (p_{ai})^{1-x}}{\sum_{i=1}^{n} q_{ai} (p_{bi})^{x} (p_{ai})^{1-x}}. \tag{30}$$

After some algebraic computations, the derivative of (30) with respect to p_{aj} is obtained as

$$\frac{\partial}{\partial p_{aj}} \boldsymbol{F}P_{xx} = \frac{U + V \cdot Z}{T} \tag{31}$$

where

$$T = \sum_{i=1}^{n} p_{bi} q_{bi} \sum_{i=1}^{n} q_{ai} p_{bi}^{x} p_{ai}^{1-x} \tag{32}$$

$$U = q_{aj} \sum_{i=1}^{n} q_{bi} p_{bi}^{x} p_{ai}^{1-x} \tag{33}$$

$$V = (1 - x) p_{bj}^{x} p_{aj}^{-x} \sum_{i=1}^{n} p_{ai} q_{ai} \tag{34}$$

$$Z = q_{bj} - \frac{\sum_{i=1}^{n} q_{bi} p_{bi}^{x} p_{ai}^{1-x}}{\sum_{i=1}^{n} q_{ai} p_{bi}^{x} p_{ai}^{1-x}} q_{aj}. \tag{35}$$

If $x = 0$ or $x = 1$, the derivative is certainly positive, for any choice of price and quantity vectors, since $\boldsymbol{F}P_{00}$ is the Paasche index and $\boldsymbol{F}P_{11}$ is the Laspeyres index (both strictly monotonic price indexes). If $0 < x < 1$, the derivative is negative when $Z < -\frac{U}{V}$ and is positive when $Z > -\frac{U}{V}$. Explicitly, condition $Z < -\frac{U}{V}$ becomes

$$\frac{q_{bj}}{q_{aj}} < \frac{\sum_{i=1}^{n} q_{bi} p_{bi}^{x} p_{ai}^{1-x}}{\sum_{i=1}^{n} q_{ai} p_{bi}^{x} p_{ai}^{1-x}} - \frac{\sum_{i=1}^{n} q_{bi} p_{bi}^{x} p_{ai}^{1-x}}{(1-x) p_{bj}^{x} p_{aj}^{-x} \sum_{i=1}^{n} p_{ai} q_{ai}}. \tag{36}$$

For any fixed $x \in (0,1)$, this inequality can be indeed satisfied by a suitable choice of $\boldsymbol{p}_a$, $\boldsymbol{p}_b$, $\boldsymbol{q}_a$ and $\boldsymbol{q}_b$ (for an explicit proof, see the Appendix). □

Finally, we observe that geo-logarithmic price indexes are in general not factor reversible (consider, for example, the Laspeyres or the Paasche indexes). It can also be proved that the only geo-logarithmic factor reversible price index is the Sato-Vartia index (Fattore 2007).

4.2 Properties of the Cofactors of Geo-Logarithmic Price Indexes

From the general theory of price indexes discussed at the beginning of the chapter, we know that the cofactor of a price index fulfills the commensurability axiom, the basis reversibility axiom and the factor reversibility axiom, if and only if the price index itself does. It then follows immediately that the cofactor of a geo-logarithmic price index P_{xy} satisfies (i) the commensurability axiom for every $x,y \in [0,1]$, (ii) the basis reversibility axiom if and only if $y = 1 - x$ and (iii) the factor reversibility axiom if and only if $x = 1$ and $y = 0$ (i.e. if and only if P_{xy} is the Sato-Vartia price index).

The following proposition shows that cofactors of geo-logarithmic price indexes satisfy also the proportionality and homogeneity axioms Martini, 1992a, 2003:

Proposition 6 *Let $P_{xy} \in \mathbf{P_{xy}}$, then Cof P_{xy} satisfies (i) the proportionality axiom and (ii) the homogeneity axiom.*

Proof.

(i) Let α be a strictly positive real number and let $\boldsymbol{q}_a = \alpha \boldsymbol{q}_b$. The vectors of virtual quantities $\boldsymbol{q}_x$ and $\boldsymbol{q}_y$, defined in (17), are simply

$$\boldsymbol{q}_x = \alpha^x \boldsymbol{q}_b, \quad \boldsymbol{q}_y = \alpha^y \boldsymbol{q}_b \tag{37}$$

and it is immediate that in this case w_{xi} and w_{yi} coincide with w_{ai} and w_{bi}, as defined in (13). Hence, we can write

$$\boldsymbol{P}_{xy}(\boldsymbol{p}_a,\boldsymbol{p}_b,\alpha\boldsymbol{q}_b,\boldsymbol{q}_b) = \prod_{i=1}^{n} \left(\frac{p_{ai}}{p_{bi}}\right)^{\frac{\tau(w_{ai},w_{bi})}{\sum_{j=1}^{n} \tau(w_{aj},w_{bj})}} \quad x,y \in [0,1]. \tag{38}$$

Substituting (38) in (3) and using (12), we obtain

$$Cof\, P_{xy}(\boldsymbol{p}_a,\boldsymbol{p}_b,\alpha\boldsymbol{q}_b,\boldsymbol{q}_b) = \frac{\prod_{i=1}^{n}\left(\frac{p_{ai}}{p_{bi}}\right)^{\frac{\tau(w_{ai},w_{bi})}{\sum_{j=1}^{n}\tau(w_{ai},w_{bi})}} \cdot \prod_{i=1}^{n}\left(\frac{\alpha q_{bi}}{q_{bi}}\right)^{\frac{\tau(w_{ai},w_{bi})}{\sum_{j=1}^{n}\tau(w_{ai},w_{bi})}}}{\prod_{i=1}^{n}\left(\frac{p_{ai}}{p_{bi}}\right)^{\frac{\tau(w_{ai},w_{bi})}{\sum_{j=1}^{n}\tau(w_{ai},w_{bi})}}} = \alpha. \tag{39}$$

(ii) Let α and β be two strictly positive real numbers. Since the functions w_{xi} and w_{yi}, defined in (18), are left unchanged by the transformations $\boldsymbol{q}_a \to \alpha\boldsymbol{q}_a$ and $\boldsymbol{q}_b \to \beta\boldsymbol{q}_b$, it follows that $P_{xy}(\boldsymbol{p}_a,\boldsymbol{p}_b,\alpha\boldsymbol{q}_a,\beta\boldsymbol{q}_b) = P_{xy}(\boldsymbol{p}_a,\boldsymbol{p}_b,\boldsymbol{q}_a,\boldsymbol{q}_b)$. Thus

$$\begin{aligned} Cof\, P_{xy}(\boldsymbol{p}_a,\boldsymbol{p}_b,\alpha\boldsymbol{q}_a,\beta\boldsymbol{q}_b) &= \frac{\alpha}{\beta}\cdot\frac{\sum_{i=1}^{n} p_{ai}q_{ai}}{P_{xy}(\boldsymbol{p}_a,\boldsymbol{p}_b,\boldsymbol{q}_a,\boldsymbol{q}_b)\cdot\sum_{i=1}^{n} p_{bi}q_{bi}} \\ &= \frac{\alpha}{\beta}\cdot Cof\, P_{xy}(\boldsymbol{p}_a,\boldsymbol{p}_b,\boldsymbol{q}_a,\boldsymbol{q}_b). \end{aligned} \tag{40}$$

Thus, we can assert that geo-logarithmic price indexes are PCH. □

Finally, we discuss the axiom of monotonicity. In general, the cofactors of geo-logarithmic indexes violate the monotonicity axiom. Consider, for example, the cofactor of the Sato-Vartia price index. Due to factor reversibility, its cofactor coincides with its correspondent which is not monotonic with respect to quantities, since the Sato-Vartia price index itself is not monotonic with respect to prices. The following proposition shows that monotonic cofactors are an exception even if we restrict ourselves to monotonic geo-logarithmic price indexes:

Proposition 7 *The only elements of the* $\mathbf{P}_{xx}$ *subfamily whose cofactors satisfy the monotonicity axiom are the Laspeyres and the Paasche indexes.*

Proof. The correspondent of a price index is monotonic with respect to quantities if and only if the price index itself is monotonic with respect to prices. The thesis now follows from Proposition 5, since the cofactor of a price index is just the correspondent of its factor antithesis.

5 Extension of the Geo-Logarithmic Family

As proved above, geo-logarithmic price indexes are PCH. Indeed there are PCH price indexes not belonging to $\mathbf{P}_{xy}$. The most important example is the celebrated Fisher index, which is not geo-logarithmic, being monotonic and basis reversible (in fact, we have previously shown that the only geo-logarithmic index satisfying these two properties is the Walsh index). The Fisher index is deeply linked to the geo-logarithmic family, being the image of $\boldsymbol{P}_{00}$ or $\boldsymbol{P}_{11}$ under the crossing operators. This suggests the possibility to generate a larger class of PCH price indexes, applying suitable transformations to $\mathbf{P}_{xy}$. To this goal, some preliminary results are needed.

5.1 PCH Preserving Transformations

In axiomatic price index theory, the antithesis and the crossing operators are routinely used to generate new price indexes out of given ones. It is immediate to check that $\boldsymbol{B}$ preserves PCH. The same is true also for $\boldsymbol{F}$; in fact, from $\boldsymbol{F} = Cor \circ Cof$, it follows that $\boldsymbol{F}P_{xy}$ shares with respect to prices the same properties that $CofP_{xy}$ satisfies with respect to quantities. Taking the cofactor of $\boldsymbol{F}P_{xy}$ we also get $Cof \circ \boldsymbol{F}P_{xy} = CorP_{xy}$, so that in turn the cofactor of $\boldsymbol{F}P_{xy}$ shares with respect to quantities the same properties that P_{xy} fulfills with respect to prices. As a consequence, also $\boldsymbol{D}$ and the crossing operators are PCH preserving, as can be checked very easily.

Finally, let us observe that the antithesis operators and the crossing operators commute. For example, if we consider $\otimes_F$ and $\boldsymbol{B}$, we have

$$(\otimes_F \circ \boldsymbol{B})P = \sqrt{\boldsymbol{B}P \cdot \boldsymbol{D}P} \tag{41}$$

and

$$(\boldsymbol{B} \circ \otimes_F)P = \boldsymbol{B}(\sqrt{P \cdot \boldsymbol{F}P}) = \sqrt{\boldsymbol{B}P \cdot \boldsymbol{B}\boldsymbol{F}P} = \sqrt{\boldsymbol{B}P \cdot \boldsymbol{D}P}, \tag{42}$$

the other commutation rules being checked in a similar way.

5.2 Closure of the Geo-Logarithmic Family

To extend the geo-logarithmic family, we can first apply to $\mathbf{P}_{xy}$ the Antithesis group and, sucessively, the Crossing monoid. This way, we first add to the geo-logarithmic family all the antitheses of its members, and successively we add all the crossings. The resulting family $\overline{\mathbf{P}}_{xy}$ is the algebraic closure of $\mathbf{P}_{xy}$ under the joint action of the Antithesis group and the Crossing monoid. It comprises only PCH indexes, since all of its members are generated from PCH price indexes (the geo-logarithmic ones) by means of PCH preserving transformations. We would have applied to $\mathbf{P}_{xy}$ the Crossing monoid first and the Antithesis group successively; in fact, we would get the same closure $\overline{\mathbf{P}}_{xy}$, thanks to commutativity of the antithesis and the crossing operators. $\overline{\mathbf{P}}_{xy}$ is by construction closed under the action of $\boldsymbol{B}, \boldsymbol{F}, \boldsymbol{D}, \otimes^B, \otimes_F$ and $\otimes_F^B$ and also under the action of any sequence of such operators.

6 Conclusion

When the joint consistency of both price and quantity comparisons is of concern, the axiomatic properties of the cofactor must be taken into account, when the price index itself is selected. In general, such properties are largely independent from the axiomatic features of the price index, so that finding out conditions assuring the axiomatic compliance of both indexes is not a trivial task. Even restricting the attention to the fundamental axioms of proportionality and homogeneity does not solve the issue, since cofactors of proportional or homogeneous price indexes need not share such properties. Here is where the geo-logarithmic family $\mathbf{P}_{xy}$ comes into play. In fact, the only known sufficient condition assuring a price index to be PCH is that the price index itself belongs to $\overline{\mathbf{P}}_{xy}$, i.e. to the image of $\mathbf{P}_{xy}$ under the joint action of the Antithesis group and the Crossing monoid. We are tempted to conjecture a more general result and to affirm that, at least under fair conditions, a price index is PCH if and only if it belongs to $\overline{\mathbf{P}}_{xy}$. At present, this is just a hypothesis and proving whether and under which conditions it is true is an interesting open problem.

Appendix: proof of inequality (36)

In this appendix, we prove that inequality (36) in the proof of Proposition 5 can indeed be satisfied, for any fixed $x \in (0,1)$, by a suitable choice of price and quantity vectors.

Let

$$W(\boldsymbol{p}_a, \boldsymbol{p}_b, \boldsymbol{q}_a, \boldsymbol{q}_b) = \frac{1}{\sum_{i=1}^{n} q_{ai} p_{bi}^{x} p_{ai}^{1-x}} - \frac{1}{(1-x) p_{bj}^{x} p_{aj}^{-x} \sum_{i=1}^{n} p_{ai} q_{ai}}, \tag{43}$$

and write (36) as

$$\frac{q_{bj}}{q_{aj}} < W(\boldsymbol{p}_a, \boldsymbol{p}_b, \boldsymbol{q}_a, \boldsymbol{q}_b) \cdot \sum_{i=1}^{n} q_{bi} p_{bi}^{x} p_{ai}^{1-x}. \tag{44}$$

Let $p_{aj} = p_{bj} = \lambda / q_{aj}$ and $p_{bi} = (1-x)^{k/x} p_{ai}$ $(i \neq j)$, where λ and k are strictly positive constants and let the other price and quantity components be fixed. For such a choice of the prices, (44) simplifies in

$$\frac{q_{bj}}{q_{aj}} < \left[\lambda \frac{q_{bj}}{q_{aj}} + \sum_{i \neq j} q_{bi} p_{ai} (1-x)^{k} \right] \cdot W^{*}(\boldsymbol{p}_a, \boldsymbol{q}_a; \lambda, k) \tag{45}$$

where

$$W^{*}(\boldsymbol{p}_a, \boldsymbol{q}_a; \lambda, k) = \frac{1}{\lambda + (1-x)^{k} \sum_{i \neq j}^{n} q_{ai} p_{ai}} - \frac{1}{(1-x)\lambda + (1-x) \sum_{i \neq j}^{n} p_{ai} q_{ai}} \tag{46}$$

does not depend upon q_{aj} and q_{bj}.

Since

$$\lim_{k\to\infty} W^*(\boldsymbol{p}_a,\boldsymbol{q}_a;\lambda,k) = \frac{1}{\lambda} - \frac{1}{(1-x)\lambda + (1-x)\sum_{i\neq j}^{n} p_{ai}q_{ai}}, \tag{47}$$

if λ_x is chosen so that

$$\lambda_x < \frac{1-x}{x}\sum_{i\neq j}^{n} p_{ai}q_{ai}$$

then $\lim_{k\to\infty} W^*(\boldsymbol{p}_a,\boldsymbol{q}_a;\lambda_x,k) > 0$. Hence, if k_x is chosen large enough, it is certainly $W^*(\boldsymbol{p}_a,\boldsymbol{q}_a;\lambda_x,k_x) > 0$. As a consequence, we have

$$L_x = \lim_{q_{bj}/q_{aj}\to 0}\left(\lambda_x\frac{q_{bj}}{q_{aj}} + \sum_{i\neq j} q_{bi}p_{ai}(1-x)^{k_x}\right)\cdot W^*(\boldsymbol{p}_a,\boldsymbol{q}_a;\lambda_x,k_x) > 0.$$

Hence if q_{bj}/q_{aj} is small enough, inequality (45) (and thus (36)) is satisfied. □

References

Balk, B. M. (2008). Price and quantity index numbers: Models for measuring aggregate change and difference. Cambridge: Cambridge University Press.

Balk, B. M. (1995). Axiomatic price index theory: A survey. *International Statistical Review, 63*, 69–93.

Bosch, S. (2003). *Algebra*. Heidelberg: Springer-Verlag.

Carlson, B. C. (1972). The logarithmic mean. *American Mathematical Monthly, 79*, 615–618.

Diewert, W. E., & Nakamura, A. O. (Eds.). (1993). *Essays in index number theory* (Vol. 1). Amsterdam: North-Holland Publishing Co.

Eichhorn, W., & Voeller, J. (1990). Axiomatic foundations of price indexes and purchasing power parities. In W. E. Diewert & C. Montmarquette (Eds.), *Price level measurement*, Amsterdam: North-Holland Publishing Co.

Eichhorn, W., & Voeller, J. (1976). Theory of the price indices, lecture notes in economics and mathematical systems, Berlin: Springer-Verlag.

Fattore, M. (2007). A characterization of the Sato-Vartia price index, Department of Statistics, University of Milano – Bicocca. Retrievd from econpapers.repec.org

Fattore, M. (2006a). *On the monotonicity of the geo-logarithmic price indexes, Società Italiana di Statistica*. Proceedings of the XLIII Scientific Meeting, Padova: Cleup.

Fattore, M. (2006b). Finite Group in Axiomatic Index Number Theory, Department of Statistics, University of Milan – Bicocca. Retrievd from econpapers.repec.org.

Fattore, M., & Quatto, P. (2004). Strutture algebriche dei numeri indice bilaterali. In *Studi in onore di Marco Martini*. Milan: Giuffré.

Fisher, I. (1922). *The making of index numbers*. Boston: Houghton Mifflin.

Krtscha, M. (1988). *Axiomatic characterization of statistical price indices*. Heidelberg: Physica-Verlag.

IMF. (2004). *Producer price index manual: Theory and practice*. International Monetary Fund, Statistics Dept.

Martini, M. (2003). *Numeri indice per il confronto nel tempo e nello spazio*. Milan: CUSL.

Martini, M. (1992a). *I numeri indice in un approccio assiomatico*. Milan: Giuffré.

Martini, M. (1992b). General function of axiomatic index numbers. *Journal of the Italian Statistical Society, 3*, 359–376.
Miller, W. (1972). *Symmetry groups and their applications*. New York: Academic Press.
Reinsdorf, M. B., & Dorfman, A. H. (1999). The Sato-Vartia index and the monotonicity axiom. *Journal of Econometrics, 90*, 45–61.
Sato, K. (1976). The ideal log-change index number. *The Review of Economics and Statistics, 58*[s](2), 223–228.
Swamy, S. (1965). Consistency of Fisher's tests. *Econometrica, 33*, 619–623.
Tung, W. (1985). *Group theory in physics*. Singapore: World Scientific Publishing Co.
Vogt, A., & Barta, J. (1997). The making of tests for index numbers, Heidelberg: Physica-Verlag.
Vartia, Y. O. (1976). Ideal log-change index numbers. *Scandinavian Journal of Statistics, 3*, 121–126.

Part V
Price Indexes in Financial Markets

Common Trends in Financial Markets

Giuseppe Cavaliere and Michele Costa

1 Introduction

Price indexes play a prominent role in financial markets, not only as a synthetic measure of changes in financial asset prices but, in particular, as benchmark levels for investors. Up to now, despite the remarkable importance of these instruments, a number of methodological issues with a substantial empirical impact have been only marginally studied. By jointly taking the time and the space dimensions into account, in this paper we aim at providing interpretative guidelines for integration and convergence processes and at contributing to the debate on the usefulness of international financial diversification.

Comparisons over time play a strategic role in financial market price index numbers. The simplest application concerns the measurement of market returns over a given time period. Although short horizons are often discussed in the financial literature, they do not exhaust all issues related to the analysis of financial markets price index numbers. On the contrary, a relevant class of topics can be analyzed over long horizons only; see, inter alia, Fama and French (1988). Integration among countries or regions, for instance, is usually slow and generally requires a long time to reach its accomplishment. Similarly, investment decisions made by some typologies of agents, such as institutional investors and households, are planned over long horizons.

The analysis of the interrelations among financial markets price index numbers over long horizons is usually carried out by means of the analysis of common trends or cointegration among a set of indexes. In the recent literature the presence of integration among financial markets is usually tested through the analysis of the existence of cointegration (Engle & Granger, 1987; Johansen, 1996) among the stock market indexes of interest; see, among others, Kasa (1992), Corhay, Rad, & Urbain, (1993) and Richards (1995). Similarly, some authors claim that financial

G. Cavaliere (✉)
Dipartimento di Scienze Statistiche, Universita' di Bologna, Bologna, Italy
e-mail: giuseppe.cavaliere@unibo.it

L. Biggeri, G. Ferrari (eds.), *Price Indexes in Time and Space*, Contributions to Statistics, DOI 10.1007/978-3-7908-2140-6_12, © Springer-Verlag Berlin Heidelberg 2010

convergence can be analyzed by testing whether the number of cointegration relations among a set of financial indexes has experienced a structural change (Rangvid, 2001; Garcia-Pascual, 2003).

Traditional tests for the presence of common trends, however, are developed under the assumption of constant volatility, while a well-known feature of financial markets price index numbers is heteroskedasticity. Furthermore, financial price index numbers are also characterized by jumps, both in level and in variability, which are not consistent with standard inferential procedures for common trends determination.

The aim of this paper is to analyze the long run interrelations among financial price index numbers by proposing a novel methodology which allows for both non stationarity and jumps in volatility.

Our purpose of obtaining results comparable to the existing literature leads us to the choice of the following four financial markets: US, Japan, Germany and United Kingdom. The choice of this set of countries leads the analysis of common trends among the corresponding price index numbers towards the direction of international diversification. In particular, the detection of common trends implies the absence of long-run benefits from diversification among the four countries of interest, while evidence of no common trends suggests profitable diversification opportunities in the long run.

2 The Data

We consider price index numbers for the four main financial markets: USA, Japan, United Kingdom and Germany. The data used are the Morgan Stanley Capital International stock price index numbers; in order to provide a greater comparability with the results published by Kasa (1992), we refer to monthly data over the period January 1974–August 1990. Furthermore, we extend the analysis to the period January 1974 - December 2007.

All series are expressed in US dollars, since, following Kasa (1992), US is considered as the reference country in the empirical analysis. In this framework we are therefore considering the case of an US investor willing to operate in international financial markets. Furthermore, we transform monetary values into real data by means of the US consumer price index, thus avoiding bias related to inflation dynamics, which could bias the results favouring the presence of a common trend. Finally, we eliminate size effects due to the different levels of the price indexes by setting the value of each series in January 1974 to 100.

Figure 1 shows a plot of each national stock price index number from 1974 to 2007, the dotted line indicating August 1990, the end of Kasa's period. The US, UK and German stock price index series share a quite stable performance during the seventies and the eighties, a sharp increase around the middle of the nineties and a maximum at the beginning of the 2000's. Also the Japanese stock price index series

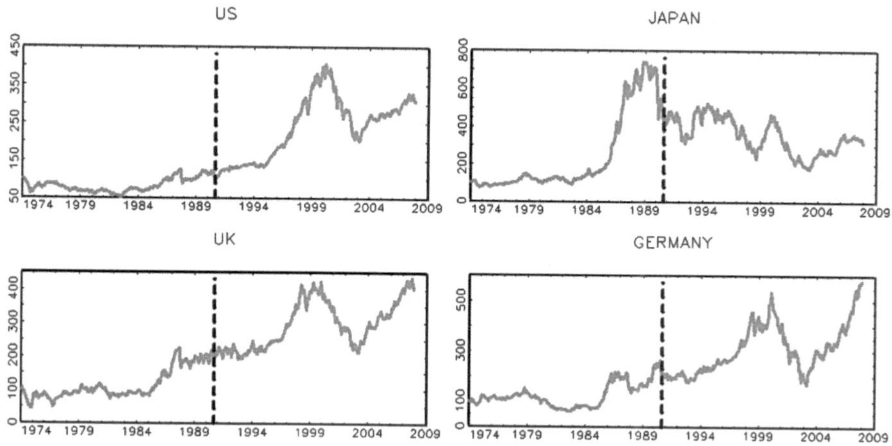

Fig. 1 National stock price index numbers, monthly data 1974–2007

initially shows a stable period, but it anticipates the upward trend to the middle of the eighties and the maximum to 1990.

Figure 1 highlights a similar profile for the US, UK and German stock price index series, which could suggest the presence of cointegration among these indexes. Also the Japanese stock price index series, although showing some relevant differences, seems to move together with the US index for long periods.

In order to provide a more detailed analysis on the long term interrelations among national stock price index numbers and, most important, with the purpose of investigating their variance structure, we calculate the index log returns, which are illustrated in Fig. 2 (the dotted line still indicating the end of Kasa's period). It is immediate to observe a well known stylized fact of financial return distributions:

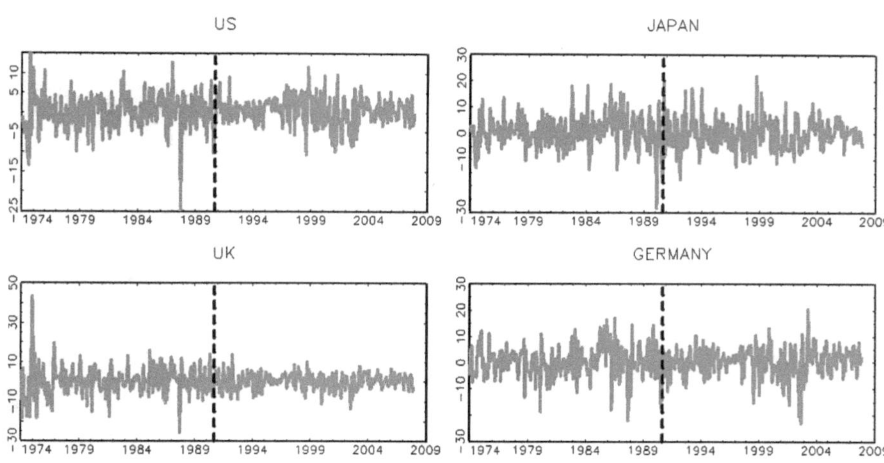

Fig. 2 National stock price index number log returns, monthly data 1974–2007

Table 1 Descriptive statistics for monthly percent log returns of stock price index numbers, monthly data 1974–2007

	USA	Japan	UK	Germany	USA	Japan	UK	Germany
	1974–1990				1974–2007			
Mean	0.03	0.75	0.36	0.40	0.28	0.28	0.34	0.43
Max	14.98	18.70	43.55	17.06	14.98	22.08	43.55	20.43
Min	–24.71	–28.35	–25.70	–21.93	–24.71	–28.35	–25.70	–22.98
Standard deviation	4.83	6.64	7.89	6.33	4.31	6.47	6.25	6.02
Skewness	–0.64	–0.20	0.53	–0.19	–0.57	0.11	0.49	–0.42
Excess kurtosis	3.20	1.44	4.37	0.92	2.86	1.17	6.48	1.42

each series is characterized by a strong heteroskedasticity. The switching between low-volatility periods and high-volatility regimes invalidates traditional cointegration test procedures and requires to develop appropriate methods in order to properly analyze the long period dynamics of stock price index numbers.

A more detailed analysis of stock price index number returns is provided in Table 1, which reports some descriptive statistics for the four return series.

Over the period 1974–1990, Japan has the highest average monthly return, which corresponds to an average annual return of 9.0%. Germany presents the second highest average monthly return, about 4.8%, followed by UK and US, with a corresponding average annual return of 4.3 and 0.36 respectively. The most volatile market is UK, with a monthly standard deviation of 7.89, followed by US, Germany and Japan, which present standard deviations of 6.64, 6.33 and 4.83, respectively.

It is possible to observe some differences between the 1974–1990 and the full sample. The stock price index series with the highest average monthly return is the German index, with a corresponding average annual return of 5.16. The UK, US and Japanese stock price index series present average annual returns of 4.08, 3.36 and 3.36, respectively. From 1974 to 2007 the most volatile market is Japan, with a monthly standard deviation of 6.47, followed by UK, Germany and US, which present standard deviations of 6.25, 6.02 and 4.31, respectively.

All series show significant excess kurtosis, thus confirming a stylized fact which characterizes the stock price index number return distribution.

In Table 2 we also provide the Pearson correlation index between stock price index number returns. All correlation measures are positive, as expected. Over Kasa's period, correlations range from 0.25 (US/Japan) to 0.52 (US/UK). Over the full period, it is possible to observe a generalized increase of the correlation coefficients, which range from 0.29 (US/Japan) to 0.56 (US/UK). The most relevant difference regards the German stock price index series, which shows a relevant increase in the correlation coefficients with the US (from 0.37 to 0.49) and the UK (from 0.40 to 0.48) series. Finally, the Japanese stock price index series shows the weakest linear correlation with respect to all other series, while the highest interrelation is observed between the US and the UK stock price index series.

Table 2 Contemporaneous correlations among monthly percentage log returns of stock price index numbers, 1974–2007

	USA	Japan	UK	Germany	USA	Japan	UK	Germany
	1974–1990				1974–2007			
USA	1.00	0.25	0.52	0.37	1.00	0.29	0.56	0.49
Japan	0.25	1.00	0.36	0.37	0.29	1.00	0.37	0.36
UK	0.52	0.36	1.00	0.40	0.56	0.37	1.00	0.48
Germany	0.37	0.37	0.40	1.00	0.49	0.36	0.48	1.00

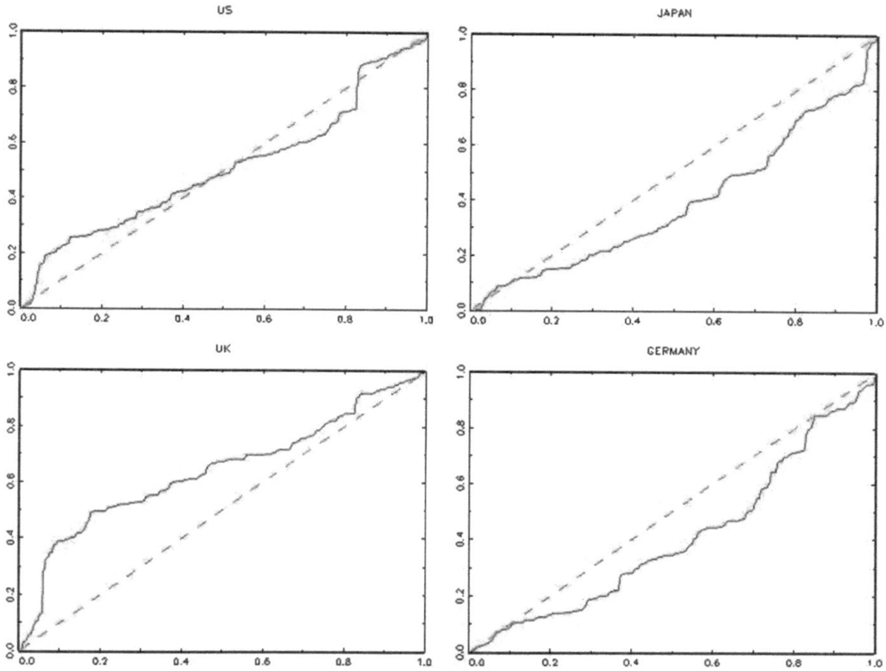

Fig. 3 Variance profiles of the national stock price index log returns, 1974–1990

Stock price index returns are typically characterized by the presence of heteroskedasticity, which has relevant methodological consequences. Figures 3 and 4, for the Kasa period and the full sample respectively, illustrate the dynamic of the variance profiles of the four stock price index log returns. These are obtained by cumulating the squared (mean adjusted) returns over all $t = 1,2,\ldots,T$. That is, as $V_i(u) = \sum_{t=1}^{\lfloor Tu \rfloor} \left(r_{it} - \hat{\mu}_i\right)^2 / \sum_{t=1}^{T} \left(r_{it} - \hat{\mu}_i\right)^2$, $u \in [0,1]$, with r_{it} denoting the log return of index i at time t, and $\hat{\mu}_i = T^{-1} \sum_{t=1}^{T} r_{it}$. It is straightforward to spot relevant departures from the homoskedastic case (diagonal line), thus confirming the importance of taking heteroskedasticity into account.

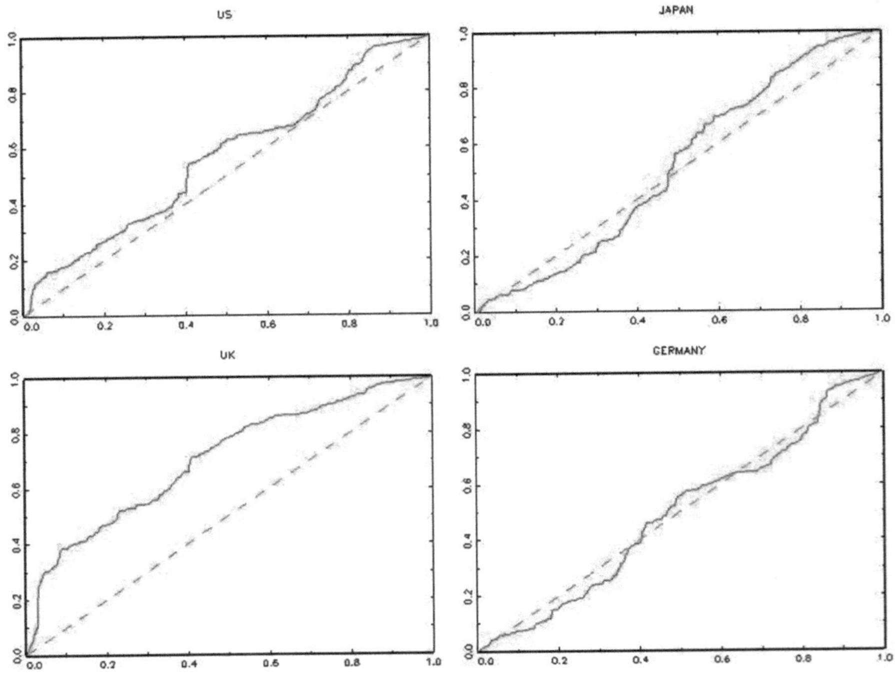

Fig. 4 Variance profiles of the national stock price index log returns, 1974–2007

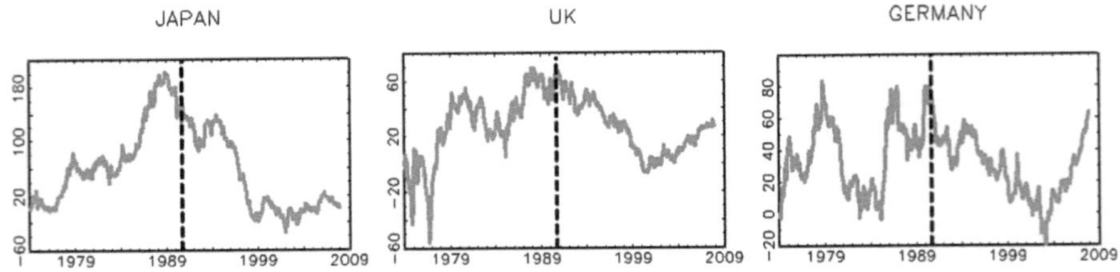

Fig. 5 National stock price index deviations from US stock index, monthly data 1974–2007

The next step of our analysis considers the deviations of each national series from the US stock price index series. Given the similar profile of stock price index series illustrated in Fig. 1, we would expect that deviations are small and perhaps stationary, at least for the UK and the German cases. However, from Fig. 5 it is clear that deviations are huge, very persistent and, most important, that they do not show stationary dynamics. The direct implications is that stock price index series do not really exhibit similar profiles.

Table 3 reports some descriptive statistics for the national stock price index numbers deviations from the US values. The Japanese stock price index series is the most different from the US series, but also the UK and the German indexes show relevant differences with respect to US stock market.

Table 3 Descriptive statistics for national stock price index deviations relative to the US, monthly data 1974–2007

	Japan	United Kingdom	Germany	Japan	United Kingdom	Germany
	1974–1990			1974–2007		
Mean	79.69	28.97	37.92	60.69	26.92	32.86
Max	202.12	69.36	84.04	202.12	69.36	84.04
Min	−1.73	−56.01	−3.96	−34.47	−56.01	−19.85
Standard Deviation	59.36	25.94	22.15	60.46	22.61	21.04
Skewness	0.61	−0.86	0.05	0.55	−0.47	0.11
Excess Kurtosis	−0.77	0.37	−1.04	−0.77	0.15	−0.72

3 Methodological Issues: The Conditionally Heteroskedastic Co-Integration Model

Procedures for determining the number of common stochastic trends among financial price index numbers are generally carried out by means of the sequential (maximum likelihood) procedure of Johansen (1996). See, among others, Kasa (1992), Richards (1995) or Ahlgren & Antell (2002). Let X_t be the $p \times 1$ vector of stock market indexes (in log form) at time t. As is standard, X_t is assumed to be integrated of order one, I(1) hereafter. Moreover, we assume that X_t evolves over time according to the following VAR(k) class of models in error correction form:

$$\Delta X_t = \Pi X_{t-1} + \sum_{i=1}^{k-1} \Gamma_i \Delta X_{t-i} + \mu_0 + \mu_1 t + \varepsilon_t, \qquad t = 1,\ldots,T \tag{1}$$

where ε_t is a $p \times 1$ sequence of iid $N(0,\Sigma)$ random variables and $\{\Gamma_i\}_{i=1}^{k-1}$ are $p \times p$ lag coefficient matrices. It is also assumed that the process is $I(1)$ and that $\Pi :=\alpha\beta'$ where α and β are full column rank $p \times r$ matrices, $r \leq p$. Hence, r is the so-called cointegrating rank, which denotes the number of cointegrating relations and $p - r$ is the number of common stochastic trends among the index numbers. Finally, linear trends (but not higher order trends) in X_t are allowed by requiring that $\mu_1 = \alpha\rho'$, ρ being a $p \times r$ matrix (cf. Johansen, 1996). We denote this model as $H(r)$.

Johansen (1996) first derives a likelihood ratio test for the model with rank r against full rank p. This is based on the likelihood ratio test statistic

$$Q_r := -2\,(\ell(r) - \ell(p))$$

where $\ell(r)$ denotes the maximized log likelihood under $H(p)$. Under the null hypothesis Q_r is well known to converge to the multivariate Dickey-Fuller distribution; the associated p-values can be computed as in Mac Kinnon et al. (1999).

In order to determine the cointegration rank, Johansen (1996) suggests to compute the sequence of likelihood ratio tests $Q_i, i = 0,1,\ldots,p-1$ and to accept $H(r)$

if, starting from $i = 0$, the models $H(i)$, $i = 0,1,\ldots,r-1$ are rejected (i.e., the associated p-values are lower than the selected significance level, say η) but $H(r)$ is not. Under the assumptions stated above, the sequential procedure is consistent, i.e.

$$P\left(\hat{r} = i\right) \to 0, \qquad i = 0,1,\ldots,r-1$$

and

$$P\left(\hat{r} = r\right) \to 1 - \eta$$

with $\hat{r}$ denoting the cointegrating rank determined sequentially.

However, in the analysis of financial price index numbers, this procedure may not deliver good results in terms of size. First of all, the approximation of the finite sample distributions delivered by the asymptotic theory is not always accurate. This occurs e.g. when the sample size is too small, or when the number of series considered is high, or when some of the autoregressive roots of the process are close to the I(2)region.[1] Second, when the independence assumption for ε_t breaks down, the size of the procedure can be far from the chosen significance level, even asymptotically. In particular, conditional heteroskedasticity can make the sequential procedure oversized both in finite samples (Cavaliere, Rahbek, & Taylor, 2008a) and, when shocks to volatility are permanent, even asymptotically (Cavaliere, Rahbek, & Taylor, 2008b). This is particular important, given that financial time series are characterized by non-independent, conditionally heteroskedastic innovations. In this case, if the size is not properly controlled, the standard (asymptotic) sequential procedure may tend to identify too many cointegrating relations.

In this paper we propose to use bootstrap methods to improve the size of the testing procedure. Bootstrap methods for testing the cointergrating rank have been discussed in the recent literature, see e.g. van Giersbergen (1996), Harris & Judge (1998), Mantalos & Shukur (2001), Trenkler (2009) and, most notably, Swensen (2006). The bootstrap approaches proposed in these papers, however, are based on resampling with replacement the residuals from the estimated model. When this resampling device is used, then (conditionally on the original data) the bootstrap innovation are iid; that is, they display no conditional heteroskedasticity. As a consequence, although being robust to certain forms of conditional heteroskedasticity, the standard residual based bootstrap methods tend to overestimate the true cointegration rank when the volatility process is non-stationary; see the discussion in (Cavaliere et al. 2008b).

To overcome problems related to the iid bootstrap under heteroskedastic shocks, in this paper we make use of a bootstrap scheme based on the so-called wild bootstrap; cf. Wu (1986), Liu (1988) and Mammen (1992). This scheme ensures that a bootstrap version of the sequential approach to determining the co-integration rank

[1] See Ahlgren and Antell (2002) for further discussions on this issue.

of Johansen (1996) will still lead to the selection of the correct co-integrating rank with probability $(1-\eta)$ in large samples, as in the i.i.d. Gaussian case. This scheme is described in the following section.

3.1 Heteroskedasticity-Robust Cointegrating Rank Determination

The bootstrap algorithm requires to estimate the VAR model (1) both unrestrictedly (i.e., setting $r = p$) and restrictedly (i.e., under the rank r, with $r = 0,1,\dots,p-1$).

Let $(\hat{\Gamma}_1,\dots,\hat{\Gamma}_{k-1})$ and $\hat{\mu}_0$ denote the ML estimates of $(\Gamma_1,\dots,\Gamma_{k-1})$ and μ_0 from the unrestricted model, $H(p)$; the corresponding unrestricted residuals are denoted by $\hat{\varepsilon}_t$, $t = 1,\dots,T$. In addition, let $\hat{\alpha}_j,\hat{\beta}_j,\hat{\rho}_j$ denote the ML estimates of α,β and ρ under the hypothesis of rank j, i.e. from the model $H(j)$. The bootstrap algorithm is based on the following steps.

ALGORITHM 1 (WILD BAOOTSTRAP SEQUENTIAL CO-INTEGRATION RANK DETERMINATION)

Step 1: Generate T wild bootstrap residuals ε_t^*, $t = 1,\dots,T$, according to the device

$$\varepsilon_t^* := \hat{\varepsilon}_t w_t$$

where $\{w_t\}_{t=1}^T$ denotes an independent $N(0,1)$ scalar sequence;

Step 2: Starting from $j = 0$, check whether the roots of the equation $\det(\hat{A}_j(z)) = 0$, where

$$\hat{A}(Z) := (1-z)I_p - \hat{\alpha}_j\hat{\beta}_j'Z - \hat{\Gamma}_1(1-Z)Z - \dots - \hat{\Gamma}_{k-1}(1-Z)Z^{k-1}$$

are equal to 1 or located outside the unit circle; the bootstrap is implemented if and only if this condition is satisfied.

Step 3: Construct the bootstrap sample $\{X_{j,t}^*\}$, $t = 1,\dots,T$, recursively from

$$\Delta X_{j,t}^* := \hat{\alpha}_j\hat{\beta}_j'X_{j,t-1}^* + \hat{\Gamma}_1\Delta X_{j,t-1}^* + \dots + \hat{\Gamma}_{k-1}\Delta X_{j,t-k+1}^* + \hat{\mu}_0 + \hat{\alpha}_j\hat{\rho}_j't + \varepsilon_t^*,\ t = 1,\dots,T,$$

with initial values, $X_{j,-k+1}^* = X_{j,-k+1},\dots,X_{j,0}^* = X_0$;

Step 4: Using the bootstrap sample, $\{X_{j,t}^*\}$, obtain the bootstrap test statistic, $Q_j^* := -2(\ell^*(j) - \ell^*(p))$, where $\ell^*(j)$ and $\ell^*(p)$ denote the bootstrap analogues of $\ell(j)$ and $\ell(p)$, respectively;

Step 5: Bootstrap p-values are then computed as, $p_{j,T}^* := 1 - G_{j,T}^*(Q_j)$, where $G_{j,T}^*(\cdot)$ denotes the conditional (on the original data) cumulative distribution function (cdf) of Q_j^*.

Step 6: If $p_{j,T}^*$ from step 5 is above the selected significance level η, set $\hat{r} := j$; otherwise, if $j < p-1$, repeat steps 2–5 with rank $j+1$ or, if $j = p-1$, set $\hat{r} = p$.

Some remarks on this algorithm are due.

First, with respect to the standard bootstrap procedures for the determination of the co-integration rank (Swensen, 2006; Trenkler, 2009), the key ingredient of the above procedure is the first step. In the standard "iid" bootstrap, the bootstrap shocks are obtained by resampling (with replacement) the residuals $\hat{\varepsilon}_1, \ldots, \hat{\varepsilon}_T$ from the unrestricted model. conversely, the wild bootstrap shocks, as defined in step 1 above, are such that they replicate the pattern of heteroskedasticity present in the original shocks since, conditionally on $\hat{\varepsilon}_t$, ε_t^* is independent over time with zero mean and variance-covariance matrix $\hat{\varepsilon}_t \hat{\varepsilon}_t'$. This is the key property for establishing that under the hypothesis of rank r and under rather general conditions on the behaviour of the conditional variance, the wild bootstrap statistic Q_r^* has the same first-order asymptotic null distribution as the standard Q_r statistic[2].

Second, the Gaussian distribution used in step 1 is not strictly necessary. As is well known in the wild bootstrap literature (see Davidson & Flachaire, 2008, for a review) in certain cases improved accuracy can be obtained by replacing the Gaussian distribution used for generating the pseudo-data by a discrete distribution. A well known example is the Rademacher two-point distribution: $P(w_t = 1) = P(w_t = -1) = 1/2$. Although Cavaliere et al. (2008b) found no discernible differences between the finite sample properties of the bootstrap unit root tests based on the Gaussian or the Rademacher distributions, in the next section we will carry out the sequential procedure using both distributions.

Finally, notice that the cdf $G_{j,T}^*(\cdot)$ required in step 5 is not known. However, as is standard in the bootstrap literature it can be approximated in the usual way through numerical simulation; cf. Hansen (1996). This is achieved by generating N (conditionally) independent bootstrap statistics, $Q_{n:j}^*$, $n = 1, \ldots, N$, computed as above but recursively from

$$\Delta X_{n:j,t}^* := \hat{\alpha}_j \hat{\beta}_j' X_{n:j,t-1}^* + \hat{\Gamma}_1 \Delta X_{n:j,t-1}^* + \ldots + \hat{\Gamma}_{k-1} \Delta X_{n:j,t-k+1}^* + \varepsilon_{n:t}^* t = 1, \ldots, T,$$

for some initial values $X_{n:j,-k+1}^*, \ldots, X_{n:j,0}^*$ and with $\{\{w_{n:t}\}_{t=1}^T\}_{n=1}^N$ a doubly independent $N(0,1)$ sequence. The simulated bootstrap p-value is then computed as $\tilde{p}_{j,T}^* := N^{-1} \sum_{n=1}^N I(Q_{n:j,r}^* > Q_j)$, and is such that $\tilde{p}_{j,T}^* \to p_{j,T}^*$ as $N \to \infty$, almost surely.

4 The Empirical Analysis

We now tackle the issue of investigating the long-term dynamics of the stock market indexes described in Sect. 2. The aim of the analyisis is to assess whether the existing empirical evidence supporting the idea that international stock markets are cointegrated.

[2]The asymptotic and small sample properties of the wild bootstrap procedure outlined in this section can be found in Cavaliere et al. (2008a,b).

In order to compare our results to those reported in the literature, we refer to the 1974–1990 period, as in most of the existing studies (cf. e.g. Kasa, 1992, and Richards, 1995). Furthermore, we extend the analysis to the 1974–2007 period, hence covering the recent evolution of the international stock markets as well.

We first estimate the VAR(k) model of Eq. (1) unrestrictedly (i.e., with full rank $r = p$). The lag length determination criteria point to two different results. The Likelihood ratio test and the Akaike information criterion provides evidence in favour of $k = 4$ lags. Conversely, the Schwartz and the Hannan-Quinn information criteria suggest setting $k = 1$, which is more in line with the well-known evidence of no significant autocorrelation in monthly stock market returns. Since the finite sample properties of cointegration tests are somewhat affected by the number of lags in the estimated VAR, we prefer to presents the tests using both $k = 1$ and $k = 4$ lags.

In Table 4 we report the results from the co-integration tests from the VAR(1) model. The first panel refers to the 1974–1990 period, while the second panel refers to the full sample (1974–2007). Together with the trace test statistics for rank r against rank $p = 4$, $r = 0,\ldots,3$, we report the corresponding 10% and the 5% asymptotic critical values as well as the asymptotic p-values, computed as suggested in Mac Kinnon, Haug, & Michelis (1999). As far as the 1974–1990 period is concerned, at the 10% significance level the sequential procedure based on the asymptotic p-values suggests rejecting the hypothesis of no cointegration ($r = 0$) and to select $r = 1$ (the asymptotic p-value for $r = 1$ is 0.088). That is, one co-integrating relation and three common stochastic trends driving the dynamics of the stock market indexes over the considered period. This result does not substantially change when the full sample is considered. The standard sequential procedure points toward $r = 1$, with an asymptotic p-value of 0.072.

Table 4 VAR(1) estimation results and p-values

Kasa sample

		Critical values		p-values		
Rank	Trace statistic	0.10	0.05	asympt.	i.i.d. bootstrap	Wild bootstrap
0	60.85	60.09	63.88	0.088	0.100	0.196
1	27.46	39.76	42.92	0.654	0.630	0.720
2	12.24	23.34	25.87	0.795	0.848	0.866
3	4.30	10.67	12.52	0.699	0.831	0.869

Full sample

		Critical values		p-values		
Rank	Trace statistic	0.10	0.05	asympt.	i.i.d. bootstrap	Wild bootstrap
0	61.98	60.09	63.88	0.072	0.069	0.279
1	26.24	39.76	42.92	0.724	0.761	0.829
2	10.26	23.34	25.87	0.913	0.935	0.934
3	3.45	10.67	12.52	0.820	0.857	0.856

As discussed earlier, in finite samples the asymptotic distributions of the co-integration tests may not deliver good approximations of the actual, finite sample distributions, even when the errors are iid. Moreover, the use of asymptotic critical values can lead to oversized tests when the data are conditionally heteroskedastic (see Cavaliere et al., 2008a, b). That is, the presence of conditional heteroskedasticity may inflate the evidence toward co-integration. In order to check the robustness of the $r = 1$ result obtained using asymptotic p-values, we compare this result to those obtained using bootstrap method.

First, we report p-values using the iid bootstrap as suggested in Swensen (2006). This bootstrap method may deliver better finite sample approximation of the distribution of the test statistics if the errors are independent and identically distributed or if the degree of conditional heteroskedasticity is low (see Cavaliere et al., 2008a). Conversely, in the presence of strong heteroskedasticity, Swensen's iid bootstrap tends to identify too many cointegrating relations. For this reason, we also present p-values obtained from the wild bootstrap procedure outlined in the previous section. As demonstrated by Cavaliere et al., (2008b), this procedure delivers p-values which are (asymptotically) unaffected by strong conditional heteroskedasticity. We report

In the last two columns of Tables 4 and 5 we report the p-values for both Swensen's i.i.d. bootstrap and the wild bootstrap proposed in the previous section. In both cases, we considered $N = 999$ bootstrap replications. The wild bootstrap p-values are computed by imposing that w_t is Gaussian; results do not change if the Rademacher distribution is used instead. These clearly show that the evidence toward the presence of cointegration may actually depend on the significant conditional heteroskedasticity which affect the stock market indexed considered.

Table 5 VAR(4) estimation results and p-values

Kasa sample						
		Critical values		p-values		
Rank	Trace statistic	0.10	0.05	asympt.	i.i.d. bootstrap	Wild bootstrap
0	71.84	60.09	63.88	0.009	0.034	0.101
1	21.12	39.76	42.92	0.937	0.959	0.974
2	11.15	23.34	25.87	0.866	0.911	0.923
3	4.48	10.67	12.52	0.672	0.692	0.720
Full sample						
		Critical values		p-values		
Rank	Trace statistic	0.10	0.05	asympt.	i.i.d. bootstrap	Wild bootstrap
0	72.30	60.09	63.88	0.008	0.019	0.113
1	24.21	39.76	42.92	0.827	0.877	0.898
2	11.34	23.34	25.87	0.855	0.883	0.895
3	4.81	10.67	12.52	0.624	0.593	0.621

The p-values from the iid bootstrap still point toward the presence of cointegration at the 10% level (the p-values associated to the test for $r = 0$ are about 0.1 for the 1974–1990 period and about 0.07 for the whole sample). However, the wild bootstrap p-values show that the hypothesis of no cointegration cannot be rejected at all conventional significance level, both on the 1974–1990 and on the 1974–2007 periods. Hence, our results confirm that, once the presence of conditional heteroskedasticity is properly accounted for, there is actually no evidence of cointegration among the international stock market indexes considered.

The results from the VAR(4), illustrated in Table 5, reinforce this view. Specifically, the asymptotic p-values still suggest to reject the hypothesis of no cointegration. The traditional procedure indicates $r = 1$ with an asymptotic p-value of 0.009 for the 1974.1990 and of 0.008 for the total period. Also the i.i.d. bootstrap p-values point toward the presence of cointegration and are lined up to the asymptotic p-values. However, when taking conditional heteroskedasticity into account, wild bootstrap p-values indicate to mantain the hypothesis of no cointegration, as for the VAR(1).

5 Conclusions

Primary focus in recent literature on common trends in financial markets has been on traditional estimation and testing procedures, without paying sufficient attention to the specific characteristics of stock price index numbers. In this paper we underline the importance to take these characteristics into account. More specifically, we make use of inference procedures that account for both heteroskedasticity and jumps in volatility, which represent a stylized fact of financial series such as stock price index numbers.

Our results contribute to the discussion on whether or not international stock indexes are cointegrated. As discussed by Richards (1995, p.632), the presence of cointegration "would imply that returns in these markets may follow different patterns in the short term, but that in the long run the levels of total return indices (i.e., prices plus reinvested dividends) are very closely linked. The existence of such a cointegrating relationship might provide an explanation for the hitherto low level of international diversification by investors, as the benefits to diversification that are implied by the relatively low short-run correlations would disappear in the long run as correlations become stronger". Our results clearly show that this is not the case. While cointegration among international price indexes is detected on the basis of traditional procedures, we found no evidence of cointegration by using p-values robust to conditional heteroskedasticity. The absence of cointegration implies that long range investment strategies based on international stock price index numbers are likely to be profitable.

Our findings also show how the analysis of the interrelations among international stock price index numbers has to be performed using specific methods, robust to the presence of conditional heteroskedasticity. This result can be generalized to the analysis of many economic and financial price indexes, whenever characterized by heteroskedasticity and jumps in volatility.

Acknowledgements Financial support from MIUR is gratefully acknowledged. We thank the editor and an anonymous referee for helpful comments on a previous draft of the paper.

References

Ahlgren, N., & Antell, J. (2002). Testing for cointegration between international stock prices. *Applied Financial Economics, 12*, 851–861.

Cavaliere, G., Rahbek, A., & Taylor, A. M. R. (2008a). Co-integration rank testing under conditional heteroskedasticity (Working paper).

Cavaliere, G., Rahbek, A., & Taylor, A. M. R. (2008b). Testing for cointegration in vector autoregressions with non-stationary volatility. *Journal of Econometrics*, forthcoming.

Corhay, A., Rad, A., & Urbain, J. (1993). Common stochastic trends in European stock markets. *Economics Letters, 42*, 385–390.

Davidson, R., & Flachaire, E. (2008). The wild boostrap, tamed at last. *Journal of Econometrics, 146*, 162–169.

Engle, R., & Granger, C. J. W. (1987). Cointegration and error-correction: Representation, estimation, and testing. *Econometrica, 55*, 251–276.

Fama, E. F., & French, K. R. (1988). Permanent and temporary components of stock prices. *Journal of Political Economy, 96*, 246–273.

Garcia Pascual, A. (2003). Assessing european stock markets (Co)integration. *Economics Letters, 78*, 197–203.

Hansen, B. E. (1996). Inference when a nuisance parameter is not identified under the null hypothesis. *Econometrica, 64*, 413–430.

Harris, R. I. D., & Judge, G. (1998). Small sample testing for cointegration using the bootstrap approach. *Economics Letters, 58*, 31–37.

Johansen, S. (1996). *Likelihood-based inference in cointegrated vector autoregressive models.* Oxford : Oxford University Press.

Kasa, K. (1992). Common stochastic trends in international stock markets. *Journal of Monetary Economics, 29*, 95–124.

Liu, R. Y. (1988). Bootstrap procedures under some non i.i.d. Models. *Annals of Statistics, 16*, 1696–1708.

Mac Kinnon, J. G., Haug, A. A., & Michelis, L. (1999). Numerical distribution functions of likelihood ratio tests for cointegration. *Journal of Applied Econometrics, 14*, 563–577.

Mammen, E. (1993). Bootstrap and wild bootstrap for high dimensional linear models. *Annals of Statistics, 21*, 255–285.

Mantalos, P., & Shukur, G. (2001). Bootstrapped Johansen tests for cointegration relatioships: A graphical analysis. *Journal of Statistical Computation and Simulation, 68*, 351–371.

Rangvid, J. (2001). Increasing convergence among European stock markets? A recursive common stochastic trends analysis. *Economics Letters, 71*, 383–389.

Richards, A. J. (1995). Co-movements in national stock market returns: Evidence of predictability, but not cointegration. *Journal of Monetary Economics, 36*, 631–654.

Swensen, A. R. (2006). Bootstrap algorithms for testing and determining the cointegration rank in VAR models. *Econometrica, 74*, 1699–1714.

Trenkler, C. (2009). Bootstrapping systems cointegration tests with a prior adjustment for deterministic terms. *Econometric Theory, 25*, 243–269.

van Giersbergen, N. P. A. (1996). Bootstrapping the trace statistics in VAR models: Monte Carlo results and applications. *Oxford Bullettin of Economics and Statistics, 58*, 391–408.

Wu, C. F. J. (1986). Jackknife, bootstrap, and other resampling methods. *Annals of Statistics, 14*, 1261–1295.

An Application of Index Numbers Theory to Interest Rates

Javier Huerga

1 Introduction

The euro area Monetary Financial Institutions (MFIs) interest rates statistics (MIR) are monthly statistics on the interest rates applied by MFIs on loans granted to and deposits received from Households (HHs) and Non-Financial Corporations (NFCs) in the euro area. MIR statistics are compiled by the Eurosystem[1] in three steps.[2] First, reporting institutions calculate weighted average interest rates of all relevant loans and deposits and submit the average interest rate and aggregated business volume to their respective national central bank (NCB). Second, each NCB compiles the national average rates and the aggregate business volumes and submits these to the European Central Bank (ECB). In the third and final step, the ECB compiles euro area average rates and aggregate business volumes for each MIR category. At each of these steps, interest rates are calculated as the weighted average, by the corresponding amounts, of its components, in order to obtain the representative rate.

One of the features of this procedure is that, if the interest rate is different across credit institutions, across creditors/debtors or across countries, a change in the business volume may also have an impact on average interest rates. The effect of changes in volume of business across countries and its impact on the euro area MIR is monitored by the ECB. Similarly, the impact of changes in volume across institutions may give rise to changes in the national MIR, and may be of interest for the national central banks in order to explain the evolution and also to the ECB when analysing the development of national MIR.

J. Huerga (✉)
European Central Bank, Frankfurt am Mein, Germany
e-mail: javier.huerga@ecb.europa.eu

[1]The Eurosystem is the monetary authority of the Eurozone. It is a system of central banks consisting of the European Central Bank and the national central banks of the member states of the European Union whose currency is the euro.

[2]For further information on the MIR categories, definitions and compilation refer to ECB (2001) and ECB (2002).

L. Biggeri, G. Ferrari (eds.), *Price Indexes in Time and Space*, Contributions to Statistics, DOI 10.1007/978-3-7908-2140-6_13, © Springer-Verlag Berlin Heidelberg 2010

The question has arisen on what tools are optimal to analyse the origin of changes in MIR and isolate, for example, changes in euro interest rates originating from changes in national interest rates from changes caused by variations in the relative weights of the countries. One possible approach to deal with this issue is to use the statistical tools developed in the field of index number theory; indeed, the traditional problem of distinguishing real from nominal growth rates in many macroeconomic variables or the calculation of price indices resembles very much the questions raised on MIR. In both cases, the objective is to distinguish changes in business volumes from changes in prices/rates. According to this view, in the words of Diewert (2002) "*the index number problem can be regarded as the problem of decomposing the change in a value aggregate,* V^1/V^0*, into the product of a part that is due to price change,* $P(p^0, p^1, q^0, q^1)$ *and a part that is due to quantity change,* $Q(p^0, p^1, q^0, q^1)$".

Nevertheless, an important difference is noted. While some economic variables for which indices are calculated do not have a relevant meaning when expressed in absolute values (e.g. price indices), interest rates certainly provide information when expressed in absolute values. Similarly a change in interest is better understood and usually communicated in absolute values rather than in percentage values. For these reasons, this paper focuses on indices that keep the absolute values of the change rather than expressing it as a percentage. These indices were first presented in the first part of the twentieth century and have recently been revisited by Diewert (2005), who called them "difference" (as opposed to "ratio") index numbers. The first section briefly describes the theoretical background to use index theory in the field of interest rates and presents the optimal solution for binary comparisons. The second section moves to multiple period indices. The third section presents some selected examples of the application of the indices to interest rates. The four section concludes.

2 Binary Indices and Interest Rates

Interest rates for the euro area are calculated as the weight average of national interest rates. The weight is provided by their national contribution to euro area total in terms of business volumes. The business volumes can be the amounts associated to new contracts (for MIR new business) or amounts outstanding on the balance sheet at the end of the period (for MIR outstanding amounts); in the following both are treated in the same manner.

Therefore the interest rate for the euro area is calculated as follows

$$\mathbf{I_t} = \sum_k \mathrm{i(k)_t}^* \ \mathrm{w(k)_t} \tag{1}$$

where

$\mathrm{I_t}$ = euro area interest rate at time t

$\mathrm{i(k)_t}$ = interest rate in country k at point in time t

$\mathrm{w(k)_t}$ = weight of the business volume of country k (compared to total euro area business volume) at time t, i.e. $\mathrm{v(k)_t}/\sum_k \mathrm{v(k)_t}$

where $\mathrm{v(k)_t}$ = business volume for country k at time t

Index theory can be easily applied to MIR by just substituting prices by interest rates and volume of transaction by weights.[3] In the case of MIR, when comparing euro area interest rates $\mathbf{I_t}$ with $\mathbf{I_{t-1}}$ it is required to find out what part of the difference corresponds to developments in national rates and what part corresponds to changes in country weights, i.e. relative changes in business volumes. The interest rate difference between two different periods is given by $\mathbf{\Delta I_{t,\,t-1}= I_{t,}- I_{t-1}}$, and it can be decomposed into difference terms that represent the pure interest rate change **Int**, the pure weight change **Wgh** and, in some cases, a mixed or composite effect **Mix**.

$$\begin{aligned}\mathbf{\Delta I}_{t,\,t-1} &= \mathrm{I}_{t,} - \mathrm{I}_{t-1} = \mathbf{Int}(\mathrm{i}_t,\mathrm{i}_{t-1},\mathrm{w}_t,\mathrm{w}_{t-1}) + \mathbf{Wgh}(\mathrm{i}_t,\mathrm{i}_{t-1}\mathrm{w}_t,\mathrm{w}_{t-1})\\ &\quad + \mathbf{Mix}(\mathrm{i}_t,\mathrm{i}_{t-1},\mathrm{w}_t,\mathrm{w}_{t-1})\end{aligned}$$

The above decomposition or sum of binary indices[4] can be performed in several different ways, and in particular by using "difference" indices inspired on the ratio index, like the Laspeyres, Paasche, Fisher, Walsh, Vartia and Marshall-Edgeworth. The latter, when expressed as a "difference" index is usually called Bennet index, and is the one that, also in this context offers the best axiomatic properties and therefore is used in the rest of the paper. This is because only the Bennet binary index complies with the time reversal and symmetry tests, while limiting the decomposition to only two terms. The Bennet decomposition in the context of MIR would be expressed as follows:

$$\begin{aligned}\Delta I_{t,t-1} &= \sum_k \Delta i(k)_{t,t-1}\left(\frac{w(k)_t + w(k)_{t-1}}{2}\right)\\ &\quad + \sum_k \Delta w(k)_{t,t-1}\left(\frac{i(k)_t + i(k)_{t-1}}{2}\right)\end{aligned} \tag{3}$$

where the first sum results in the interest rate effect **Int** $(\mathrm{i}_{t,}, \mathrm{i}_{t-1}, \mathrm{w}_{t,}, \mathrm{w}_{t-1})$ and the second sum provides the **Wgh** $(\mathrm{i}_{t,}, \mathrm{i}_{t-1}, \mathrm{w}_{t,}, \mathrm{w}_{t-1})$, with no mixed effect.

While this formula already serves for the purposes of MIR analysis a further fine-tuning can be obtained by taking into account that the net changes in weights is to be zero $\sum_k \Delta \mathrm{w}(\mathrm{k})_{t,\,t-1} = 0$.

[3]It is noted that "weights" in the context of interest rates has a slightly different meaning than "weights" in the context of index theory. In the latter "weight" is usually calculated as the division of transactions in one product (prices by quantities) by total transactions; in MIR it is simply the percentage of the value (in euro) of loans/deposits over total loans/deposits, therefore comparable to quantities in usual index theory but not to the usual meaning of "weight" in index number theory.

[4]The term "binary" is used for the comparisons between two consecutive periods as in Stuvel (1989).

Therefore (3) can also be expressed as

$$\Delta I_{t,t-1} = \sum_k \Delta i(k)_{t,t-1} \left(\frac{w(k)_t + w(k)_{t-1}}{2}\right) + \sum_k \Delta w(k)_{t,t-1} \left(\frac{(i(k)_t - I_t)_t + (i(k)_{t-1} - I_{t-1})}{2}\right) \tag{4}$$

This "extended" version of the Bennet decomposition produces the same aggregate results that the previous version.[5] However, the contribution of each country (k) to the weight effect changes, measured as

$$\Delta w(k)_{t-t-1} \left(\frac{(i(k)_t - I_t)_t + (i(k)_{t-1} - I_{t-1})}{2}\right) \tag{5}$$

now is expressed in terms of the impact in respect of the average interest rate. Therefore the latter is most relevant, intuitive and easier to interpret when measuring contributions to the aggregate by each country, while it results in the same results as the former in aggregate terms.

3 From Binary Index to Multiple Period Index

Similarly to the ratio index theory, the difference indices can also be expanded to constitute a series of data that permits different comparisons of one of the components across time. For doing that, two different aspects must be considered. Firstly, it must be decided what precise formula will be used for the comparison. This issue has already being discussed in the previous section and therefore, the Bennet decomposition is used. In the second place, the question arises, when having multiple periods, whether to use a direct index or a chain index. For illustration purposes this alternative can, at least in theory be expanded to a re-basable index and a Divisa index.

A direct Bennet for the interest rate component on the basis of the Bennet decomposition would be as follows:

$$Index\_I = Index_t^{DB} = \sum_k \Delta i(k)_{t,0} \left(\frac{w(k)_t + w(k)_0}{2}\right) \tag{6}$$

where DB indicates "direct Bennet-type" (index).

[5] The "extended" decomposition was initially proposed by Coene (2004) on the basis of previous work by Berthier (2001).

A chain index in this context is built so that each link is chained to the previous one by adding them. Therefore, the difference chain index of interest rates on the basis of the Bennet-type decomposition would be as follows:

$$Index\_II = Index_t^{CB} = \sum_t \sum_k \Delta i(k)_{t,t-1} \left(\frac{w(k)_t + w(k)_{t-1}}{2} \right)$$

where CB stands for "chain Bennet-type" (index).

In theory, a chain index can be translated to a continuous time modelisation. With continuous time, the sum appearing in the chain index formula would be substituted by an integral, as follows:

$$Index_t^{DIB} = \oint_t w(k)_t{}^* \frac{di(k)_t{}^*}{dt} dt$$

where DIB stands for Divisia – Bennet – type index.

However, this index is not applicable in practice, because we have neither a function that explains weights in terms of time nor a function of interest rates in terms of time. Therefore additional input would be needed to model and estimate the appropriate functions.

A possible compromise between the direct index and the chain index would be a re-baseble direct index, consisting in applying the direct index for a limited number of periods and as from one point in time referring the comparisons to a different point in time. The frequency of the change in the reference point can be different; for monthly series, it would possibly be defined in terms of years, e.g. every year or every 5 years. The index formula would be as follows:

$Index\_III = Index_t^{DRB} = \sum\limits_k \Delta i(k)_{t,s} \left(\frac{w(k)_t + w(k)_s}{2} \right) + Index_s^{DRME}$ for periods where the last change in reference period before t happened in s (starting with s=0). In the case of yearly rebasing, s = 0,12,24, ..., in case of quarterly rebasing s = 0,3,6,9 ... DRB stands for "difference re-basable Marshall-Edgeworth-type" (index).

The main difference between a chain index and a direct index is that a chain index implicitly reflects in its value the way followed from period 0 to period t, giving different results depending on the road followed from one point to another. A direct index provides the same results for a point in time regardless of the path of intermediate data, because it only considers the weights at the first and last period.

A clear advantage of a chain index is that the weights are continuously updated. That results in that, when comparing the index at any two points in time, only the weights for the chosen interval have an impact on the changes in the index. It is also noted that, contrary to other data sets for which relative weights may not always be available with the same frequency as prices, that problem does not occur in MIR. On the contrary, a direct index implies that the initial weighting has a very strong bearing on the whole index. When trying to find out the evolution of interest rates in isolation of weights between two periods different from the initial period, it seems somehow strange that the weights of the initial period have a bearing on the final result regardless of how distant from the initial period the two periods examined are, and how much may have changed in the meantime. In addition, some of the possible advantages of direct indices in usual index theory do not apply to difference indices. In particular, some direct ratio indices (e.g. Laspeyres or Paasche, but not Fisher) can be interpreted as a ratio of expenditures together with a mean of price relatives. This interpretation is not applicable (or easily adaptable) to difference indices, regardless on whether they are direct or chain indices.

Nevertheless, a possible disadvantage of a chain index is that it is unclear whether considering the intermediate steps when comparing euro area interest rates between two points in time would help an analyst, who is possibly more interested in knowing whether the difference in rates between the two separate periods is attributable to changes in national rates or to changes in country weights. On the other hand, the evolution of MIR data is monitored in the ECB on a monthly basis, and the chain index would better help to link the monthly analysis with the longer period comparisons.

Finally, the advantages of a chain index would be re-enforced if the different indices do not differ very much in practice. In that case, it could be said that in normal circumstances all indices provide similar results, but in case there is any important (isolated) change in weights, it would be taken into account in the index only for the span of periods included in the comparison.

4 Application of Indices to MIR - January 2003 to October 2008

In order to analyse the possible relevance of the three indices (Index I, Index II, Index III) proposed in section 2, they have been applied to each of the 43 euro area MIR monthy data for the period January 2003 to October 2008. The 43 euro area indicators refer to different categories of loans (e.g. new loans to households (HHs) for house purchase, with initial rate fixation up to 1 year, or outstanding amounts (OA) of deposits of non-financial corporations with maturity up to 2 years). Each index is compared with the accumulated change of the actual euro area aggregate interest rate starting in January 2003.[6]

In general the results show that the evolution of the accumulated changes in actual euro area aggregate interest rate does not differ much from the different indices of interest rates, except for a few of the 43 interest rate categories. That indicates that indicates that the changes in weight across countries are generally small, having a very limited impact on euro area aggregates for long periods, except

[6]A particularity of MIR has been taken into account in the calculations. Whenever no operation has taken place on new business or no outstanding amounts remains for a single category in a country, no figure is reported to the ECB for that country. If this absence of interest rate figure were treated as zero it would result in a spurious impact in the interest rate component. To avoid this, whenever no interest rate was reported for a specific category and month, the latest previously reported interest rate is used to calculate the interest rate effect, resulting in no impact on the interest rate component.

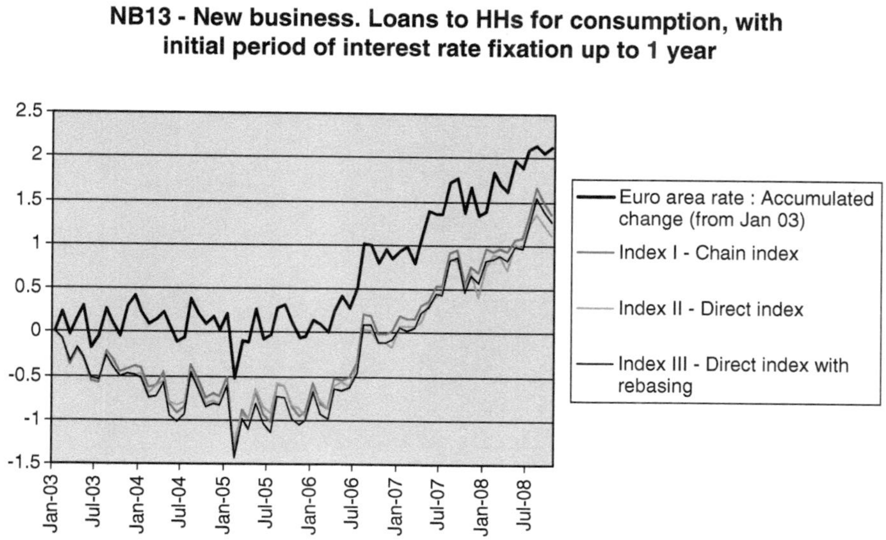

Chart 1 Indices applied to NB 13 for the period Jan03 to Oct08

on a few cases. The most clearer case is the new loans to households for consumption purpose. Chart 1 shows the differences between the accumulated changes in the euro area rate and the indices, particularly at beginning of the period analysed and also at the very end. Therefore the index can be used in the economic analysis to make explicit that, while in the period 2003–2005 the euro area average stayed around its initial value in January 2003, this was influenced by changes in weights, in particular with an increase of the weight of countries with relative higher rates and a parallel decrease of the countries with lower interest rates.

Focusing on the behaviour of the indices as compared with one another, in most of the cases the indices are quite close to each other. Differences are particularly prominent in the new deposits from non-financial corporations (NFCs) with maturity over 2 years (NB10). In Chart 2, the different behaviour of the indices is analysed for a selected period. It is noted that the chain index immediately reacts to the decrease in the interest rates in May 2005, while the other indices operate with a certain lag. This example also speaks in favour of the use of a chain index.

Regarding the series that shows the higher differences in the development of the indices, the decomposition (binary indices) is an important tool on its own, in addition to being a building block of the multiple period index. New deposits from households redeemable at notice up to 3 months notice (NB5) is possibly the most illustrative case on the link between indices and decomposition. For this MIR category, a weight effect occurred in June 2005 (Chart 3); this effect caused a drop of the Indices in June while the original series increases (Chart 4). For the subsequent periods, the indices and original series behave very much the same way, with just a difference in level coming from that particular weight effect. In fact, the weight

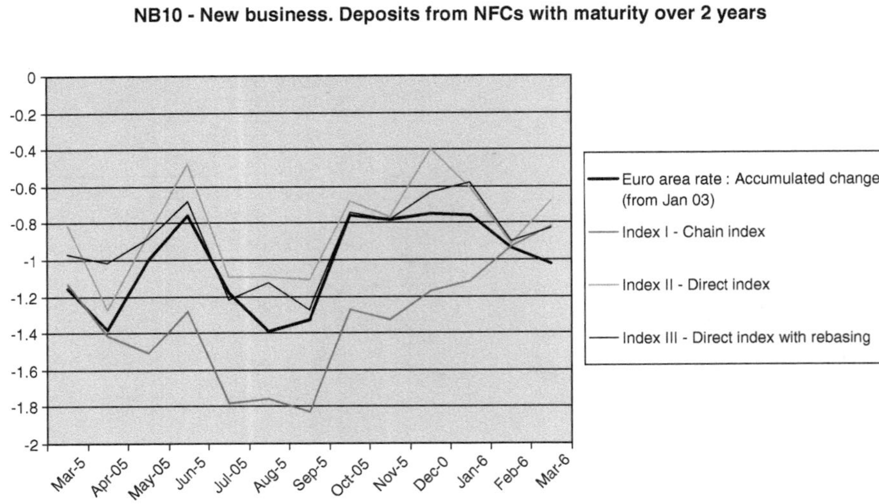

Chart 2 Indices applied to NB 10 for a selected period (Mar05 to Mar06)

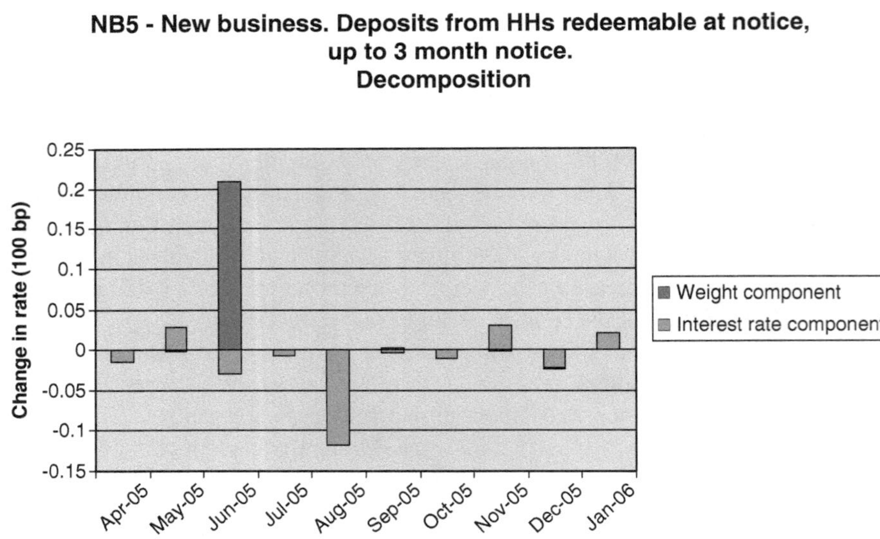

Chart 3 Decomposition to NB 5 for a selected period (Apr05 to Mar06)

effect in June 2005 is caused by a change in the statistical classification of a particular financial instrument in one member state. Therefore the index correctly discounts the weight change caused by this statistical re-classification.[7]

[7]Other ways of avoiding this type of statistical break, like for example the reporting of pre-break values, are beyond the scope of this paper.

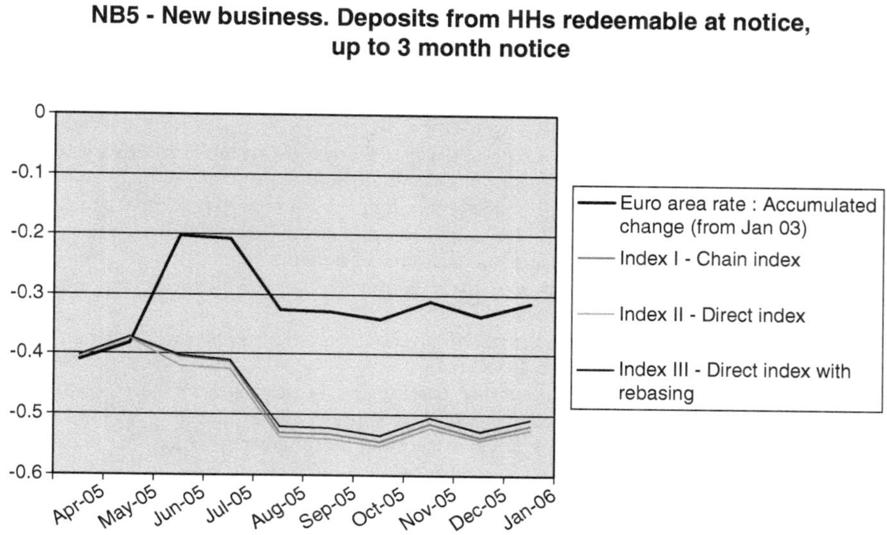

Chart 4 Indices applied to NB 5 for a selected period (Apr05 to Mar06)

5 Conclusion

This paper is an application of index number theory to MFI interest rates (MIR) statistics, on the basis of the work by Diewert (2005) using differences rather than ratios. This approach seems more appropriate for interest rates, for which changes are usually measured in absolute rather than in relative terms. The application of index number theory to MIR shows that the regular calculation and publication of month-to-month decomposition (binary index) at euro area aggregate level could help analysts to interpret monthly changes in the euro area rates. In addition a chain difference index, which would accumulate all month-to-month decompositions from the starting point of MIR statistics (Jan03) could also be useful. In that sense, Index I (chain index) may have some advantages as it solves the so-called "index number problem", while for the analysed set of data seems to offer better results. The index would permit to assess the evolution in longer periods in terms of changes in interest rates. Finally, an index will also exclude the weight impact of new countries entering the euro area.

Acknowledgments This contribution is based on an original idea of Steven Keuning. That idea was developed and discussed in an ECB Working Paper prepared by J Huerga and L Steklacova. I am very grateful to Olivier Coene from the National Bank of Belgium, who contributed with comments and actually suggested the "extended Bennet" formulation, proposed in this paper. I would also like to thank Julia Weber for her in-depth review and suggestions on previous work, Holger Neuhaus, Ruth Magono, Roswitha Hutter and Jean-Marc Israël for their help and support, and the members of the Working Group on Monetary and Financial Statistics (WG MFS) and the participants in the International Workshop on Price Indices organised by the University of Florence on 29–30 September 2008 for their comments on previous discussions. The views expressed in this note are those of the author and do not necessarily reflect those of the ECB or the Eurosystem.

References

Allen, R. G. D. (1975). *Index numbers in theory and practice*, New York: MacMillan – Palgrave.

Balk, B. M., & Diewert, W. E. (2001). A characterisation of the Törnqvist price index. *Economics Letters, 72*, 279–281.

Berthier, J. P. (1997, June). Calcul des contributions aux écarts entre évolution du PIB à prix 80 et évolution aux prix de l'année précedente, Note ISEE, Paris.

Coene, O. (2004a, November 12). Some remarks concerning the difference between the method of Berthier and the method of the ECB: Decomposition into factors – discussion document, dated 12 November 2004, Brussels: Internal National Bank of Belgium.

Coene, O. (2004b). *Methodology for analysis of adjustments of weighted averages*, Brussels: Internal National Bank of Belgium.

Diewert, W. E. (1995, September). *On the stochastic approach to index numbers* (Discussion Paper No. DP95-31). University of British Columbia.

Diewert, W. E. (2002, March). *Harmonized indexes of consumer prices: Their conceptual foundations* (Working Paper No. 130). Frankfurt: European Central Bank.

Diewert, W. E. (2005, January). Index number theory using differences rather than ratios, *The American Journal of Economics and Sociology, 64*(1), 311–360.

ECB. (2001). Regulation of the European Central Bank of 20 December 2001 concerning statistics on interest rates applied by monetary financial institutions to deposits and loans vis-à-vis households and non-financial corporations (ECB/2001/18), OJ L 318, 27.11.1998, p8.

ECB. (2002, October). *Manual on MFI Interest rate statistics: Regulation ECB/2001/18*. Frankfurt: ECB.

Eichhorn, W., & Voeller, J.(1976). *Theory of the price index*, Berlin: Springer.

Huerga, J., & Steklacova, L. (2008, September). *An application of index numbers theory to interest rates*, (ECB Working Paper). Frankfurt.

Hulten, C. R. (2003). Divisia index numbers, *Econometrica, 41* (6) November 1973.

National Research Council. (2002). *At what price? Conceptualizing and measuring cost-of-living and price indexes*. Washington, DC: National Academy Press.

Stuvel, G. (1989). *The index-number problem and its solution*, London: MacMillan.

Törnqvist, L. (1936). The Bank of Finland's consumption price index, *Bank of Finland Monthly Bulletin, 10*, 1–8.

Von der Lippe, P. (2001). Chain indices: A study in price index theory, *Statistisches Budesamt, Spectrum of Federal Statistics, 16*, Wiesbaden.

Sector Price Indexes in Financial Markets: Methodological Issues

Michele Costa and Luca De Angelis

1 Introduction

Price index numbers play a prominent role in financial markets both as synthetic measure of changes and as reference for risk diversification. Despite the importance of these instruments, some methodological issues with a relevant empirical impact have been only marginally analyzed. The purpose of this paper is to contribute to the analysis of two aspects.

Firstly we examine stock price indexes related to specific economic sectors and propose a new stock's classification methodology able to obtain homogenous sectors under the risk-return profile. In our opinion traditional sector definition is not always successful in correctly discriminate homogeneous sectors. Furthermore, it is strongly static, since it is rarely updated, and considers only the main business, while stock companies frequently operate in different sectors. On the contrary, the new classification can be updated if the stock's risk-return profile changes over time. We provide a framework for the application of latent class models to the financial market analysis and we introduce a methodological dimension into the stock's classification. In particular we suggest a method able to define the composition of the basket underlying sector price index numbers.

The new sectors obtained on the basis of our proposal allow a relevant improvement in terms of the coverage and the representativeness of the risk-return profile.

The second issue refers to the effects of the weighting structure on the volatility and on the correlation between indexes. That is a key point, since volatility and correlation directly determine the risk diversification processes. We compare different index numbers proposals, from the simplest version to the methodology currently used and based on the concept of floating shares. Furthermore, we apply the different index numbers to the traditional and the new stock's classification in order to find evidence about the interrelations between aggregation processes and

M. Costa (✉)
Dipartimento di Scienze Statistiche, Universita' di Bologna, Bolgana, Italy
e-mail: michele.costa@unibo.it

L. Biggeri, G. Ferrari (eds.), *Price Indexes in Time and Space*, Contributions to Statistics, DOI 10.1007/978-3-7908-2140-6_14, © Springer-Verlag Berlin Heidelberg 2010

classification methods. Our final step refers to the implications of classification, basket composition, and weighting structure for portfolio analysis.

2 Methodological Issues

This section briefly summarizes the framework for the application of latent class (LC) models to the stock's classification (Costa & De Angelis, 2008). LC models, originally introduced as a method for sociological researches (Lazarsfeld, 1950; Hagenaars, 1990), are characterized by a strong ability to classify different objects into homogenous latent classes on the basis of the responses to a set of observed indicators (Moustaki & Papageorgiu, 2005). The use of latent tools is quite successful in the field of financial markets, where the evaluation processes are developed over two variables, the risk and the expected return, which are not directly observable. Our purpose is to introduce a new sector classification obtained by exploiting the potential of LC models for classifying stock companies into homogenous groups under risk-return profile. The new sector classification loses its economic trait based on the company's main business but it acquires a strong financial value. The direct effect of the new classification is to provide an efficient and straightforward method in order to define the composition of the basket underlying the sector price indexes. We also contribute to the debate on the classification of financial time series, which is receiving growing attention in the statistical literature (Otranto, 2008; Lisi & Otranto, 2008).

Besides the basket composition, also the weighting structure of the sector price index numbers represents a topic of great interest. On the analogy of price index numbers of the most important economic variables, there has been the tendency to adopt the total quantities of the different financial assets as a weighting structure. With respect to such choice, it has been argued that alternative formulations, developed on the basis of traded quantities, or potentially tradable quantities, can lead to a more correct measure of price changes. Different weighting structures can lead to very different outcomes (Lisi & Mortandello, 2004) about the variability and the risk measures of financial price index numbers. Our purpose is to evaluate the effects of alternative proposals on both new and traditional classification as well as on portfolio diversification processes.

2.1 A New Stock's Classification

We suggest a LC model in which some observed characteristics of the stock's return distribution, expressed in the form of categorical variables, are treated as indicators. More specifically, we resort to three indicators (the mean, M, the standard deviation, S, and the first percentile, P, of the return distribution) in order to obtain a (categorical) latent variable denoted by X which is able to explain the association among the manifest variables and to allocate the stocks in the new sector classification. Furthermore, we use the traditional sectors C as covariates.

The LC model is specified as

$$\pi_{mspc} = \sum_{x=1}^{K} \pi_{xmspc} = \sum_{x=1}^{K} \pi_{x|c}\pi_{m|x}\pi_{s|x}\pi_{p|x} \tag{1}$$

where m, s, p, c, and x are the generic indices of M, S, P, C, and X respectively, while K denotes the number of latent classes. The probability π_{mspc} indicates the proportion of stocks with mean $M = m$, standard deviation $S = s$, first percentile $P = p$, and which belong to sector c. The probability π_{xmspc} denotes the proportion of stocks in the five-way contingency table whereas $\pi_{x|c}$ is the probability of belonging to latent class x (prior probability given the covariate c). The other πs indicate the conditional probabilities: for instance, $\pi_{m|x}$ is the conditional probability of being in category m of variable M, given that the stock belongs to latent class x.

In this framework, all the relationship among the indicator variables M, S, and P is explained by the latent variable X which is influenced by the covariate C[1]. As shown in the Fig. (1), no relationship exists between C and the indicators.

The parameter estimation is obtained by maximizing the log-likelihood function using the EM algorithm (Dempster, Laird, & Rubin 1977).

LC analysis allows to determine the smallest number of latent classes K which is sufficient to explain the associations among the indicators. K can be determined on the basis of likelihood ratio test L^2 and of some information criterion such as the AIC or the BIC statistic. Since the latent classes are interpreted as the sectors of the new classification, the test for the choice of K represents a crucial step and allows to determine the number of the sectors on the basis of a straightforward methodology.

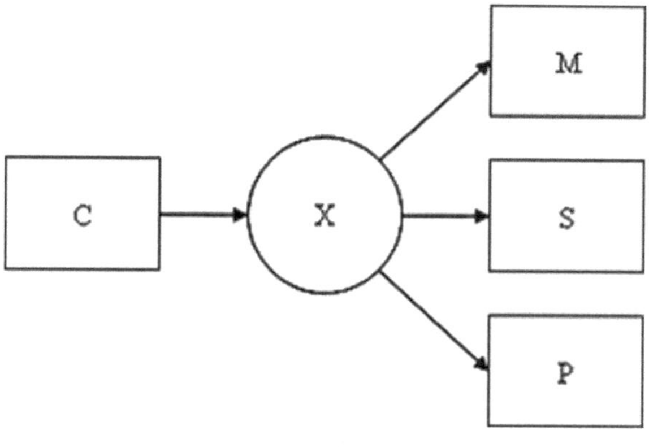

Fig. 1 LC model graphical representation

[1]We report Eq. (1) in a simplified form, where does not explicitly appear the probability of each traditional sector π_c; the complete expression for π_{xmspc} is: $\pi_{xmspc} = \pi_c\pi_{x|c}\pi_{m|x}\pi_{s|x}\pi_{p|x}$.

The final step of LC analysis is to classify the stocks into the appropriate latent class. Stocks are assigned to the class for which the posterior membership probability is the highest. This approach is usually known as LC cluster model because the goal of classification into K homogenous groups is identical to that of cluster analysis (Magidson & Vermunt, 2001, 2004). Estimates for the posterior membership probability can be calculated using the Bayes' theorem:

$$\hat{\pi}_{x|mspc} = \frac{\hat{\pi}_{xmspc}}{\sum_{x=1}^{K} \hat{\pi}_{xmspc}} \tag{2}$$

where $\hat{\pi}_{xmspc}$ is obtained by substituting the model parameter estimates in place of the corresponding parameters in Eq. (1) for $x = 1, ..., K$. The allocation rule refers to the posterior membership probabilities in Eq. (2), thus one stock will be classified into latent class 1 if

$$\hat{\pi}_{x=1|mspc} > \hat{\pi}_{x=i|mspc} \tag{3}$$

with $i = 2, ..., K$. By means of Expression (3), we achieve the composition of the basket underlying the new sector price index numbers.

2.2 Sector Price Index Numbers

Our second proposal refers to the development of a synthetic indicator for each sector. We suggest different alternatives in order to evaluate the effects of the weighting structure on the index characteristics. Furthermore we want to compare the indexes obtained within the new classification to the traditional references.

The most direct and immediate option is the arithmetic average of the simple indexes p_{it}/p_{i0}:

$$I1 = \frac{1}{n} \sum_{i=1}^{n} \frac{p_{it}}{p_{i0}} \tag{4}$$

where p_{it} and p_{i0} are the prices of the i-th stock at time t and at time 0 respectively, while n is the number of stocks included in the basket. Notwithstanding its extreme simplicity, index $I1$ can provide useful information and, since 1882, it is adopted for the calculation of the Dow Jones Indexes.

The second proposal follows the rules traditionally used in financial markets, which duplicate the Laspeyres-based methodology developed for price index numbers related to the main economic variables. In this framework, the market capitalization at the base time is employed as weighting structure:

$$I2 = \frac{\sum_{i=1}^{n} \frac{p_{it}}{p_{i0}} p_{i0} q_{i0}}{\sum_{i=1}^{n} p_{i0} q_{i0}} \tag{5}$$

where q_{i0} is the number of shares at time 0 for the i-th company. Most of financial price index numbers are usually obtained following Expression (5).

The number of shares represents the fundamental element in the our next index proposal, where the weights are given by q_{it}:

$$I3 = \frac{\sum_{i=1}^{n} \frac{p_{it}}{p_{i0}} q_{it}}{\sum_{i=1}^{n} q_{it}} \tag{6}$$

Furthermore, it is also possible to resort to the volumes V_{it} traded at time t as weighting structure:

$$I4 = \frac{\sum_{i=1}^{n} \frac{p_{it}}{p_{i0}} V_{it}}{\sum_{i=1}^{n} V_{it}} \tag{7}$$

Our last proposal refers to the methodology currently adopted in the S&PMIB index, where are acknowledged some critical observations about the use of standard price index numbers $I2$. The main criticism refers to the representativeness of q_i, the total number of shares. When the majority or controlling shareholders own a relevant number of shares, it could be appropriate to exclude this quantity from q_i and from the index computation, since it does not participate to the regular trading activity.

A reference which could ensure an higher representativeness is defined on the basis of the fraction of free floating shares f_i. The measurement of the fraction of free floating shares is still considered as an open problem in the financial community because it does not exist a common procedure in order to compute the quantities f_i. In the following we resort to the Standard and Poor's methodology and we suggest to use both the simple number of free floating shares:

$$I5 = \frac{\sum_{i=1}^{n} \frac{p_{it}}{p_{i0}} q_{it} f_{it}}{\sum_{i=1}^{n} q_{it} f_{it}} \tag{8}$$

and the free floating capitalization:

$$I6 = \frac{\sum_{i=1}^{n} \frac{p_{it}}{p_{i0}} q_{it} f_{it} p_{i0}}{\sum_{i=1}^{n} q_{it} f_{it} p_{i0}}. \tag{9}$$

If the choice of the base period represents a traditionally crucial point in price indexes theory, in financial price indexes the reference base plays a even more relevant role.

First, index revisions frequently modify the original basket by introducing or eliminating some companies. Therefore, at time t, some stocks initially included in the index could be no more present, while some other stocks could have been added to the basket and a direct comparison with the base period would be not appropriate. Second, capital operations modify both the stock prices and the number of shares, thus introducing a gap with respect to the base period. Finally, also extraordinary or ordinary payments (such as dividends) modify the stock's value and, as the capital operations, introduce a gap with respect to p_0.

Therefore, in financial market indexes, it is necessary to adjust the values of prices p_0 and quantities q_0 related to the base period. The expressions of the indexes $I1$ to $I6$ represent a general reference, but their empirical computation requires to transform p_0 and q_0 by resorting to an adjustment factor k_t which includes the effects of basket modifications, capital operations and payments occurred from the base period to time t.

Since a correct comparison cannot be directly operated between time t and time 0, in the following we calculate the previous indexes with respect to time $t-1$, but including the adjustment factor k_t:

$$I = \frac{\sum_{i=1}^{n} \frac{p_{it}}{p_{i0}} k_t W_{it}}{\sum_{i=1}^{n} W_{it}} \tag{10}$$

where W_{it} represent the generic weighting structure.

3 The Data

The data object of an empirical analysis consists of the monthly return from January 2002 to December 2007 of 136 stocks quoted at the Italian Stock Market. The stocks belong to 5 different sectors (number of stocks per sector is reported in brackets): energy (5), consumer discretionary (50), financials (58), utilities (11), materials (12). Sectors are defined on the basis of Global Industry Classification Standard (GICS) methodology developed by MSCI and Standard and Poors in 1999 and currently used within the S&PMIB index.

Table (1) reports some descriptive statistics for the stocks belonging to the 5 GICS sectors analyzed while Fig. (2) illustrates the mean and the standard deviation of the return distributions.

As shown in Fig. (2), the stocks belonging to the GICS consumer discretionary sector (depicted as circles) are quite different in both standard deviation and mean return, thus suggesting the presence of a strongly heterogeneous sector. Indeed, from

Table 1 Descriptive statistics for the GICS sectors

	Number of stocks	Mean return	Standard deviation
Energy	5	1.755	7.325
Consumer discretionary	50	0.500	9.351
Finance	58	0.870	7.875
Utilities	11	0.915	6.381
Materials	12	0.707	7.988

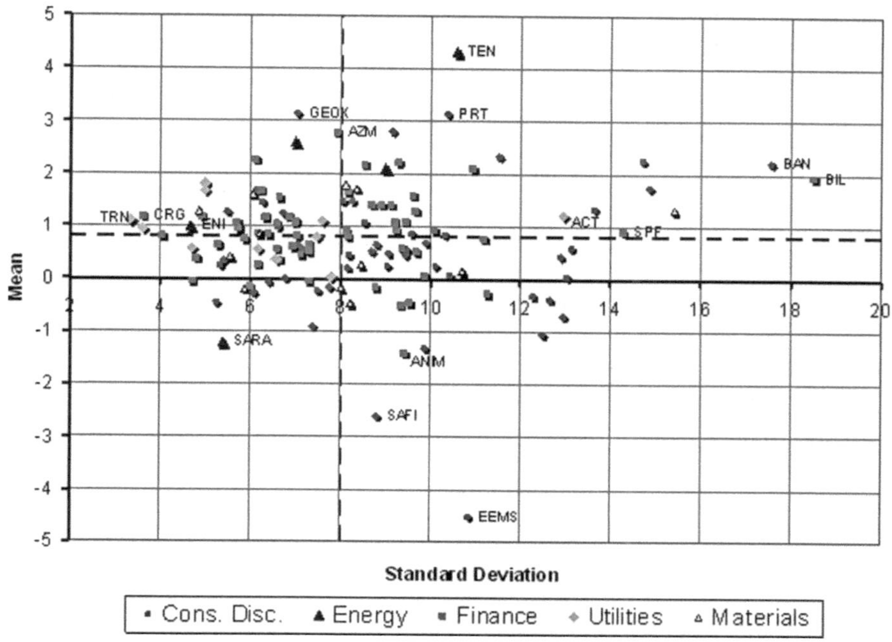

Fig. 2 Mean and standard deviation of 136 Italian stocks return distribution by GICS sector, monthly data 2002–2007

Table (1) arises that consumer discretionary sector has the highest value of standard deviation and the lowest value of mean return. An analogous profile is also detectable in GICS finance sector, which groups stocks dissimilar in both return and volatility: the squares depicted in Fig. (2) are quite dispersed as also confirmed by the high standard deviation value reported in Table (1). Furthermore, Table (1) reveals that GICS energy sector has the lowest volatility and the highest mean return but the black triangles in Fig. (2) clearly show how the stocks belonging to this sector have totally different values of mean return. On the contrary, GICS sectors materials and utilities are characterized by high volatility and quite homogenous mean returns. The logical conclusion which can be drawn from this evidence is that traditional sectors are not able to detect similar groups in mean return and standard deviation.

In order to suggest a new stock's classification and according to our proposal developed in Sect. 2.1, we consider 3 (ordinal) indicator variables with 2 categories each (Low and High): the mean (M), the standard deviation (S), and the first percentile (P) of the return distribution, which are illustrated in Fig. (3).

The five GICS sectors are generally present in both categories of each indicators, thus confirming their heterogeneity. The three indicators M, S, and P represent the observed variables for the LC model, while we use the GICS sectors as covariates. In the next section are presented the estimation results.

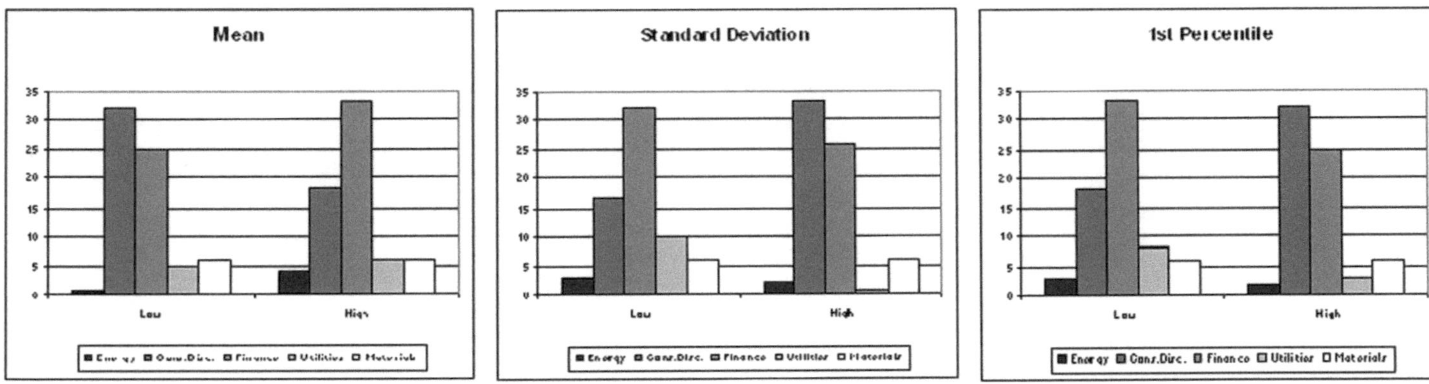

Fig. 3 Indicator variables for the latent class model

4 The Results

Our first results are related to the composition of the basket underlying the sector price indexes, which is achieved by means of the LC model illustrated in Sect. 2.1. In the following we show the results of the LC model estimation and the consequent stock's classification which defines the new sectors.

Furthermore, we contribute to the debate on the weighting structure of financial price index numbers by calculating the indexes listed in Sect. 2.2 with respect to the traditional and the new sector definition as basket compositions. We analyze the effects of both index weighting structure and basket composition on index's performance in the perspective of financial portfolio diversification.

4.1 The Estimation of the Latent Class Model

LC analysis usually begins by fitting a model with one latent class and then evaluating its goodness of fit by performing a likelihood ratio test. If the null hypothesis is rejected, the analysis continues by adding a second latent class, thus estimating a 2-class model. Again, a test is performed and if the value is unacceptable then a further latent class must be added to the model. This procedure continues until the LC model reaches an adequate fit. An alternative method in order to choose the best model is to compare the values of some information criterion. In the following we resort to Akaike Information Criterion based on the log-likelihood.

Table (2) reports the results from latent class model estimation with different number of classes. The 1-class model is rejected on the basis of the likelihood ratio test L^2 ($L^2 = 110.63, df = 32, p < 0.001$). The 2-class model provides a significant reduction of L^2 statistic, however its value is still too high ($L^2 = 40.65$, $df = 27$, $p = 0.04$). According to both likelihood ratio test and AIC information criterion, the 3-class model is the best estimated model.

The results related to the 3-class model estimation are shown in Table (3) which illustrates prior and conditional probabilities in addition to the means for each indicator. Prior probabilities $\hat{\pi}_x$ indicate that 43% of stocks are estimated to be in class1, 29% in class2, and 28% in class3. The interpretation of the characteristics of each latent class can be obtain on the basis of the conditional probabilities and the means

Table 2 Results from LC models estimation with different number of classes: log-likelihood *LL*, AIC information criterion and likelihood ratio test L^2

Model	LL	NPar	L^2	df	p-value	AIC(LL)
1-class	–282.789	3	110.628	32	1.4E-10	571.579
2-class	–247.798	8	40.646	27	0.04	511.596
3-class	–242.125	13	29.300	22	0.14	510.251
4-class	–239.544	18	24.137	17	0.12	515.088

Table 3 Results related to the 3-class model

Profile	Class1	Class2	Class3
Prior prob. $\hat{\pi}_x$	0.4332	0.2904	0.2763
Conditional prob.			
Indicator M			
$\hat{\pi}_{m=1\|x}$	0.5752	0.0280	0.9053
$\hat{\pi}_{m=2\|x}$	0.4248	0.9720	0.0947
Mean	1.4248	1.9720	1.0947
Indicator S			
$\hat{\pi}_{s=1\|x}$	0.0089	0.7675	0.9872
$\hat{\pi}_{s=2\|x}$	0.9911	0.2325	0.0128
Mean	1.9911	1.2325	1.0128
Indicator P			
$\hat{\pi}_{p=1\|x}$	0.0555	0.9573	0.7146
$\hat{\pi}_{p=2\|x}$	0.9445	0.0427	0.2854
Mean	1.9445	1.0427	1.2854

of the indicators. From Table (3) it arises that class1 has the highest values of indicator S and P means (1.99 and 1.94 respectively) and a medium value of indicator M mean (1.42), therefore class1 is characterized by high risk and medium mean return. Class2 has low values of indicator S and P means (1.23 and 1.04 respectively) and the highest value of indicator M mean (1.97): it allows the best investment opportunities since an high return matches a low risk. Finally, class3 has low values of indicator S and P means (1.01 and 1.29 respectively) and the lowest value of indicator M mean (1.09): it associates low return to low risk.

The estimation of the 3-class model also allows to classify the 136 stocks into each latent class using the posterior membership probabilities expressed in Eq. (2), thus achieving the new sector definition. The new sectors are constituted by 58, 43, and 35 stocks assigned to class1, class2, and class3 respectively. The traditional sectors are allocated into the latent classes as shown in Table (4).

The probabilities in Table (4) denote the weight that each GICS sector has within each latent class and they enable us to compare the difference between the traditional sector definition and the stock's allocation achieved by the LC model. It can be observed that class1 collects 63% of the stocks belonging to consumer discretionary sector and no stock from the utilities sector. On the contrary, stocks from

Table 4 New and traditional stocks classification

Traditional Sector	Class1	Class2	Class3
Energy	0.3902	0.4020	0.2078
Consumer discretionary	0.6293	0.1353	0.2354
Finance	0.3517	0.3745	0.2738
Utilities	0.0002	0.4754	0.5243
Materials	0.4257	0.3139	0.2604

energy, finance, and materials sectors are quite equally divided into the three classes. This can be taken as an evidence that traditional sectors are quite heterogenous and cannot properly discriminate stocks under the risk-return profile.

4.2 The Synthetic Price Index Numbers

In order to analyze the effects of different weighting structures, Table (5) reports the mean and the standard deviation of the synthetic price index numbers related to the stocks belonging to the 3 latent classes and the 5 traditional GICS sectors.

In Fig. (4) the values of the various price index numbers reported in Table (5) are depicted in the mean-standard deviation space.

From Table (5) and Fig. (4), it is possible to observe some relevant differences between the price index numbers.

First, free floating capitalization indexes (*I*6) perform quite similar to traditional fixed capitalization Laspeyres index numbers (*I*2). This evidence contributes to the debate on the role and importance of the free floating approach. The question which arises in the light of these results is if it is worth using *I*6 as the foremost stock market index number instead of *I*2 which is both less complicated to define and does not require a constant and expensive data activity.

Second, indexes with weighting structure defined on the basis of the number of shares (*I*3) and the free floating shares (*I*5) show the lowest mean values: they likely suffer from the performance of the companies with many shares emitted even though their importance (i.e., their prices) in the stock market is moderate. This may lead to anomalous results such as the negative values of mean return for consumer discretionary sector. Because of this flaw, the weighting structure based only on the number of shares is rarely used in financial index numbers.

Third, the indexes based on the volumes (*I*4) assume, in most of the cases, the highest values of both standard deviation and mean return.

In order to evaluate the effects of extreme values on the new and the traditional stock's classification, Table (6) reports the first and the fifth percentiles of the different synthetic indexes. Fig. (5) shows the impact that different weighting structure

Table 5 Mean and standard deviation of the indexes related to the new and traditional sectors

	*I*1		*I*2		*I*3		*I*4		*I*5		*I*6	
	m	*s*	*m*	*s*	*m*	*s*	*m*	*s*	*m*	*s*	*m*	*s*
Class1	0.65	5.66	1.31	6.90	0.88	6.34	1.60	7.15	1.07	6.44	1.36	6.97
Class2	1.39	3.64	1.23	3.22	1.20	3.55	1.47	3.69	1.23	3.32	1.17	3.64
Class3	0.25	3.73	0.55	4.26	0.31	4.36	0.36	4.71	0.32	4.43	0.55	4.48
Energy	2.26	6.05	1.35	4.85	1.64	5.08	1.41	4.93	1.53	4.78	1.26	4.71
Cons. Disc.	0.62	4.92	0.76	5.33	−0.01	6.34	0.64	7.31	−0.10	6.54	0.69	5.48
Finance	0.82	4.54	0.88	5.19	0.89	5.38	1.48	6.44	0.96	5.37	0.90	5.15
Utilities	0.75	4.43	0.71	4.15	0.72	3.91	0.72	4.04	0.69	3.92	0.68	4.12
Materials	0.73	4.70	1.13	5.40	0.55	5.57	2.18	8.15	0.45	5.97	1.13	5.51

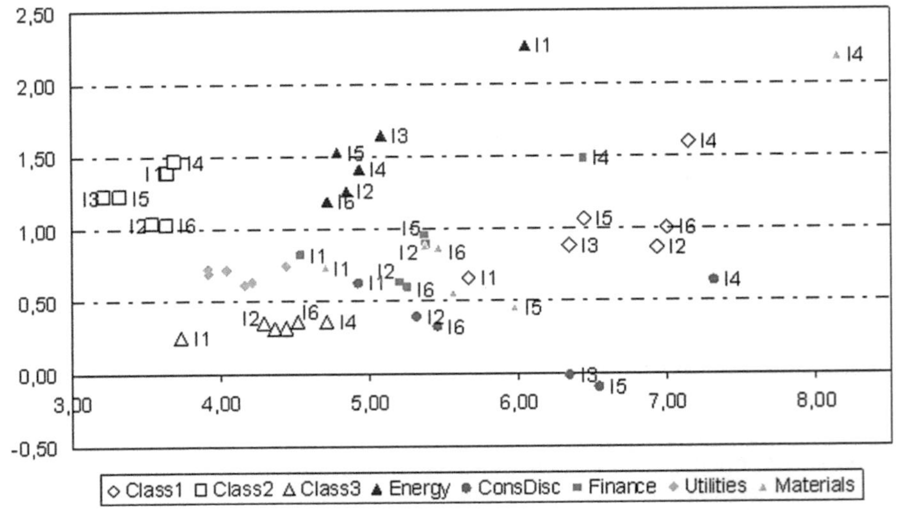

Fig. 4 Synthetic price index numbers related to the new and traditional sectors

has on price index's volatility: this bar chart illustrates sector's standard deviation and first percentile for 4 synthetic index numbers. Price index *I*4 confirms to be often characterized by the highest values, whereas price indexes *I*2 and *I*6 lead to the very similar results of standard deviation and first percentile. From Fig. (5), it also arises that the arithmetic average price indexes (*I*1) are generally characterized by lower values of standard deviation and 1st percentile than the other price index numbers, except for those sectors which are constituted of few stocks, such as energy and utilities.

Finally, from the analysis of the results reported in Tables (5) and (6), and depicted in Fig. (4) it seems that, in the framework of the traditional classification, the differences among the synthetic index numbers are more marked. We interpret the greater homogeneity of the indexes related to the latent classes as a result

Table 6 First and fifth percentiles of the price index numbers related to the new and traditional sectors

	*I*1		*I*2		*I*3		*I*4		*I*5		*I*6	
	*P*1	*P*5	*P*1	*P*5	*P*1	*P*5	*P*1	*P*5	*P*1	*P*5	*P*1	*P*5
Class1	−14.2	−9.6	−20.9	−9.2	−17.2	−9.0	−16.8	−10.0	−19.3	−9.4	−22.0	−9.0
Class2	−10.9	−6.7	−15.8	−6.6	−18.2	−7.2	−19.2	−8.3	−18.1	−6.9	−15.4	−6.6
Class3	−9.5	−6.1	−12.1	−6.3	−12.1	−5.9	−11.8	−6.6	−11.8	−5.4	−12.6	−6.7
Energy	−13.3	−8.0	−9.6	−7.3	−10.4	−7.0	−9.6	−7.7	−10.1	−6.6	−9.5	−7.3
Cons. Disc.	−12.3	−8.7	−12.4	−9.6	−14.7	−9.4	−16.5	−11.1	−15.7	−9.7	−13.2	−9.3
Finance	−10.9	−6.7	−17.1	−7.2	−18.2	−7.2	−19.2	−8.2	−18.1	−6.9	−16.6	−7.6
Utilities	−11.7	−7.9	−10.7	−7.1	−10.3	−6.7	−10.2	−7.0	−10.5	−6.6	−10.7	−7.0
Materials	−10.7	−8.0	−12.6	−7.7	−11.9	−8.4	−14.2	−9.1	−12.4	−9.1	−12.9	−8.0

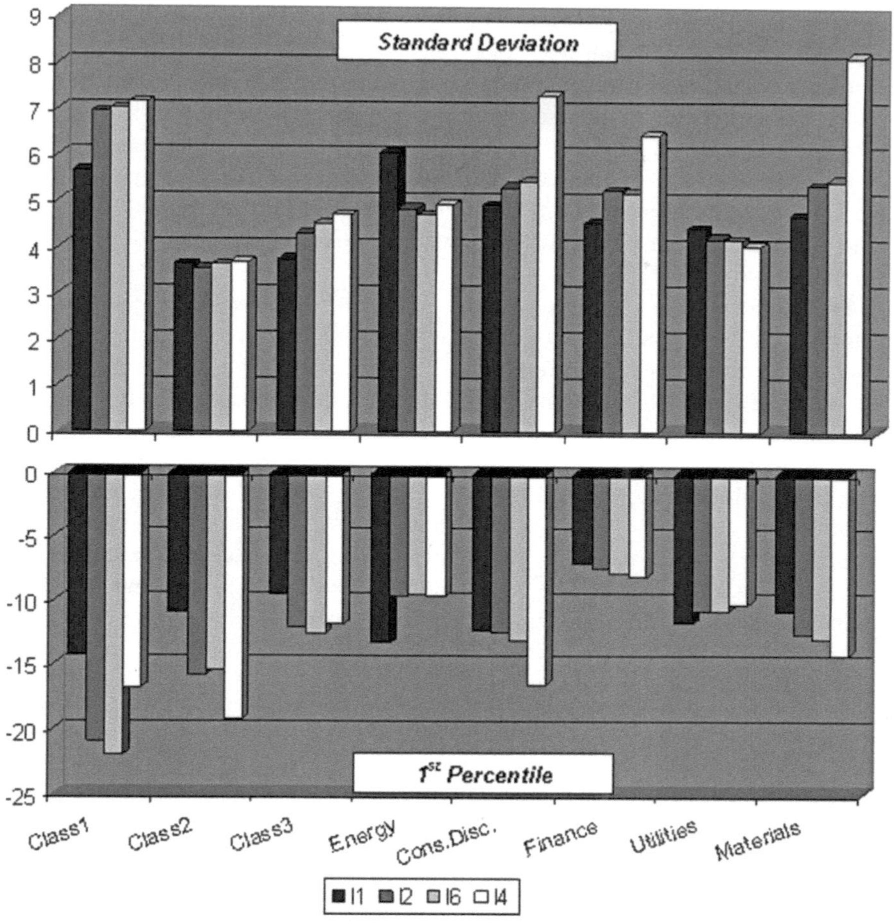

Fig. 5 Effects of different weighting structure on price index's volatility: standard deviation and 1st percentile of price index *I*1, *I*2, *I*6, and *I*4

of the stronger power of the new classification in detecting sectors with similar risk-return profile.

4.3 The Implications for Portfolio Analysis

In the framework of the standard portfolio theory, the mean and the standard deviation of the stock's return distribution are the main reference in order to derive efficient financial portfolios, that is minimum risk portfolios for a given level of mean return. The solution of the minimum problem achieved by the method of Lagrange multipliers implies the use of the correlation matrix, thus employing also the information about the interrelation structure. The set of efficient portfolios is called efficient frontier and, by adding the risk-free asset, it leads to the Capital Market Line (CML): an half-line in the mean-stardard deviation space which summarizes the best investment opportunities. By comparing the efficient frontier based on the traditional sectors to the efficient frontier related to the new classification, it is possible to evaluate the effects of sector definition on the portfolio analysis.

In particular, we consider the portfolio combinations of the traditional and the new sectors adding the 3-month Italian Treasury Bill as the risk-free asset, thus obtaining the two CMLs which allow an easier comparison between the two classifications.

In Fig. (6) are illustrated the efficient frontiers obtained within the new stock's classification (solid line) and within the traditional framework (dashed line). In order to observe the effects of the weighting structure on the investment opportunities we calculate the CML on the basis of different indexes proposals. Fig. (6) panel *a* illustrates the results obtained by using equally weighted indexes *I*1. Figure (6) panel *b* refers to total capitalization indexes *I*2, while Fig. (6) panel *c* reports the case of free floating capitalization. Finally, Fig. (6) panel *d* summarizes the three previous proposals.

Efficient frontiers obtained by using latent classes dominate in all cases the efficient frontiers calculated on the traditional GICS sectors. This striking feature of the new stock's classification allows much more interesting investment opportunities, with a relevant reduction of the risk level. Furthermore, this property is robust with respect to the weighting structure.

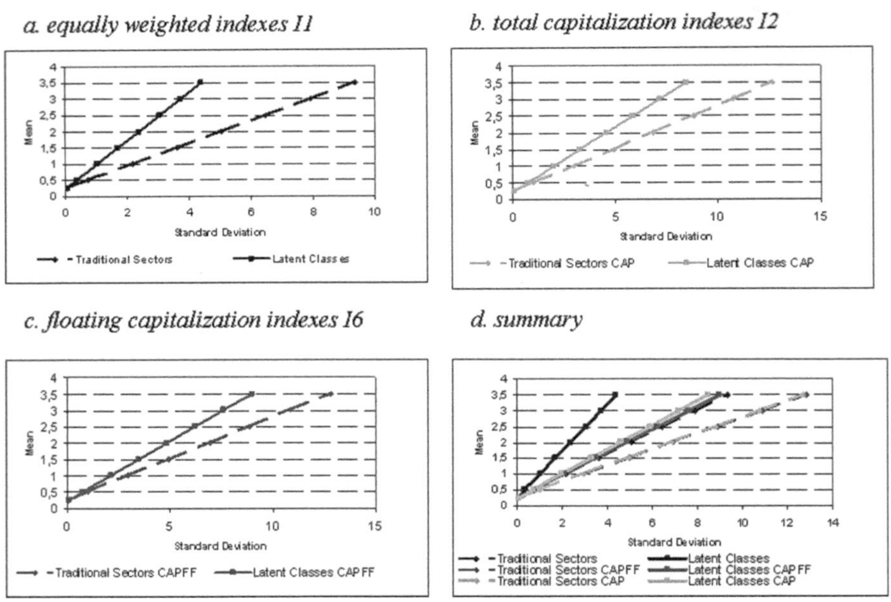

Fig. 6 Capital market lines within the traditional and the new stock's classification

It is also worth noting that equally weighted indexes $I1$ lead to the better performing efficient frontiers, while the results obtained by using $I2$ and $I6$ indexes are quite similar.

5 Conclusions

In this paper we face two methodological issues regarding sector price index numbers in financial markets.

First, we show how LC models represent an appropriate method in order to define a new classification in which the stocks are allocated into homogenous sectors under risk-return profiles. We find evidence of a 3-class latent model which allows to classify stocks into three new sectors: one which allows the best investment opportunities because it collects the stocks which perform the best, a second group of stocks characterized by low return and low volatility, and a third one which includes stocks with higher risk and low return.

Our proposal allows to overcome some problems related to traditional sector definition and to indicate a methodologically correct solution. The use of latent classes leads to an improvement in the quality of the new sectors, in particular with respect to the coverage and the representativeness of the risk-return profile. Furthermore, we define the composition of the basket underlying sector price indexes following a strictly correct methodological process. Our results are consistent with the standard portfolio theory and outperform the stock's classification based on the traditional sectors, thus giving new and improved investment opportunities.

Second, we construct price index numbers with different weighting structure for both the traditional and the new sector definitions. We find out that the results related to the new sector definition are less heterogenous, thus underlying that the LC methodology applied in order to define the new stock's classification is robust with respect to the index weighting structure.

We also show that indexes based on the free floating capitalization, which are having a great success in financial markets in the last years, perform quite similar with respect to the less complicated Laspeyres-type price index numbers based on the market capitalization. The analysis of the effects on volatility and extreme values of the synthetic indexes highlights how the plainest index structure, the simple arithmetic average, performs pretty well in most of the cases. This evidence also arises considering the efficient frontiers achieved in the standard portfolio theory framework.

A free floating capitalization weighting structure represents an important methodological and theoretical achievement but it also requires an heavy data adjustment activity. The introduction and the use of the new price index numbers should be therefore carefully evaluated by balancing a higher cost and an informative content which seems quite similar to the results of the traditional price index numbers.

References

Costa, M., & De Angelis, L. (2008, June 11–13). Sector classification in stock markets: A latent class approach. In *Book of Short Papers, Meeting of the Classification and Data Analysis Group of the Italian Statistical Society, Caserta*. Heidelberg : Springer.

Dempster, A. P., Laird, N. M., & Rubin, D. B. (1977). Maximum likelihood from incomplete data via the EM algorithm. *Journal of the Royal Statistical Society B, 39*, 1–38.

Hagenaars, J. A. (1990). *Categorical longitudinal data - loglinear analysis of panel, trend and cohort data*. Sage : Newbury Park.

Lazarsfeld, P. F. (1950). The logical and mathematical foundation of latent structure analysis. In S. Stouffer et al. (Ed.), *Measurement and prediction*. New York : Wiley.

Lazarsfeld, P. F., & Henry, N. W. (1968). Latent structure analysis. Boston: Houghton Mill.

Lisi, F., & Mortandello, F. (2004). Numeri indici di borsa: flottante e volatility. *Statistica Applicata, 1*, 17–37.

Lisi F., & Otranto, E. (2008). *Clustering mutual funds by return and risk levels* (Working Paper CRENoS 200813) Sardinia: Centre for North South Economic Research, University of Cagliari and Sassari.

Magidson, J., & Vermunt, J. K. (2001). Latent class factor and cluster models, Bi-plots and related graphics displays. *Sociological Methodology, 31*, 223–264.

Magidson, J., & Vermunt, J. K. (2004). Latent class models. In D. Kaplan (Ed.), *The sage handbook of quantitative methodology for the social sciences*. Thousand Oaks: Sage Publications.

Moustaki, I., & Papageorgiu, I. (2005). Latent class models for mixed variables with applications in archaeometry. *Computational Statistics and Data Analysis, 48*, 659–675.

Otranto, E. (2008). Clustering heteroskedastic time series by model-based procedures. *Computational Statistics and Data Analysis, 52*, 4685–4698.

FSC
www.fsc.org
MIX
Papier | Fördert gute Waldnutzung
FSC® C083411

Kolibri 360 GmbH
Ferdinand-Jühlke-Straße 7
99095 Erfurt, Deutschland
produktsicherheit@kolibri360.de